Adhesion of Polymers

I0826692

Adhesion of Polymers

Roman A. Veselovsky
Vladimir N. Kestelman

McGraw-Hill
New York Chicago San Francisco Lisbon London Madrid
Mexico City Milan New Delhi San Juan Seoul
Singapore Sydney Toronto

Library of Congress Cataloging-in-Publication Data

Veselovsky, Roman A.
Adhesion of polymers / Roman A. Veselovsky, Vladimir N. Kestelman
p. cm.
Includes index.
ISBN 0-07-137045-5
1. Adhesives. 2. Polymers. I. Kestelman, Vladimir N. II. Title.
TP968.K42 2001 2001026703
668′3—dc21

McGraw-Hill 
A Division of The McGraw-Hill Companies

1 2 3 4 5 6 7 8 9 0 DOC/DOC 0 8 7 6 5 4 3 2 1

ISBN 978-0-07-173792-0

The sponsoring editor for this book was Kenneth P. McCombs, the editing superviser was David E. Fogarty, and the production supervisor was Pamela A. Pelton. It was set in Century Schoolbook by Keyword Publishing Services Ltd.

This book was printed on recycled, acid-free paper containing a minimum of 50% recycled, de-inked fiber.

Contents

Preface

The development of contemporary technology and industry is closely related to the creation of new polymeric materials, among which adhesives are playing an increasing role. Their production is being increased at higher rates than that of other polymeric materials. Adhesives find wide application in novel fields of technology. Such enhanced interest in adhesives can be attributed to several factors:

1. Modern technology involves new types of materials that cannot be joined by means of traditional mechanical methods such as welds, rivets, screws, and bolts. These materials include different types of ceramics, glass ceramics, alloys, composites, etc.
2. Newly developed adhesives characterized by strength, heat resistance, and noncombustibility better meet the requirements of the technology.
3. Adhesion is frequently the most effective way of joining very different materials in ways that can be achieved using relatively simple equipment. The range of materials that can be cemented is practically unlimited.
4. Application of adhesives results in valuable properties of the article produced, such as improved strength, waterproofness, resistance to vibration, and decreased weight.

The problem of improving adhesion strength is paramount not only for adhesive-bonded joints. Filled and reinforced polymers are of primary significance among new polymeric materials. These include glass-reinforced plastics, laminated plastics, coatings, woodchip boards, and compounded and reinforced rubbers. The properties of these materials are determined mainly by interaction of the polymer with the filling and reinforcing materials.

At present, there are many hypotheses in the theory of adhesive phenomena but they cannot be practically applied for developing new adhesives insofar as these hypotheses mostly explain different phenomena that occur in the course of cementing and fracture of adhesive-bonded joints. This is essentially related to the fact that the process of formation of an adhesive-bonded joint is a complex set of closely interrelated phenomena. It is necessary to clearly differentiate two concepts—adhesion and adhesion strength. Generally, adhesion strength, defined as the adhesion work determined by the experimental data on mechanical failure of the adhesive-bonded joint, differs considerably from the adhesion work determined by means of the thermodynamic equations or by the interaction energy between surface layers of atoms of the adhesive and the substrate. One of the reasons for such incongruence is the fact that the formation of the adhesive-bonded joint involves a great number of factors, which in the course of loading of the joint facilitate its premature failure. Among these factors must be included the formation of weak layers of different types between the adhesive and the substrate, and of the internal stresses in the adhesive layer.

All of this complicates the study of adhesion phenomena, hinders the scientific approach to the problem of controlling the adhesive properties, and is one of the reasons why, despite the great scientific and practical importance of research on creation of new, efficient adhesive compounds, progress in this field has been achieved mainly empirically.

This book generalizes the results of studies performed in the Department of Adhesion and Adhesives of the Institute of Macromolecular Chemistry of the Ukrainian National Academy of Sciences. It considers some regularities of the formation of adhesive-bonded joints, presents thermodynamic and physical-chemical substantiation of new principles of controlling the adhesion strength and other important properties of polymeric adhesives, and describes application of these principles in the course of developing adhesives for various fields of engineering and medicine.

One of the basic principles of controlling the properties of adhesives considered here is inclusion of surface-active substances (surfactants) capable of chemical interaction with the adhesive components and entering into the adhesives' composition. Such reactive surface-active (RS) substances differ radically from chemically indifferent surface-active (IS) substances. In the course of polymerization of oligomers containing IS substances there is a decrease of the critical concentration for micelle formation (CCMF) and formation of substantial quantities of large micelles of surfactant, which results in weak layers on the boundary between the adhesive and substrate and in decrease of the adhesion strength.

RS substances react chemically with molecules of the polymerizing oligomers to form macromolecules that contain both oligomers and the surface-active substance, which permits the decrease of CCMF and breakdown of RS micelles without damage to the adsorption layer on the substrate surface. Thus, application of RS substances provides for controlling the properties of the polymer boundary layers without initiating deleterious side effects. Use of RS substances allows an increase in the adhesion strength and water resistance of adhesive-bonded joints, making adhesives capable of cementing metals and other materials in water and petroleum products.

Of great interest is the application as adhesive compounds of polymeric blends, such as thermodynamically incompatible polymers one of which has a high modulus of elasticity and the other a low modulus. When the adhesive cures, the second polymer is liberated as a separate finely dispersed phase. Such separation of the blend into high- and low-modulus phases provides for controlling the relaxation properties of the adhesive while maintaining its high strength.

A special type of polymeric blends is interpenetrating networks (IPN), which represent a system formed in the course of building up one crosslinked polymer inside the ready-made network-matrix of another under conditions of no chemical reaction between the networks. Such IPN-based adhesive compounds are noted for considerable long-term strength, which is explained by features of the deformation processes that occur in the IPN layer when it is loaded.

The application of these principles of increasing the adhesion strength and of controlling other properties of adhesives provides for development of polymeric compounds with a number of valuable features: for example, structural adhesives with high short-term and long-term strength even when cementing untreated surfaces in various liquids; sealing adhesives that combine high adhesion strength and elasticity of the adhesive layer; foaming adhesives; medical-purpose adhesives capable of bonding biological tissues in an environment of tissue fluids, of being infiltrated by living tissues, and of being excreted from the organism at prescribed times; photopolymeric compounds; binders for forming glass-reinforced plastics in liquids; and others. Compounds can be cured both at high and at subzero temperatures; their adhesion strength is of low dependence on air humidity or pressure during cementing, on adhesive layer thickness, or on treatment of the surfaces to be bonded. These stipulations for high efficiency of adhesive compounds permit their application in fields of engineering where adhesives have not so far been used, for example, when repairing underwater oil and gas pipelines, oil tanks, ships on the high seas, and so on.

Chapter

1

The Process of Adhesive-Bonded Joint Formation

The strength and serviceability of adhesive-bonded joints are mainly the result of the play of forces of intermolecular interaction between the adhesive and the substrate. The forces of interaction between two condensed bodies at distances on the atom size scale can produce high adhesion strength [1, 2]. For example, calculations performed by Joraelachvili and Tabor [3] showed that when cementing materials with high surface energy the adhesion strength must be 170.0 MPa. With a gap between the contacting surfaces of 4×10^{-10} m [4] the force of interaction between them is 115.3 MPa for a Teflon–polyamide couple, 113.9 MPa for a polyamide–polyamide couple, 434.0 MPa for a Teflon–metal couple, and 569.0 MPa for a polyamide–metal couple. The values presented substantially exceed the usual adhesion and cohesion strengths determined for polymers, although only dispersion forces between contacting surfaces were taken into account in the calculations. Taking high values of the interaction forces into account, some researchers consider that failure of adhesive-bonded joints must always be of the cohesion type [5].

The reason for the lack of concordance between the theoretically and experimentally determined values of the adhesion strength lies in the fact that the process of achieving high adhesion strength is hindered by a number of phenomena that accompany the formation of adhesive-bonded joints. These factors that decrease the strength of adhesive-bonded joints can be divided into two groups:

1. Weak layers on the boundary between the adhesive and the substrate.

2. Internal stresses of the adhesive-bonded joint.

Weak layers on the substrate surface are formed in the case of incomplete wetting by the adhesive as well as when there are foreign impurities on that surface. These impurities may come from the atmosphere, the substrate, and/or the adhesive. A solid surface is always contaminated with "foreign" substances. There are no strict quantitative criteria of surface cleanness: the surface is considered to be clean within a permissible amount of contamination [6]. Gases, vapors, and various greaselike substances are adsorbed onto the substrate surface from the atmosphere [7]. The surface of a metal is almost always coated with an oxide film. The first stage of interaction between the metal surface and the gas is the formation of a gas monolayer on the surface; the sorption rate is so high that at room temperature it cannot be measured [8]. At high cohesion strength and with good adhesion to the metal, the oxide film does not produce any considerable effect upon the adhesion strength, while in the case of poor adhesion to the metal it will influence the adhesion strength of an adhesive-bonded joint.

Adsorption or the presence of oxide layers on the substrate surface can be demonstrated by the fact that surfaces of metals are considerably reactive only at the moment of formation of the above layers and can initiate polymerization reactions and start chemosorption interaction with polymers.

One of the sources of dirtiness of the substrate surface is low-molecular weight substances in the substrate itself that gradually diffuse to the surface and accumulate there. In the case of polymers, such substances might be plasticizers, softeners, stabilizers, residual monomer, or various additives.

Low-molecular weight impurities in the adhesive can markedly decrease its adhesiveness. Fractional precipitation of polyethylene results in removal of low-molecular weight impurities, which therefore significantly increases its adhesion to various materials [9, 10]. In a polymeric system that contains plasticizer, it is adsorbed mainly by the surface of the solid body and, in the course of time, the concentration of the plasticizer at the boundary increases up to an equilibrium. Low-molecular weight fractions in the epoxy resin are characterized by high absorptivity. Their presence results in considerable decrease of adhesion strength of the bonding. Removal of these fractions from the resin by means of multiple adsorption operations (from 76% solution in toluene in SiO_2) results in 20% increase of the adhesion strength.

The adsorption of impurities in the composition of an adhesive onto the substrate surface can be judged by the change of the system interphase tension. To study the interphase tension on the boundary between oligomers and metal, a model system was used with purified and distilled mercury as the metal. The interphase tension was determined using the sessile drop method [11] by measuring the parameters of a mercury drop in a transparent adhesive with a polarographic system. Removal of low-molecular weight fractions from ED-20 resin resulted in increase of the interphase tension (Fig. 1.1). The epoxy resin was purified by vacuum distillation and the basic product was collected at boiling point of 232°C (1 Pa) and contained about 25% epoxy groups. Figure 1.1 shows that the unpurified ED-20 resin has minimum value of the equilibrium interphase tension, while the purified resin has the maximum value. Addition of alcohol to the unpurified resin results in some increase of the interphase tension.

With purified resin, alcohol decreases the interphase tension, with the exception of OP-10 (allyl phenol oxyethylated ester) for which it produces some increase (by 1 mN/m) [12]. These findings can be explained by the fact that the alcohol facilitates desorption of the low-molecular weight resin fractions with surface-active properties from the boundary between the resin and the mercury by increasing their compatibility with the bulk resin. The free energy advantage for alcohol adsorption on the mercury surface is less than that for low-molecular weight fractions, which is why it results in increase of the interphase tension.

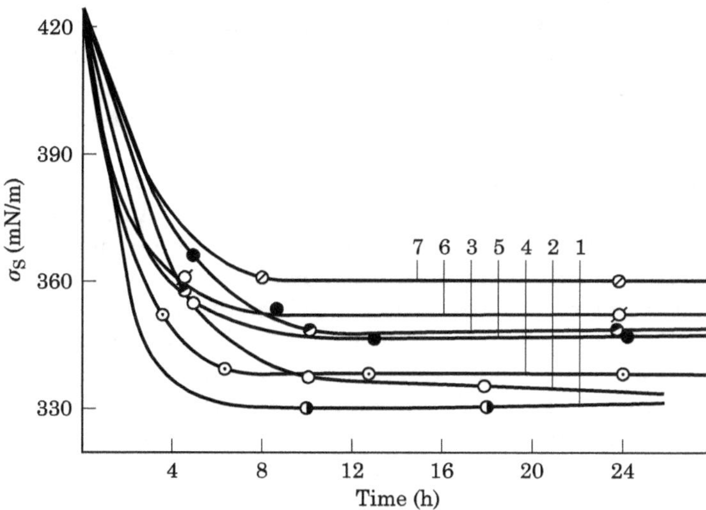

Figure 1.1 Change over time of interphase tension at the interface of mercury and ED-20 epoxy resin with different additives: (1) undistilled ED-20 without additives; distilled ED-20 with (2) 1% ethanol, (3) 1% butanol, (4) 1% hexane, (5) 1% octane, (6) 1% decane; (7) distilled ED-20 without additives.

TABLE 1.1 Effect of Addition of Alcohol on Adhesion Strength of Epoxy Adhesive

Adhesive	Alcohol	Adhesion strength (MPa)
Unpurified resin: ED-20 + PEPA	No alcohol	18
	Ethanol	23.6
	Decanol	20.6
Purified resin: ED-20 + PEPA	No alcohol	25.4
	Ethanol	23.6
	Decanol	21.0

Table 1.1 presents information on the effect of additives upon the adhesion strength of an adhesive based on ED-20 resin cured by polyethylene polyamine (PEPA). It is evident that the lowest adhesion strength is exhibited by the adhesive based on the unpurified resin. The sorption of low-molecular weight fractions of an epoxy resin from the interphase boundary caused by addition of alcohols to this adhesive, i.e. resin purification, results in increase of the adhesion strength. Adding further alcohol to a purified resin results in formation of a weak layer on the interphase boundary, producing some decrease of the adhesion strength.

In the course of formation of polymers, some monomer remains in the cured adhesive. This monomer can diffuse to the boundary, plasticizing faulty surface layers of the polymer and adsorbing on the substrate surface [13]. Monomer at the boundary between the adhesive and the substrate inevitably decreases the adhesion strength. For example, by electron microscopy one can observe single crystals of caprolactam on the surface of PC-4 polycaprolactam [14]. Removal of the polymer surface layer usually results in noticeable increase of the adhesion strength in the course of bonding [15].

For many types of adhesives the adhesion strength depends on the type and quantity of catalyst(s) used to cure the adhesive and of polymerization initiators. This can be understood as being due to different extents of cure of the adhesive on the interphase boundary. One study investigated the effect of the residual methyl methacrylate (MMA) present in the adhesive layer upon the internal stresses and adhesion properties of adhesives based on a 40% solution of polybutylmethacrylate (PBMA) in MMA or in butylmethacrylate (BMA). The adhesives were cured at room temperature with initiating systems such as benzoyl peroxide (BPO)–dimethylaniline (DMA), methyl ethyl ketone peroxide (MEKP)–DMA, and n-toluenesulfonic acid–DMA. The quantity of BPO, MEKP and n-toluenesulfonic acid was varied from 0.5% to 2%; in all cases the quantity of DMA was equal to 1%. The quantity of unreacted monomer in the adhesive interlayer was determined by a polarographic method using a P-60 polarograph with dropping mercury cathode and bottom mercury as anode. The polarograms were recorded against 0.2 mol/l of tetrabutylammonium iodide in dimethylformamide (DMFA).

The concentration X of the residual monomer was determined from

$$X = \frac{hcw}{h_1 + (h_1 - h)V} \tag{1.1}$$

where h_1 is the total height of the polarographic wave after adding the standard solution; h is the height of the wave that corresponds to the concentration X; V and c are the volumes of the studied and standard solutions; and w is the concentration of the standard solution. (Here and hereafter, the weight fraction of substances is given in percent.)

A layer of the adhesive was dissolved in DMF immediately after failure of the adhesive-bonded joint.

Figure 1.2 shows the dependence of internal stresses and the adhesion strength of the specimens on content of the residual monomer. The quantity and type of the initiator affect the adhesion strength in the same way as the polymerization reaction. No change of the cohesion strength of the polymers was observed with variation of the content of monomer within the range examined.

With vacuum treatment of the specimens (at 200°C, 103 Pa) for 10 days, the content of the monomer in the adhesive layer decreased and the adhesion strength increased. The character of adhesion failure of the joint changed to cohesive type. The data allow a more complete explanation of the so-called latent period—the period of achieving the maximum adhesion strength. It is evident that the strengthening of some adhesive-bonded joints in the course of time is caused by volati-

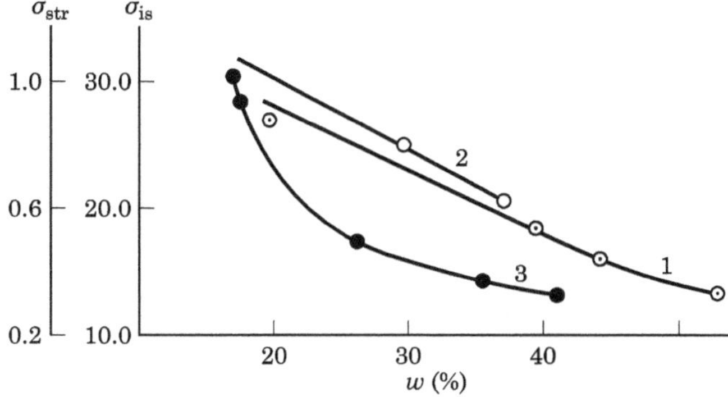

Figure 1.2 Effect of residual MMA on the strength σ_{str} of adhesive-bonded joints (1) before and (2) after vacuum treatment and (3) on the internal stresses σ_{is} in the adhesive layer.

lization or by more complete bonding, the monomer being capable of migration to the substrate surface in the adhesive layer. Additional heterogeneity at the boundary between the adhesive and the substrate may result from the influence of the substrate surface on the chemical reactions in the adhesive. Selective adsorption of different components of the adhesive by the substrate surface results in a change in the conditions of the reaction insofar as it causes a different distribution of components of the cured system within the boundary layer. As a result, not only the kinetic but also the chemical conditions of the reaction change with violation of the stoichiometry of the process. Let us consider the effect of substrates of various surface energy upon the process of formation of the boundary layer of an epoxy polymer [16]. ED-20 epoxy resin cured by PEPA was used for the study. Reversed-phase gas chromatography was used to study the properties of the polymer boundary layers, using an LKhM-72 chromatograph with air thermostat and flame ionization detection. Hydrophobicized glass, glass, and iron were used as support materials, which in this case served as substrate models. The hydrophobicized glass was obtained by treating glass balls with a solution of dimethyldichlorosilane in toluene, with the formation of a polysilicon film on the glass surface.

Substrates of different types influence the glass vitrification temperature of the polymer. The glass-transition temperature T_g was determined from the turning point of the dependence of the logarithm of the retention volume on the inverse temperature, $\log V_q$ vs. $1/T$. As Fig. 1.3 shows, the effect of the substrates of low (hydrophobicized

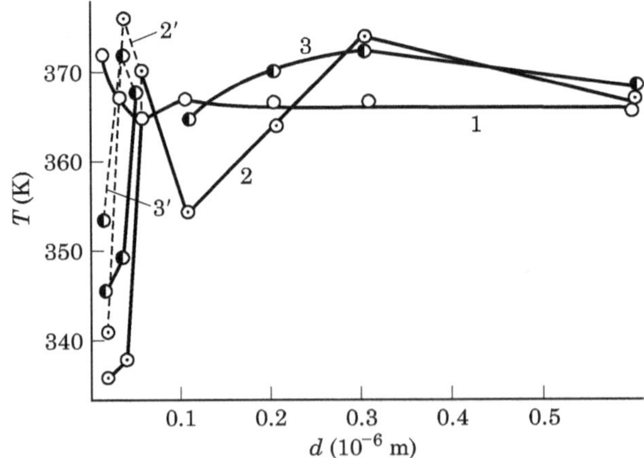

Figure 1.3 Dependence of the glass-transition temperature T_g on film thickness d for ED-20-PEPA on hydrophobicized glass (1), on iron (2, 2′), and on glass (3, 3′); (2) and (3), heating for 5 h; (2′) and (3′), heating for 10 h at 423 K.

glass) and high (iron, glass) energy on the polymer T_g on the boundary with the substrate differs widely. For low-energy surfaces the polymer vitrification temperature does not significantly depend on the polymer layer thickness. The increase of T_g for a film 0.01 μm thick is related in this case to limitation of the mobility of the polymeric chains close to the solid surface. The dependence of T_g for high-energy surfaces illustrates the complex structure of the boundary layer. For a film 0.01–0.03 μm thick, there is no kink characterizing the glass vitrification temperature in the $\log V_q–1/T$ curve within the range of temperatures studied. This suggests that the boundary layer of the epoxy compound up to 0.03 μm thickness does not undergo the transition to the polymeric state under cold cure, although it may have sufficient mechanical strength due to the energy field of the surface. In fact, as shown in [17], a high-energy surface can selectively sorb the epoxy resin, as a result of which the adhesive layer is depleted by the curing agent and the stoichiometry of the compound is disturbed. In this case, the polymer layer enriched by epoxy resin has substantial extent. In addition, undercure of the adhesive boundary layer may be caused by decrease of the polymer chain mobility due to the energy of interaction with a solid surface, by limiting the conformation set, or by blocking of the active groups of components of the compound by the solid surface.

Heating the specimens for 5 h at 423 K results in a kink in the $\log V_q–1/T$ curve that corresponds to the polymer glass vitrification temperature in a film of 0.01–0.03 μm thickness. In this case, the effect of the iron surface on the glass vitrification temperature is greater than that of the glass surface. Heating at the same temperature for 10 h results in an increase of T_g; further heating of specimens for 15 h does not cause any futher increase of T_g. As Fig. 1.3 shows, even heating does not increase the glass vitrification temperature of specimens 0.01 μm thick to the T_g value of the block polymer. This suggests that the primary reason for adhesive undercure on the boundary with a high-energy substrate is depletion of the compound boundary layer by the curing agent. The increase of the glass-transition temperature of a 0.05 μm thick film of polymer can be explained by this layer containing excess of the curing agent, which reacts with the resin under prolonged heat treatment and results in increased extent of crosslinking. The subsequent layer of 0.1–0.2 μm thickness with lower T_g seems to be explained by the structural looseness of the polymer within this zone [18]. Beginning at 0.3 μm thickness, the properties of the polymer boundary layer approach those of the polymer in bulk.

These data allow determination of characteristic changes of values of the retention volume V_q depending on the film thickness.

Insignificant change in V_q with varying film thickness is observed for polymer on the hydrophobicized surface.

For the high-energy surfaces of the 0.03 μm thick film there is abrupt increase of V_q that can be explained by decrease of the polymer structure density [19]. Heat treatment of the film results in V_q decreasing, i.e. in structure identification. Figure 1.4 shows the dependence of the parameter $\chi_{1,2}$ of thermodynamic interaction between the polymer and the solvent on the film thickness. It is evident that values are similar for the 0.03 μm thick epoxy films made both with and without a curing agent, which indicates the latter case to be one of incomplete gel-formation. Heating the 0.03 μm thick film on iron or glass substrates causes an increase in the parameter $\chi_{1,2}$ that is related to the adhesive post-cure reaction that occurs in the course of heating. As the film thickness increases, the influence of the polar substrate on the magnitude of $\chi_{1,2}$ has less effect, although it is observed for thicknesses up to 0.6 μm.

The process of formation of the epoxy polymer in bulk was studied using IR spectroscopy, and on the boundary with the KRS-5 element was studied by the method of disturbed total internal reflection [20]. In 24 h the degree of conversion of the epoxy groups was 28% in bulk and was zero at the boundary. In 8 h of heating the conversion level was 70% both for bulk and at the boundary. Formation of a surfactant

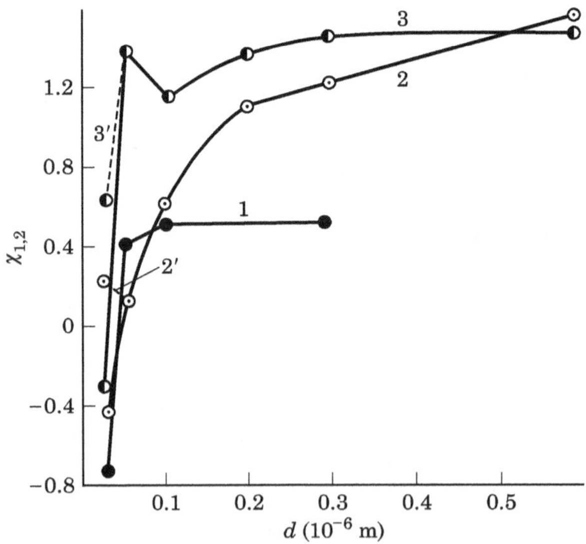

Figure 1.4 Dependence of the parameter $\chi_{1,2}$ of interaction of polymer solvent on the film thickness for ED-20 (1) and ED-20-PEPA (2, 2′, 3, 3′). Designations are the same as for Fig. 1.3.

monolayer on the substrate surface results in an increase in the level of conversion of the epoxy groups in the boundary layer to 20% for an unheated specimen. Undercure of the adhesive on the boundary with the substrate resulting from the influence of the solid surface on polymer formation causes a decrease of adhesion strength of adhesive-bonded joints; heating the specimens increases the conversion of the epoxy groups and enhances the adhesion strength. Thus, when cementing steel specimens the adhesion strength of the thermally untreated adhesive is 10.3 MPa at normal break-off; for specimens heated at 423 K for 5 h it is 20.9 MPa; and for 10 and 15 h it is 30 MPa. Useful increase of the adhesion strength can be achieved through formation of the surfactant monolayer on the substrate. To clarify the mechanism, a monolayer of stearic acid from a solution in chloroform was applied to the surface of iron and glass support materials. This decreased the effect of the surface energy field on the polymer formation process. Thus, for a 0.03 μm thick polymer film applied to glass and iron surfaces with a surfactant monolayer, T_g was 369 K and 364 K, respectively, and the heating produced essentially no change.

These results indicate that a layer of undercured polymer can be formed when epoxy adhesives are used on high-energy substrates. When cementing surfaces with low surface energy, no such layer was observed, but in this case the achievement of high adhesion strength is hindered by poor substrate wetting by the adhesive.

Formation of weak boundary layers is confirmed by a study of the molecular mobility of filled epoxy polymers. The availability of the solid surface results in a decrease of the molecular mobility in the boundary layer [21] as a result of limiting the conformation set and adsorption interactions of the polymer molecules with a solid body at the boundary. The nature of the filler surface has little effect on the molecular mobility of the epoxy polymer and on the change of mobility of its side-groups and segments. It has been concluded [21] that the primary role in the change of mobility is played by geometric limitation of the number of possible conformations of macromolecules close to the surface of the particles, i.e., by the entropy factor rather than by energetic interactions of the surfaces.

Such information provides the basis for more detailed consideration of the physical-chemical processes that occur in the course of curing of epoxy systems and that do not affect their relaxation behavior.

Dielectric relaxation of ED-20 epoxy resin (molecular weight = 450), both unfilled and filled with marshallite and cured by PEPA, was studied. The dielectric relaxation was measured using an E8 4-digit capacity meter at an applied frequency of 1 kHz and the temperature range of measurements was 123–424 K.

Epoxy polymers manufactured from commercial resins are known [12] to contain low-molecular weight impurities, which can adsorb on the interface between the polymer and the solid substrate, and form a weak interlayer. In epoxy resin cured at room temperature and not thermally treated, there are two maxima in the dielectric losses observed in the low-temperature range (Fig. 1.5): at 153 K a dipole-grouped process of relaxation is manifested, while at a higher temperature of 231 K there is the onset of a relaxation process governed by the mobility of kinetic units larger than those responsible for the dipole-grouped motion but smaller than chain segments. The maximum in tan δ that corresponds to 153 K can apparently be explained

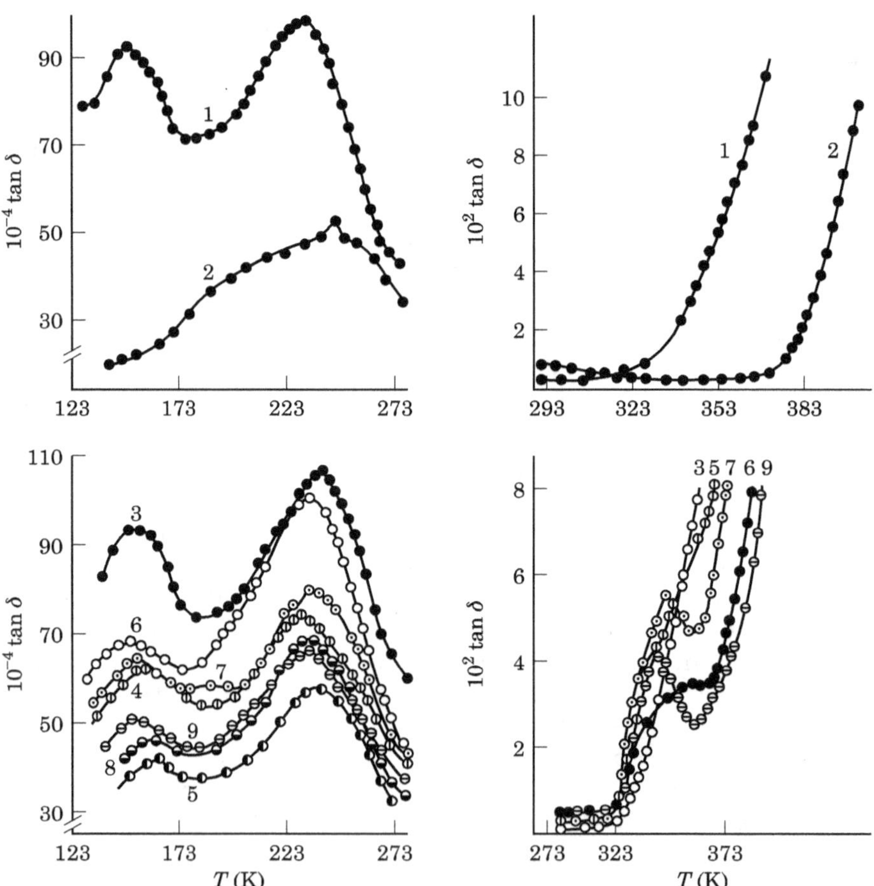

Figure 1.5 Dependence of tan δ of the cured epoxy polymer on temperature: (1) unthermostated; (2) thermostated at 423 K for 2 h; (3) unfilled; (4) 10%, (5) 20%, (6) 50%, (7) 100%, (8) 150%, and (9) 200% marshallite.

by increase of the polymer molecular mobility caused by low-molecular weight impurities therein. After the heat treatment of the polymer, the relaxation process observed at 153 K disappears and the maximum of the molecular mobility at 231 K shifts into the region of higher temperatures. Such a temperature shift of the maximum in the dielectric loss is caused by complete cure of the polymer system and by volatilization of the low-molecular weight impurities, which bring about the decrease of molecular mobility. Increase of the polymer moisture content and holding it for 2 days at room temperature results in shift of the $\tan\delta$ maximum to lower temperatures and increase of the absolute value of $\tan\delta$.

As Fig. 1.5 shows, the maximum for some systems does not show up at high temperatures. This can be explained by the increase of the $\tan\delta$ value caused by low-molecular weight impurities of the polymer serving as centers of conductivity. The change in the $\tan\delta$ values for the epoxy system in the range of 333–383 K measured after 2 h thermal treatment is explained by the increase of the crosslink density of the space-network polymer and by the volatilization of the low-molecular weight impurities.

The amount of filler in the system needed to obtain the polymer interlayer of a certain thickness was:

Content of marshallite in the polymer (%)	10	20	50	100	150	200
Distance between particles of the filler (μm)	6.354	4.155	2.035	0.928	0.325	0.154

At low temperatures, there are two regions of temperature transitions: 163–153 K and 238–231 K. The first characterizes the dipole-grouped relaxation process and the second characterizes the dipole-segmented process. With decrease of the polymer layer thickness in the system, the relaxation processes shift toward higher temperatures. The decrease of the dipole-grouped and dipole-segmented mobility at low filler content (up to 50%) is apparently explained by depletion of the low-molecular weight impurities in the polymer due to their adsorption on the filler surface. With increase of the filler content (over 50%), a considerable portion of the polymer makes a transition to the state of the boundary layer. Thus, if the thickness of the faulty boundary layer characterized by polymer undercure in the given system is about 0.1–0.2 μm, addition of 50% filler to the system means that the thickness of the polymer layer on its surface is 1 μm, i.e., 10% of the polymer is in the boundary layer state. With increase of the filler content, the maximum shifts to lower temperatures as a result of enhancement of molecular mobility of the polymer due to decrease of the extent of conversion of the system. At a filler content over 100%, a considerable proportion of polymer enters the boundary layer state and further increase of the filler quantity has no noticeable effect on the molecular mobility. The decrease of mobility within the range of 100–200% of filler is apparently caused by entropy and energetic interactions of the polymer with the surface of filler particles [21].

These results show that the relaxation behavior of filled epoxy polymer is affected by two factors: the presence of the low-molecular weight impurities and the formation of the undercured boundary layer.

Thus, for polymer formation in the course of reaction on the surface, the principal difference of interaction between the polymer and the high-energy substrate lies in the fact that an intermediate undercured layer is formed on the surface due to the absorption of reagents by that surface. The layer of completely cured polymer follows the intermediate layer. Such a two-layer structure may be valuable from the point of view of adhesion strength.

The presence of the solid surface affects the formation of crosslinked and linear polymers in different ways [22]. In the case of crosslinked polymers, sufficiently large branched molecules are formed at comparatively early stages of the process. In the formation of linear polymers, the critical molecular weight of macromolecules at which the surface becomes capable of effecting its conformation set is higher than for crosslinked polymers.

Study of the process of swelling of the polymeric coatings and filled polymers found that the crosslinked concentration when the network forms with surface present is significantly less than in bulk. In addition, instances were observed of nonmonotonic dependence of the molecular weight of the polymer between crosslinks in the network on the filler concentration, which indicates that the surface can order the arrangement and growth of chains [23]. The nature of the polymer and the type of surface determine the course of the process.

The solid surface has equivocal effects on different stages of the polymerization process. Adding filling agents results in decrease of induction period of the polymerization of the methacrylic esters and in acceleration of polymerization in the reaction region before and at the point of gel formation. After the gel formation point, the abrupt decrease of mobility of the elements of the polymerization system causes a drop in the polymerization rate, which practically ceases depending on the concentration and the character of the filler [24]. The example of polycaprolactam [25] shows that that the number-average molecular weight at the polymer surface can be less than in bulk.

In a study of the influence of glass fiber on the formation of the glass-reinforced plastic, it was found [26] that the presence of the glass fiber determines both the rate and the depth of the bond cure, which in turn brings changes to the elastic properties and the bond stressed state around the fiber. Close to the fiber the organic silicon bond cures much more slowly than in bulk, i.e., it becomes weaker, which results in cohesion failure along the bond layer when glass-reinforced plastic is loaded.

The chemical bonding of the polymer with the surface of the fiber glass by means of finishing substances has been investigated [27]. In the majority of cases IR spectroscopy revealed the absence of chemical interaction of the organic silicon finishing agent containing vinyl groups with the polyester resin in the course of cure, although such interaction has been observed [28].

Thus, the solid surface can hinder the course of polymerization or polycondensation. It is to be expected that formation of the low-molecular weight interlayer between the adhesive bond and the substrate cannot but affect the adhesion strength.

The effect of the solid surface on polyurethane formation is specific. The surface does not hinder formation of the polyurethane itself, although polyaddition on the surface frequently proceeds at a different rate from that in bulk. This is related to the effect of the adsorption ordering the boundary layer on the kinetics of the interaction [29]. The dependence of polyurethane formation on the solid surface present is apparently explained by a number of factors. The isocyanate groups are capable of strong interactions with various surfaces (metals, glass), and they can react both with one another and with a great number of other compounds such as water, alcohols, amines, and unsaturated compounds. Many substances (salts of metals, amines, phosphines, and others) catalyze reactions of isocyanate.

Let us consider the formation of polyurethane in the presence of magnesium chloride. This salt was selected because metals of group 2 form comparatively strong coordinate bonds with the oxygen-containing groups of the reactive system [30]. These groups provide for a certain interaction between the growing chains and the solid surface and this must influence the kinetics of the polyurethane formation. In addition, the metals of this group are frequently components of compounds used as the filling agents for plastics.

The system of polytetramethylene glycol (PTMG) of molecular weight 1000 and 4,4-diphenylmethane diisocyanate (DPMDI) was investigated. The ratio of NCO and OH groups was 1:1. Chlorobenzene was the solvent. The total concentration of the reacting substances was 0.2 mol/l. Magnesium chloride was added to the solution as a powder with the particle size $\sim 5\,\mu m$. The powder was dried at 150°C (10 Pa) for 3 h. Analysis showed that the reagent contained 79% $MgCl_2$, 4.7% MgO, and 16.3% H_2O. Experiment determined that adding MgO in the same quantity as that of the salt, and even somewhat more, has no effect on the course of the process. We related the acquired result to the actual quantity of $MgCl_2$. DPMDI was vacuum distilled at 156°C (10 Pa). PTMG was dried at 80°C (2×10^2 Pa) to residual moisture content of 0.01%. The reaction was carried out in a dry argon environment. The reaction kinetics were studied by sampling with subsequent titration of the unreacted isocyanate groups.

The spontaneous reaction of polyurethane formation in solutions follows a second-order rate equation in the early stages. Toward the end of the reaction, the polymer formation rate constant increases. The rate constant of polyurethane formation from DPMDI and PTMG calculated by the second-order equation for 60°C up to 76% conversion is 9.0×10^{-4} l/(mol.s), and from 76% conversion to the end of the reaction is 15.5×10^{-4} l/(mol.s). Some increase of the reaction rate constant at 76% conversion seems to be related to catalysis of the reaction by the urethane groups formed or by secondary reactions in the system.

The addition of magnesium chloride causes a slight increase of the rate constant at the reaction first stage (up to 70% conversion). As soon as the magnesium chloride content in the system reaches 9.43 g per mole of hydroxyl groups, the rate constant rises to 13×10^{-4} l/(mol.s). Further increase of the salt content in the reactive system brings about no further elevation of the rate constant. At the first stage when the salt is added, the activation energy changes insignificantly, from 38.1 kJ/mol for spontaneous reaction to 57 ± 2 kJ/mol for the system with 6.30 g of magnesium chloride added per mole of hydroxyl groups (the constants were measured within the temperature range 40–70°C). At the second stage (after 70% conversion) the rate constant rises abruptly in line with the increase of the quantity of added salt.

It should be noted that the increase in the reaction rate constant is not related to $MgCl_2$ solubility in the reactive system. Polarographic determination of $MgCl_2$ solubility in chlorobenzene at 60°C gives a value of about 3×10^{-3} g/100 ml. Addition of such a quantity of salt into the reactive system does not change the rate constant.

The polymer formed at the second stage of the reaction with magnesium chloride differs qualitatively from the polymers that are formed at the first stage and in the course of spontaneous reaction. Microscopic studies indicate that intensive formation of the polymer directly on the salt particles begins at the second stage. The content of nitrogen in this polymer exceeds that estimated theoretically in polyurethane. Thus a polymer is not soluble in the solvents that are usually used for such polyurethanes. At 270°C the polymer does not melt and begins to decompose. An intense absorption band at $\sim 1660\,\text{cm}^{-1}$ appears in the IR spectrum of the polymer in line with the bands typical of the urethane groups. The intensity of this band increases with the increase of the quantity of the magnesium chloride added to the reactive system. This band is related to the urea bonds, which can be formed on account of water in the salt. Additionally, the extent of the carbonyl band splitting increases at $\sim 1720\,\text{cm}^{-1}$ in line with the increase of the quantity of magnesium chloride. This splitting can be related to formation of allophanic bonds that disintegrate under heating (after holding the specimen at 120°C for 6 h, the splitting of the carbonyl band disappears).

The appearance of the accelerating effect of the magnesium chloride only after 70% conversion (i.e. at sufficiently high concentration of the urethane groups in the growing polymeric chains) and formation of the polymer on the surface of the salt particles in the second stage of the reaction suggest that availability of the urethane groups facilitates adsorption of the growing polymer chains from the solution onto the surface of the salt solid particles. In line with this, it was shown that the polymer contains large amounts of allophanic bonds, which is why the accelerating effect of the $MgCl_2$ particles can be explained on the one hand by the accelerating effect of the surface due to its ordering influence, and on the other by the acceleration of formation of allophanic and urea groups. Apparently the reaction of the isocyanate groups with water contained in $MgCl_2$ becomes possible only when they are in direct contact with the surface (i.e., in case of adsorption), so that the band that characterizes the urea bonds does not appear at once but becomes more intense at the second stage of the reaction. In the IR spectrum of the polymer obtained with anhydrous MgO, there was no $1660\,\text{cm}^{-1}$ band. The same can be assumed with respect to the allophanic bonds, which may appear with contact of the isocyanate and urethane groups in the sorbed layer.

Thus, the reactions of branching and formation of the network polymer take place on the surface of $MgCl_2$. The orienting effect of the boundary surface with a solid body in the polyurethane formation reaction may not only accelerate the principal process but also facilitate the course of side-reactions (in the case considered, these are the reactions of formation of branching through the allophanic groups).

The solid surface can cause breakdown of the adsorbed polymer. For example, immediately after adsorption of the polyesters on quartz glass, aluminum oxide, and other hydrated surfaces, hydrolysis is observed [31]. Such behavior is understandable for these substances given that upon adsorption the polar group of the ester must be in direct contact with the solid body if there is no steric hindance. It is also possible that solvating water molecules located on the surface are more hydrolytically active due to the orientation than are those in bulk phase. Products of ester hydrolysis adsorb on the surface and block it, breaking the hydrolysis reaction. In this connection, the bulk concentration of the hydrolyzed ester is very low and cannot be registered by common methods of analysis.

Thus, the adhesive contacts the substrate via a layer of substances that frequently differ from the adhesive in composition. If the cohesion strength of these substances is less than that of the adhesive, this will determine the failure stress of the adhesive-bonded joint. The adhesive, which has the same composition as that of the adhesive in bulk, can form a weak zone in the substrate surface. Adhesives are polymers and the particular nature of a polymer must have effects at all stages of formation and operation of an adhesive-bonded joint.

The effect of the solid body surface propagates a considerable distance from it; that is, the influence of the surface on the chains that are in direct contact with it propagates via other chains into the bulk of the material. The range of action of the surface forces is a consequence of changes in the intermolecular interactions between chains that are directly adjacent to those in contact with the surface. Two factors limit the molecular mobility of chains close to the boundary: adsorption reactions of macromolecules with the surface and decrease of their entropy. Close to the boundary, the macromolecule cannot adopt the same number of conformations as in bulk, so that the surface limits the geometry of the molecule. As a result, the number of states available to the molecule in the surface layer decreases. These limitations on conformation are the primary reason for the decrease of molecular mobility close to the boundary [32].

A result of the range of action of the solid body surface is that an area of adhesive of considerable depth and close to the boundary is involved. It governs the details of the structure of the adhesive surface layer as well as the change in many physical-chemical properties of the adhesive and necessarily influences the adhesion strength.

The effect of the solid body surface on the structure of the polymer boundary layers has been considered in detail in a great number of publications and we will not consider this issue in detail. We point out only that a number of publications have reported correlation between the structure of the boundary layer and the adhesion strength for couples such as metal–polycaproamide coating [33], fluoroplastic–steel [34], and epoxy rubber polymers–metals [35].

In the course of formation of the adhesive-bonded joint, internal stresses appear in the adhesive layer. These stresses can change the process of formation of the polymer boundary layer and cause the formation of faults. With increase of the internal stresses in polystyrene, the softening temperature of the oriented films decreases. The same phenomenon has been observed with other polymers [36].

The faulty polymer layers on the boundary with the substrate can have low adhesion strength and serve as a source of disintegration of the adhesive-bonded joint. It was shown [37, 38] that in this case the polmer remains on the substrate as a very fine film or as separate spots, its quantity increasing with the time and temperature of the contact of the adhesive and the substrate.

In many cases precision methods allowed detection of the thin layer of adhesive on the substrate surface [37–39]. For example, using disturbed total internal reflection it was shown [40] that when polyethylene was peeled off the quartz a thin layer of the polymer remained on the latter. The remnants of the polyethylene are completely washed off the quartz by xylene, which is proof of the absence of interaction between the quartz and the polyethylene. Such disintegration of adhesive-bonded joints is determined both by the adhesion interaction between the polymer and the substrate and by differences between the behavior of the polymer when in thin layers on the substrate and when in the free state [85–89].

From the results described, one can conclude that a significant change of the mechanical properties of a polymeric material on a substrate or with a solid filler surface present can be expected in rather thin layers or in the thin interlayer of the polymeric matrix in filled polymers when the thickness becomes comparable with the thickness of the boundary layer.

Studies of the dynamic mechanical properties of polymeric objects characterized by the adhesion bond with a solid surface found that the gradient of the segmental mobility and of the indices of the mechanical properties was close to the interphase boundary during withdrawal from the solid surface.

Study of the mechanical properties of the boundary layers of reactive plastics, epoxy polymers in particular, is of special interest in that undercure of the compound is observed on the surface of the solid body; the primary reason for this is selective adsorption of components of the reactive system, which results in violation of stoichiometry of the initial products close to the solid body surface.

Let us consider the elastic properties of epoxy polymer thin films on substrates with various surface energies. The elasticity of the boundary layers and its contribution to the elasticity of the film on the substrate were assessed by the dependence of the modulus of elasticity on the thickness of the polymeric coating formed on high-energy (aluminum) and low-energy (polyethylene terephthalate) surfaces. ED-20 epoxy resin (molecular weight 420, epoxy number 21.6) was selected for study; it was cured onto the substrate with PEPA (molecular weight 120); the weight ratio of components was 10:1.

Analysis of the known methods for determining the modulus of elasticity of fine films on substrates, especially high-modulus films, showed that either they are of insufficient precision or they require the construction of special apparatus because of the complexity of processing the results obtained [91–93]. In this case, the modulus of elasticity was determined under conditions of tension at constant rate by the difference of results of tests on the epoxy coating with the substrate and on the substrate itself. The procedural problems are related to the fact that to obtain authentic results it is necessary that the forces that appear in the high-modulus metallic substrate and in the low-modulus polymeric coating when being deformed are comparable. This makes it necessary to use a very fine metallic foil as substrate, the manufacture of which is very intricate. For this reason the modulus of elasticity of the epoxy coating was determined using a combined substrate that comprised a film made of polyethylene terephthalate coated with a super-thin aluminum layer 0.05 μm thick by vacuum condensation. Both sides of such a substrate were used as models of the high-energy (aluminum) and low-energy (polyethylene terephthalate) surfaces for the application and formation of the coating under study.

To obtain thin coatings, the epoxy resin was diluted with acetone. The coatings, applied onto the substrate by special doctor blades, were cured for 20 h at 296 K and then for 15 h at 333 K. The thickness of the cured coatings was determined by gravimetry.

For mechanical tests, strip-shaped specimens of width 10 mm and test length 100 mm were used. The experiment was carried out using the FU-1000e (produced in former East Germany) general-purpose tensile testing machine, which performed automatic recording of the extension curves at a deformation rate of 5 mm/min. To enhance the accuracy of determination of the modulus of elasticity, it was calculated from extension diagrams as the ratio of the stress increment to the relative deformation increment when the latter changed from 0.2% to 0.5%. A series of 30 similar specimens were tested on the high-energy surface, but for the thinner coatings the picture was reversed. With a 5 μm thick coating on the polyethylene terephthalate, its modulus of elasticity was 1.34 times higher than when the coating was applied to aluminum; at 1 μm thickness the film on aluminum became of higher modulus, the difference between moduli reaching 38%.

It should be noted that for thin coatings the dependence under consideration is approximately linear. Extrapolation to thinner coatings results in a modulus of elasticity of $(60–70) \times 10^3$ MPa for the coating on the aluminum substrate and of $(26–29) \times 10^3$ MPa for the coating on the polyethylene terephthalate substrate at zero thickness; in other words, these values are close to the modulus of elasticity of the respective substrates onto which they are applied.

If such an extrapolation is reasonable, it means that the modulus of elasticity of thin films formed on various substrates depends on that of the substrate itself, i.e. the mechanical behavior of the low-modulus polymeric film for slight deformations approaches the behavior of the higher-modulus material of the substrate. The dependence obtained for the coating modulus of elasticity on the coating thickness in the region of slight thickness concurs with the existence of the gradient of the structural and the segmental mobility in polymers close to the boundary with the solid surface. The increase of the modulus of elasticity of the epoxy coating when applied to a substrate in line with the decrease of its thickness can be related not only to the energy and entropy effects of the substrate on the process of the structure formation and on the final structure of the polymer boundary layers but also to the effect of the adhesion interaction on the process of their deformation together with the substrate [85–89], which results in effects of combined strengthening and increase of deformability of polymeric films. In this case the lower values of the modulus of elasticity of the epoxy coatings on the high-energy substrate compared with the moduli for coatings on the low-energy substrate at coating thicknesses of 2–20 μm can be explained by a small contribution to the elasticity of the coating of the weak boundary layers [94, 95] formed, in line with formation of the polymer on the high-energy surface.

A screening effect is also possible, which decreases the range of transmission of the externally fed deformation energy from the solid surface of the higher-modulus substrate to the boundary layer and farther to more remote layers of the polymeric coating via adhesion bonds. Statistical processing of the results of the mechanical tests showed that the relative error in determining the modulus of elasticity of the coatings on the substrate does not exceed 10% for the confidence coefficient of 0.95. The experiments obtained the dependence of the modulus of elasticity of the epoxy coating on its thickness within limits of 1–100 μm as presented in Fig. 1.6. The tested coating was applied to the aluminum and polyethylene terephthalate surfaces of the substrate.

It is evident that decreasing the coating thickness down to about 10 μm has a slight effect on the value of the modulus of elasticity of the epoxy coating, independently of the nature and the energy field of the substrate surface. Apparently, the modulus of elasticity of the polymeric coating in this range of thickness corresponds to that for this material in bulk, where the contribution of the boundary layer is not practically exhibited.

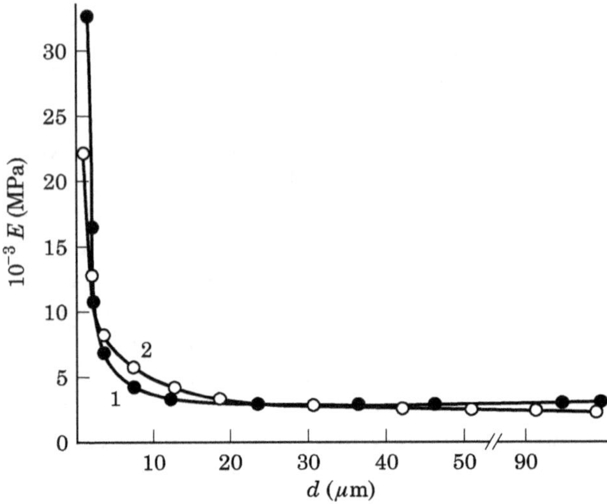

Figure 1.6 Dependence of the elastic modulus of the epoxy coating on thickness on (1) aluminum and (2) polyethylene terephthalate substrate.

Subsequent thinning of the coating on both substrates results in regular increase of the coating modulus of elasticity that is especially abrupt at thicknesses less than 3–4 μm in addition to that for the range of film thicknesses of 2–20 μm where the elasticity of the epoxy coatings on the low-energy surface appears to be higher. To determine the quantitative contribution of the adhesion interaction of the epoxy coating with a substrate to the value of the modulus of elasticity, epoxy coatings in the shape of free films 100 μm thick, formed under similar conditions on both sides of the combined substrate as well as on aluminum foil, were tested. The epoxy coating was separated by dilution of the PET film with phenol at 313 K and sulfuric acid was applied to the aluminum for a short period. An experimental check of the short-term effect of these reagents on the thick epoxy films did not detect any noticeable change in their elasticity (Table 1.2).

The results show that the modulus of elasticity of the thickest of the studied films on the high-energy substrate appeared to be 20% and for the low-energy substrate 13% higher than that of the films tested without substrates. This is evidence that, even for large thicknesses of coatings in which the boundary layers have no effect on the mechanical properties, the adhesion bond of the coating with the substrate has a noticeable effect on the coating modulus of elasticity. However, for the above experimental conditions the effect cannot be considered significant. Obtaining quantitative data on the role of such a factor with

TABLE 1.2 Dependence of the Elastic Modulus of Epoxy Films on Method of Production

Coating	Modulus of elasticity, $10^{-3}E$ (MPa)
On the surface of the combined substrate made of	
Aluminum	3.02
Polymer	2.49
Separated from the surface of the substrate made of	
Aluminum	2.51
Polyethylene terephthalate	2.20
Separated from the aluminum foil	2.43

deformation of the thinner films on the substrates is of significant interest for understanding the mechanism of deformation and failure of the coatings and adhesive-bonded joints, but it presents serious procedural problems, which were not overcome in the present studies. There are grounds for assuming that, with further decrease of thickness of the coating film on a substrate, the contribution of this effect will increase, and that for very fine films the elastic behavior of the polymer–substrate system will be determined by the combined influence of the adhesion interaction forces on the phase boundary and the particular features of the cohesion behavior of the boundary layers.

Thus, despite unsoundness of the structure of the polymer boundary layers, their mechanical properties can be high. The zone of failure of the adhesive-bonded joint in this case will depend on the correlation of the weakening and strengthening effects of the substrate on the polymer layer in contact with it.

Chapter

2

Adhesive Properties Control by Surface-Active Substances

2.1 Alteration of Properties of Polymeric Composites under the Influence of Surface-Active Substances

Control of adhesion interaction by the addition to adhesives of surface-active substances (surfactants) is of great theoretical and practical interest. The particular effects of surfactants lie in their ability to decrease the surface tension of the solution due to positive adsorption on the surface. Coating the surface of solid bodies and of liquids with the finest layer of a surfactant added to the system in very small quantities permits changes of the conditions of phase interaction and the progress of the physical-chemical processes.

Increase of adhesion strength is observed when there is adsorption of a monolayer of surfactant, for example, of a fatty acid, between the adhesive and the substrate. The monolayer can be applied by the Langmuir method, by a mechanical-chemical procedure that consists in polishing the metal surface using pastes based on the particular surfactant, or by dipping the specimens into the surfactant solution. The results obtained are explained by the fact that in the case of adsorption of diphilic molecules there appear areas of close-packed molecules that prevent formation of the weak boundary layers. Using tagged stearic acid it was determined that the failure of polyethylene–metal adhesive-bonded joints with the metal coated with a monolayer of stearic acid is always accompanied by cohesion failure of the polyethylene. In the region close to the substrate modified by surfactant a high degree of order of the polymer is observed.

Microscopic study of the cross-section of polyethylene from which aluminum had been separated (by dissolving in alkali) showed that crystalline formations grouped perpendicularly to the surface were located in the area previously in contact with the metal surface. This layer is 50 μm thick; it was not observed on the reverse side of the film. Apparently, the hydrophobic parts of the molecules of the adsorbed surfactant oriented perpendicular to the metal surface stimulated a corresponding orientation of the macromolecules of the polymer melt later applied. Treating the substrate surface with multilayer adsorption films decreases the adhesion. This seems to occur not because of decrease of interaction of the diphilic molecules with the substrate surface but because of insufficient interaction between separate monolayers. Surfactant monolayers on solid surfaces have high physical-mechanical properties. For example, a monolayer of stearic acid on a steel surface has a modulus of elasticity of 70.0 MPa under axial compression. Liquidlike properties are inherent to the surfactant polymolecular layer, but the layer can be of high viscosity. Measurement of the bonding forces between the solid particles coated with surfactant layers illustrated the multiple decrease of strength of the contacts with the surfactant present that is considered to be the result of screening of the cohesion interaction of the particles by the surfactant layers; this can be considered a direct display of the mechanical properties (strength) of these layers, i.e., of their ability to serve as structural-mechanical barriers.

Adsorption modification of the surface of fillers and pigments to improve the quality of polymeric coatings has been widely used. With adsorption modification by surfactant, the nature of the solid phase surface and consequently the character of its interaction with the polymer change. These adsorption layers lyophilizing the surface of particles of the pigment or the filler bring together the molecular properties of the polymer and the filler. Regularities of the effects of surfactant used as adsorption modifiers of fillers upon the polymer adsorption and their control have been investigated by applying the corresponding surfactant. The surface of filler "polymerophilized" by surfactant initiates adsorption of the polymers from solution. The adsorption (initiated by dichlorostearic acid) of perchlorovinyl resin from the dichloroethane is assumed to be caused by stretching out of the globules by the activated surface of the adsorbent; when the surface of the filler is partially filled with the surfactant molecules, the polymer is able to chemosorb onto the unmodified portions of the surfaces.

Surfactants are often added into paint materials to enable the painting of wet surfaces. Optimal concentrations of surfactant are not the same for different binders. One should always use the minimum possible concentration at which good spreading of paint over the wet surface is provided because excess of surfactant decreases the strength and water resistance of the coating.

The ability of surfactants to force spoiling off surfaces and of penetrating into pores and capillaries is widely exploited in the application of coatings and in cementing bodies of porous and vesicular structure (wood, concrete, etc.).

The use of curing agents for polymeric adhesives and coatings with surface-active properties is of considerable interest. These include curing agents for epoxy resins like oligoamineamides (OAA) or versamides. Such substances change the surface tension of the epoxy resins and their wetting capacity is enhanced by the fatty alkyl chain and by the polar amine groups; in structure they are similar to detergents. The curing of epoxy resins by liquid polyamide resins gives products that have good wetting properties and better adhesion and elasticity compared with epoxy resins cured by polyamines. They can be applied on wet surfaces because, in the course of applying the coating, water on the surface of the solid body emulsifies in the coating and is removed during drying.

Nevertheless, it should be pointed out that there is comparatively little evidence of adhesion enhancement when surfactant is added to adhesives and paints. In the majority of cases, addition of surfactant to polymer results in decrease of adhesion—in practice this is frequently used to obtain coatings that hinder aircraft icing, compounds for removing paints, and so on. Modification of the oxidized surface of aluminum by the majority of tested surfactants (such as octadecylamine; stearic acid; ethyltriethoxysilane; γ-aminopropyltriethoxysilane, AMG-9) decreases the short-term strength of the adhesive-bonded joints, and the more modifier is adsorbed the greater is the decrease. The drop in strength of the bond of the filler particles with the binders is used to reduce the internal stresses in coatings when the filler particles are modified by surfactant.

Frequently, surfactants increase the adhesion of polymers only within a narrow range of concentrations, below which the surfactant has practically no effect and above which the adhesion decreases. The effect of surfactant is well illustrated by the data presented in [62]. The adhesion strength was determined by the method of peeling polyethylene containing different types of added surfactant off aluminum. Typical results were:

	Content of surfactant (%)					
Surfactant	0	0.05	0.1	0.2	0.3	0.5
OS-20	50	–	80	200	100	0
DS-10	50	80	200	340	200	20

It is seen that the effect of the surfactant on the adhesion strength of polymer is related not only to the surface but also to the bulk properties of surfactant. Let us now pass to considering the details of surfactant behavior in oligomeric and polymeric solutions.

2.2 Colloid-Chemical Properties of Surfactants in Heterochain Oligomers

Consider the properties of reactive surface-active (RS) substances (alkyl phenol oxyethylated ester (OP-10), dodecyl nitrile (DDN)) and of chemically indifferent surface-active substances (IS) like the oligoamidoamines L-18, L-19, L-20 capable of reacting with such oligomeric solvents as the ED-20 and DEG-1 epoxy resins and polyoxypropylenetriol (POPT) of molecular weight 750. These oligomers are unassociated low-polarity liquids, as confirmed by the data of Table 2.1.

The total surface energy E_n, the surface molar entropy S_m, and the parachor P were calculated by the equations

$$E_n = \gamma - T\frac{d\gamma}{dT} \tag{2.1}$$

$$P = V\gamma^{1/4} \tag{2.2}$$

$$S_m = f\,V^{3/2}N_A^{1/2}\frac{d\gamma}{dT} \tag{2.3}$$

where V is the molar volume, N_A is the Avogadro number, f is the packing factor, γ is the surface tension, and T is temperature. The

TABLE 2.1 Thermodynamic Properties of Oligomeric Solvents

	DEG-1 Temperature (°C)		POPT Temperature (°C)		ED-20 Temperature (°C)	
Property	20	60	20	60	20	60
Surface tension, γ (mN/m)	44.7	40.66	36.13	30.50	48.9	46.2
Density, $10^{-3}\rho$ (kg/m^3)	1.157	1.1367	1.0111	0.972	1.145	1.133
Total surface energy, E_n (mN/m)	74.0	74.0	76.48	76.30	91.3	91.2
Parachor P[a]	560 (550)	557 (550)	1920 (1923)	1915 (1923)	1010 (1006)	1010 (1006)
Molar surface entropy, S_m (J/mol · K)	32.9	33.7	106.0	109.5	65.4	36.4

[a] Theoretical values of the parachor are given in parentheses.

theoretical values of the parachor are given in parentheses in Table 2.1. The surface tension of the oligomers at different temperatures and its dependence on the surfactant concentration were determined by the Wilhelmy method.

Absence of association in the oligomer is indicated by the fact that the values of the parachor (calculated from the experimental data and from the summation of atoms, bonds, and groups in a molecule according to the additive rule) coincide and hardly depend on the temperature. A similar picture was observed for the surface molar entropy. This indicates that the extent of macromolecular association does not change within the studied range of temperatures. The entropy of formation of the surface per cm^2

$$\Delta S = \frac{-d\gamma}{dT} \tag{2.4}$$

also shows little dependence on temperature.

The selected surfactants are surface-active with respect to the ED-20 and DEG-1 solvents but not toward POPT. In this case L-19 can be considered indifferent in that at 20°C its rate of interaction with DEG-1 is very low.

Figure 2.1 shows that on all the curves at a certain concentration of surfactant there is a distinct kink after which the surface tension of the solution becomes independent of the surfactant concentration. This is likely related to the formation of micelles of the surfactant molecules that occurs at the surfactant concentration equal to CCMF (the critical concentration for micelle formation). Thus, the surfactant solutions in the oligomers are subject to the same regularities as those in other solvents. For the OP-10 solutions in DEG-1 two kinks are observed on the isotherm; the kink at the lower surfactant concentration is usually related to preassociation.

Table 2.2 presents CCMF (C_k), areas occupied by the molecules in the extremely saturated adsorption layer (S_{max}), and surface activity (G) for OP-10 and L-19 in epoxy resins. CCMF was determined by the kink in the curves (Fig. 2.1). The areas occupied by molecules were calculated by the formula

$$S_{max} = \frac{1}{\Gamma_{max} N_A} \tag{2.5}$$

where Γ_{max} is the maximum adsorption determined by the equation

$$\Gamma = \frac{c_k}{RT} \frac{\delta\gamma}{\delta c_k} \tag{2.6}$$

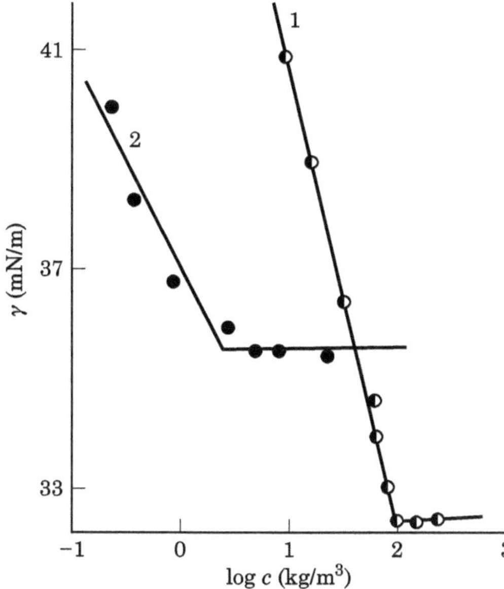

Figure 2.1 Isotherms of surface tension of DDM (1) and L-19 (2) solutions in DEG-1 at 20°C.

Surface activities were determined by

$$G = \frac{\gamma - \gamma_k}{c_k} \tag{2.7}$$

where γ is the surface tension of the oligomer, and γ_k is the same but at surfactant concentration equal to CCMF.

As Table 2.2 indicates, the areas occupied by the molecules of OP-10 and L-19 differ more than 1.5-fold, although these substances are of similar structure. One can assume that the OP-10 molecules form associations with the DEG-1 molecules that result in considerable increase of the area occupied by a molecule on the surface.

Let us calculate the area that should be occupied by OP-10 molecules in an extremely saturated adsorption layer. The effective area of the normal cross-section of the hydrocarbon chains for benzene derivatives (according to structural analysis) is $23.8 \times 10^{-10}\,m^2$, and the distance between the axes of two oriented chains in the saturated layers is not less than $(4.5–6) \times 10^{-10}\,m$. Consequently the effective cross-section for the OP-10 molecule with two hydrocarbon chains is about $(52.1–53.6) \times 10^{-10}\,m^2$. This value is less than that determined experimentally and practically coincides with the area occupied by the L-19 molecule in the adsorption layer. Thus, L-19 does not form asso-

TABLE 2.2 Some Properties of Surfactant Adsorption Layers in Oligomers

	DEG-1–L-19			DEG-1–OP-10			ED-20–OP-10		
Temperature (°C)	c_k (kg/m^3)	$10^{-10}S_{max}$ (m^2)	G (mN·m^2/kg)	c_k (kg/m^3)	$10^{-10}S_{max}$ (m^2)	G (mN·m^2/kg)	c_k (kg/m^3)	$10^{-10}S_{max}$ (m^2)	G (mN·m^2/kg)
20	2.82	52.8	3.34	24	84.4	0.346	–[a]	–	–
40	–[a]	–	–	26.3	135	0.195	5.63	87.2	1.37
60	–[a]	–	–	41.6	177	0.076	14.1	23.0	0.44

[a] Not applicable—see text.

ciations with the oligomer, although it includes both proton-donor and proton-acceptor groups. It can be assumed that L-19 molecules under close packing in the adsorption layers form hydrogen bonds with one another.

Let us now consider the dependence of the surfactant surface activity on temperature. The DEG-1–L-19 system was studied only at 20°C because at higher temperature the reaction between its components becomes noticeable. Taking account of high viscosity, ED-20 was investigated at 40°C and 60°C only. As Table 2.2 shows, elevation of the temperature results in a decrease of the surface activity and an increase of OP-10 CCMF in the oligomers that is related to the increase of the surfactant solubility.

Alteration of the surface tension of the DEG-1–OP-10 system with temperature is greatly influenced by the surfactant concentration (Fig. 2.2). At 11 kg/m^3 of OP-10, the surface tension does not depend on temperature, and at high surfactant concentrations even increases. The OP-10 associations with DEG-1 must be of increased solubility and must easily desorb from the boundary surface in the course of raising the temperature. Thus, temperature increase may result both in decrease of the surface tension on account of increase of the thermal motion of the molecules and in its increase due to the desorption of surfactant molecules, and this increase has a direct dependence on the surfactant concentration. Superposition of these factors for certain surfactant concentrations can bring about independence of the system surface tension on temperature, confirmed by experiment. In the case of surfactant desorption, the system entropy increased due

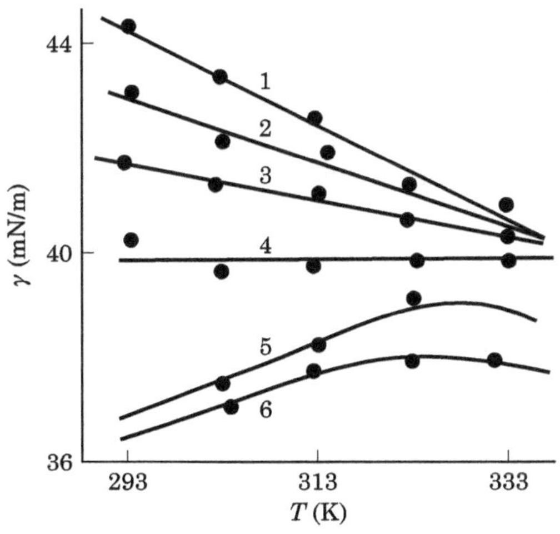

Figure 2.2 Dependence on temperature of the surface tension of OP-10 solutions in DEG-1 of various concentrations: (1) 0, (2) 0.4, (3) 2.0, (4) 4.0, (5) 20.0, (6) 40.0 g/l.

to more uniform distribution of surfactant molecules in the oligomer; when the adsorption increases, the system entropy decreases, as illustrated by the decline of the straight lines in Fig. 2.2.

Let us consider the effect of the formation of surfactant micelles on the thermodynamic compatibility of the system. The ED-20–OP-10 system was used for study. Using reversed-phase gas chromatography, thermodynamic parameters such as the excess free energy, enthalpy, and entropy of mixing were determined. In addition, the experimental values of the parameters of the Flory–Huggins thermodynamic interaction χ_{ij} were found. The molecular weights of the oligomers were determined by means of the ED-68 ebulliograph. The studies were performed with the LKhM-72 chromatograph with a flame ionization detector and an air thermostat. Nitrogen served as support gas. The AW 60/80 chromosorbent was used as a substrate. The films were applied from solution in methylene chloride.

The measurements were performed with extrapolation to zero dilution of the sorbate (heptene, benzene), The following equation was used to calculate the mixing thermodynamic parameters:

$$\ln V_{\mathrm{M}} = \sum x_i \ln V_{\mathrm{M}i} - \frac{\Delta G_{\mathrm{M}}}{RT} \tag{2.8}$$

where x_i is the mole fraction of the ith component of the blend that is used as stationary phase; V_{M} and $V_{\mathrm{M}i}$ are retention volumes of the sorbate by a mixed phase and by an individual ith component; and ΔG_{M} is the excess free energy of the mixed phase.

The blending excess enthalpy was calculated from

$$\Delta H_{\mathrm{CM}} = \sum \chi_i \Delta H_i - \Delta H_{\mathrm{M}} \tag{2.9}$$

where ΔH_{CM} and ΔH_i are the enthalpies of mixing the sorbate with the blend and with the ith component of the blend; ΔH_{M} is the excess enthalpy of the polymer–polymer blend.

The value of ΔH_{M} was determined from

$$\Delta H_{\mathrm{M}} = R \frac{\delta}{\delta(1/T)} \left[-\ln V_{\mathrm{q}} p_1 - \frac{p_1}{RT}(B_{1,1} - V_1) \right] \tag{2.10}$$

where p_1, $B_{1,1}$, and V_1 are the pressure of the saturated vapor, the second coefficient, and the sorbate molar volume at the column temperature, V_{q} is as defined previously.

The excess entropy of the blending was determined from

$$\Delta S = \frac{\Delta H_{\mathrm{M}} - \Delta G_{\mathrm{M}}}{T} \tag{2.11}$$

The parameter of the thermodynamic interaction between the sorbate and the blend was calculated in the following way:

$$\begin{aligned}\chi_{1,2,3} &= \left(\frac{\chi_{1,2}}{V_1}\varphi_2 + \frac{\chi_{1,3}}{V_1}\varphi_3 - \frac{\chi_{2,3}}{V_2}\varphi_2\varphi_3\right)V_1 \\ &= \ln\frac{273R(w_2v_2 + w_3v_3)}{p_1V_1V_{\mathrm{q}}} - \left(1 - \frac{V_1}{V_2}\right)\varphi_2 \\ &\quad - \left(1 - \frac{V_1}{V_3}\right)\varphi_3 - \frac{p_1}{RT}(B_{1,1} - V_1)\end{aligned} \tag{2.12}$$

Here $\chi_{1,2}$ and $\chi_{1,3}$ are the parameters of the thermodynamic interaction between the sorbate and the individual components of the blend; $\chi_{2,3}$ is that between the components of the blend; the φ_i are the volume fraction of the components; the V_1 are the molar volumes of the sorbate and the blend components; w_2, w_3 and v_2, v_3 are the weight fractions and the specific volumes of the blend components. For the individual components χ was determined on separate columns. Numerical values were calculated from

$$\chi = \ln\frac{273RV_2}{V_{\mathrm{q}}p_1V_1} - \left(1 - \frac{V_1}{M_2V_2}\right) - \frac{p_1}{RT}(B_{1,1} - V_1) \tag{2.13}$$

where M_2 is the molecular weight of the stationary medium.

When surfactant is added to the oligomer, the specific holding volume of the oligomer begins to decrease; it subsequently increases, but up to 30% of surfactant content it remains less than the volume calculated under conditions of additive dependence on the blend composition.

The data presented indicate the complex thermodynamic behavior of the system. Actually, as Fig 2.3 shows, the Flory–Huggins thermodynamic interaction parameter between the system components has a value lower than the critical value in the region of surfactant concentrations up to 30% that characterizes the thermodynamic stability of the system. The $\chi^*_{2,3}$ critical parameter for the given oligomeric system was calculated using the equation

$$\chi^*_{2,3} = \frac{1}{2}\left[\left(\frac{V_1}{V_2}\right)^{1/2} + \left(\frac{V_1}{V_3}\right)^{1/2}\right]^2 \tag{2.14}$$

and was equal to 0.452.

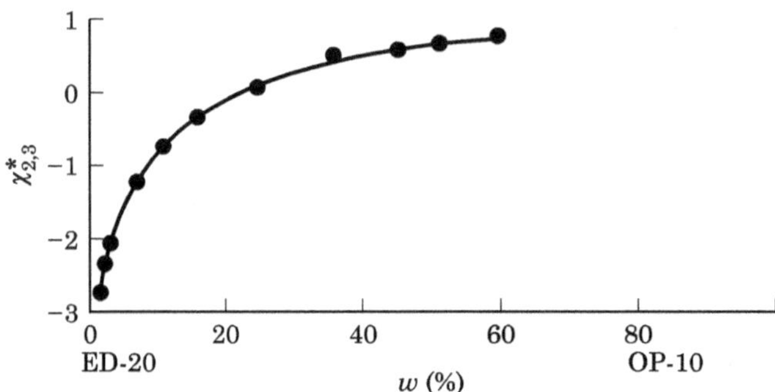

Figure 2.3 Dependence of $\chi^*_{2,3}$ on composition for ED-20–OP-10 system at 60°C.

The maximum thermodynamic stability is displayed by the system when the surfactant concentration is equal to CCMF. With increase of the surfactant concentration, the interaction parameter increases and the system becomes thermodynamically incompatible.

The enthalpy within the studied range of relationship of the blend components has negative value; blending ED-20 and OP-10 results in release of heat, although from the enthalpy value one cannot unequivocally judge the state of the system. The system is characterized more completely by the change of the excess free energy ($\Delta G_M = \Delta H_M - T\Delta S_M$) in the course of blending the components because this accounts for both the energy and entropy state of the system.

Increasing the surfactant concentration in the system results in a decrease of the components" compatibility that is characterized by the increase of the excess energy of mixing, although the system becomes thermodynamically incompatible only when the surfactant content is about 30%.

Thus, change of the oligomeric properties under the influence of surfactant does not in principle differ from that when aqueous or hydrocarbon solvents are used. Formation of associations of surfactant with oligomer abruptly increases the area occupied by a surfactant molecule in the extremely saturated adsorption layer and results in decrease of the surfactant surface activity and in the anomalous dependence of surface tension on temperature.

The study of thermodynamic parameters of mixing showed that the oligomer–surfactant system is thermodynamically stable even when the surfactant concentration is considerably above CCMF. This is a surprise and seems to be related to the fact that the surfactant molecules aggregate into the micelles in such a way that those parts of the surfactant molecules are directed at the oligomeric solvent that have maximum affinity to it.

Now let us consider the colloid-chemical properties of the oligomer–surfactant system under conditions in which the system components are capable of chemical interaction.

2.3 Surface Tension of Heterochain Oligomers with Surfactant Additives

The surface activity of a surfactant is determined first by the relationship between its hydrophilic and hydrophobic parts, or the hydrophilic–oleophilic balance. In the course of interaction of a surfactant with an oligomer, the surfactant's hydrophilicity and hydrophobicity, its structure and hydrophilic–oleophilic balance will change depending on its polarity and where in the surfactant molecule the active groups are located. But independently of this, such an interaction always results in increase of the surfactant solubility in the oligomer and in decrease of its surface activity.

The interaction between epoxy resin and OAA (oligoamide amine) occurs as follows:

$$
\begin{array}{l}
CH_3(CH_2)_5-CH-CH-CH{=}CH(CH_2)_7CONH(CH_2NH)H \\
\qquad\qquad\quad | \qquad\; | \\
\qquad\qquad\quad | \qquad CH(CH_2)_7CONH(CH_2CH_2NH)H \\
\qquad\qquad\quad | \qquad\; | \\
\quad CH_3(CH_2)_5CH \quad CH
\end{array}
$$

$$
+\,2\,\underset{\diagdown\; O\;\diagup}{CH_2-CH}-R-\underset{\diagdown\; O\;\diagup}{CH-CH_2} \longrightarrow
$$

$$
\begin{array}{l}
CH_3(CH_2)_5-CH-CH-CH{=}CH(CH_2)_7CONH(CH_2CH_2NH)CH_2-CH-OH \\
\qquad\qquad\quad | \qquad\; | \qquad\qquad\qquad\qquad\qquad\qquad\qquad\qquad\quad | \\
\qquad\qquad\quad | \qquad CH(CH_2)_7CONH(CH_2CH_2NH)CH_2CH \qquad\qquad R-\underset{\diagdown\; O\;\diagup}{CH-CH_2} \\
\qquad\qquad\quad | \qquad\;\diagdown \qquad\qquad\qquad\qquad\qquad\quad \diagup\;\diagdown \\
CH_3(CH_2)_5-C-CH{=}CH \qquad\qquad\qquad\quad OH \quad R-\underset{\diagdown\; O\;\diagup}{CH-CH_2}
\end{array}
$$

Figure 2.4 presents the isotherms of the surface tension of the DEG-1–surfactant system at various time intervals after mixing. The reaction between DEG and surfactant was carried out at 80°C, with subsequent cooling of the system down to 20°C for determination of the surface tension. At 20°C the rate of reaction between DEG-1 and OAA is low and the changes in surface tension in the course of measurement (for less than 0.5 h) are within the limits of the experimental error. Figure 2.4 shows that with the increase of the interval from the time of blending up to the measurement of the surface tension the kink on the isotherms shifts in the direction of higher surfactant

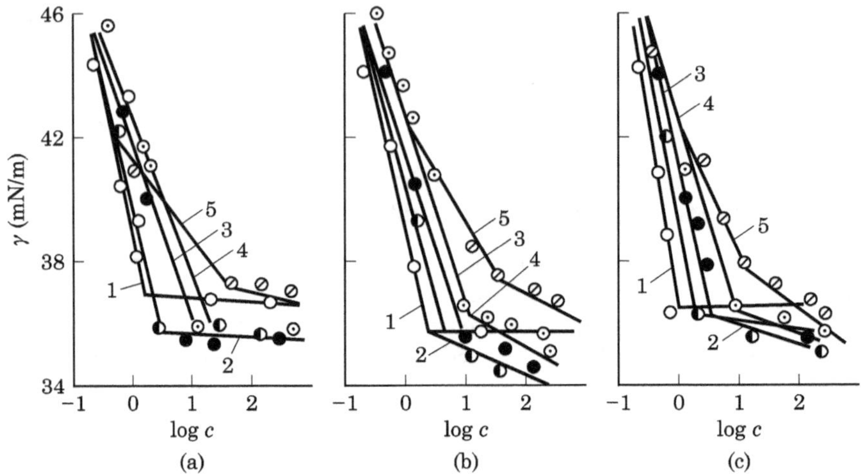

Figure 2.4 Isotherms of the surface tension of oligoamines (a) L-20, (b) L-19, and (c) L-18 at various time points: (1) after blending; with heating at 800°C for (2) 1 h, (3) 3 h, (4) 5 h, and (5) after completion of the reaction.

concentrations, i.e., CCMF of the surfactant increases in line with the increase of the system degree of conversion.

Let us consider the reasons for this. In addition to the increase of surfactant affinity to the solvent, CCMF may increase due to the growth of the micellar mass and to the decrease of the tendency to aggregation because of increase of branching of the surfactant molecules. The contribution of the micellar mass change can be calculated under the condition of invariance of other factors of OAA and products of reaction (PR) of OAA with resin by Equation (2.15) [67]:

$$\mathrm{CCMF}_{\mathrm{PR}} = \mathrm{CCMF}_{\mathrm{OAA}} \frac{M_{\mathrm{PR}}}{M_{\mathrm{OAA}}} \tag{2.15}$$

where M_{PR} and M_{OAA} are the molecular weights of the products of reaction and of the oligoamide amines.

In calculating the molecular weight of the reaction products one can assume that with high probability there is no interaction between the reaction products and OAA, because of the great excess of the reactive epoxy groups in the system. In our case the molar ratio of OAA to DEG-1 did not exceed 16.7×10^{-3}. The contribution of the growth of branching of the surfactant molecules to the increase in CCMF can be accounted for if one takes into consideration that CCMF increases 1.5–2 times when there is surfactant branching.

Table 2.3 presents the colloid-chemical characteristics of OAA and of PR; it is evident that, although PR themselves are surfactants, their micellar mass is substantially less than that of OAA. The aggregation number for PR is less than 2, i.e., the largest aggregates consist of two surfactant molecules. Evidently the PR aggregates cannot be considered as micelle formations, which is confirmed by the fact that the lower branches of the curves in Fig. 2.4 are not parallel to the concentration axis. Of particular interest is the fact that sphericity of molecular associations does not change in the course of the reaction. This seems to be related to the branching of the polar part of the surfactant molecule and to its location in such a way that the nonpolar parts are inside the protomicelles. This conclusion follows from the invariance of the effective volumes of 1 g of OAA and PR calculated by the Gut and Simkh equation when varying the surfactant concentration.

The data in Table 2.3 suggest that the primary contribution to increase of the system CCMF is made by the increase of the surfactant solubility. In fact, $CCMF_{PR}$ estimated by the micellar mass for conditions where surfactant molecule branching is more extensive than that of OAA, is 3-fold lower than CCMF determined experimentally. Thus, at least 2/3 of the CCMF growth is caused by increase of the surfactant affinity for the solvent.

The decrease of the aggregation of surfactant particles in the course of reaction is confirmed by light scattering studies. A study using the turbidity spectrum showed (Fig. 2.5) that the decrease of the colloid particle concentration occurs in line with the decrease of the colloid particle size. The decrease of the particle size is also confirmed by the data from viscosity–diffusion (Table 2.3).

Let us review the information acquired. As shown above, the solubility of PR in the system is higher than that of OAA. At the beginning of the reaction when the PR concentration is low, the PR are in the molecular-dispersed state, which is why they do not affect the determined size of the colloid particles in the system. As Fig. 2.5 shows, in the early stages of the reaction there is an abrupt decrease of the particle concentration while their size does not change significantly. Once the concentration of the PR molecules formed exceeds CCMF, a decrease of the particle size is observed such that the size of the PR aggregates is smaller than that of OAA.

The data permit an easy explanation of the regularities of the change in surface tension during the course of DEC-1 reaction with various quantities of OAA (Fig. 2.6). The surface tension is seen to increase with time at low content of OAA in the system; at higher OAA concentration the surface tension increases less strongly, and at OAA concentration of 5% change of the surface tension is not observed. In the initial stage of the reaction there is a decrease of

TABLE 2.3 Surfactant Colloid-Chemical Characteristics

	Before reaction				Products of reaction				Estimated $CCMF_{PR}$ (mol/m^3)	
OAA branching	Molecular weight[a]	CCMF (mol/m^3)	Micellar weight[b]	Aggregation number	Molecular weight	$CCMF_{PR}$ (mol/m^3)	Micellar weight	Aggregation number	By micellar weight	By
L-18	770	1.16	–	–	1070	6.61	–	–	–	–
L-19	860	1.63	2.73	31.7	1940	9.75	0.180 to 0.213	1.40 to 1.66	0.127	3.26
L-20	940	1.2	–	–	1690	7.81	–	–	–	–

[a] Defined from IR-spectroscopy data.
[b] Defined by a method of viscosity–diffusion.

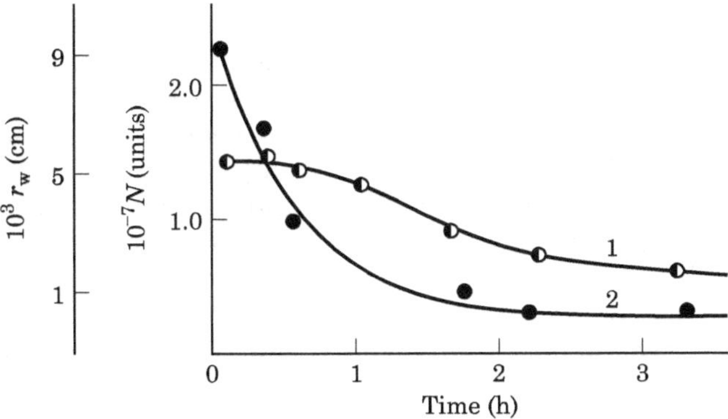

Figure 2.5 Change with time of size r_w (1) and number concentration N (2) of colloid particles in the course of reaction (at 800°C) for a 7.5% solution of L-19 in DEG-1.

the surface tension, this stage occurring for longer time at higher surfactant concentration.

As noted above, the colloid-chemical behavior of RS substances does not differ from that of IS substances when there is no chemical interaction of RS substances with oligomer. Thus, the change in OAA properties in the process of reaction is related only to the change in their structure. The interaction of OAA with the oligomer results in

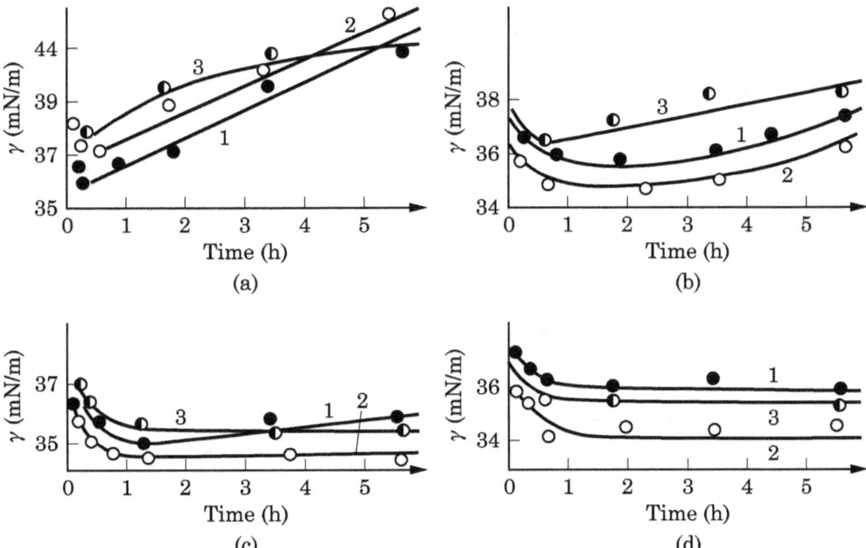

Figure 2.6 Change with time of the surface tension of surfactant solutions [(1) L-18; (2) L-19; (3) L-20] in DEG-1 in the course of reaction at various surfactant concentrations: (a) 0.1%, (b) 0.5%, (c) 1%, and (d) 5%.

increase of the surfactant solubility. In this case its surface activity estimated by Equation (2.7) decreases 150–170 times. This results in a decrease of the surfactant excess adsorption that causes increase of the solution surface tension. When the surfactant concentration in the system is higher than CCMF, the disintegrating micelles of OAA serve as a source of surface-active molecules. Although the increase of the PR solubility results in the decrease of their relative adsorption (see Table 2.4), the absolute adsorption attainable also increases because of the increase of CCMF; that is, by increasing the PR concentration in the system one can achieve greater decrease of the surface tension than with OAA despite the much lower surface activity of PR.

In OAA interactions with the oligomer there is formation of large branched molecules whose cross-section S is much greater than that of the initial OAA (see Table 2.4)

The surface tension of OAA solutions in diethylene glycol, a non-reactive analogue of DEG-1, decreases linearly with increasing temperature for OAA content up to 5%. The increase of PR solubility compared with OAA means that as the system temperature is raised PR desorb from the boundary surface. This initiates the surface tension rise that is superimposed on the surface tension decrease due to increase of molecular thermal motion. The dependence of the surface tension on temperature for various PR concentrations is well described by a straight-line equation.

2.4 Surface Tension of Curing Oligomers

The chemical composition as well as molecular weight and structure changes when the oligomer is being cured. In the course of reaction of the epoxy resin with amines, the epoxy groups are replaced by hydroxyl groups; when polyesters react with isocyanate, the hydroxyl and isocyanate groups are replaced by urethanes, and so on. Such a change in the solvent necessarily affects the surfactant's affinity for it, which in turn results in change of the micellar mass and CCMF

TABLE 2.4 Adsorption Properties of Surfactants

	OAA		PR	
Initial surfactant	$10^{-14}\Gamma$ (mol/m^2)	$10^{10}S$ (m^2)	$10^{-14}\Gamma$ (mol/m^2)	$10^{10}S$ (m^2)
L-18	3.66	45.6	2.0	83.5
L-19	3.04	52.8	1.6	104.0
L-20	3.04	52.8	1.98	83.5

[70, 97, 98]. Consequently, the surface tension of the polymerizing system with added surfactant is affected both by changes of the structure and the chemical composition of the polymerizing oligomer and by the change of surfactant colloid-chemical properties. To obviate difficulties in separation of these two factors, DEG-1 epoxy resin was selected as an oligomeric solvent. The surface tension of DEG-1 when reacting with PEPA does not change up to a considerable degree of conversion. This allows us to relate the observed changes to the surfactant colloid properties only. Figure 2.7 presents isotherms of the surface tension of DEG-1 with DP-10 for various degrees of conversion of epoxy groups, which was controlled by the quantity of PEPA added.

Completion of the reaction and the degree of conversion of the epoxy groups were determined by IR spectroscopy. To determine the equilibrium value of the surface tension, a quantity of PEPA was added such that the extent of epoxy group conversion would not exceed 30–35%. In this case the resin remained in the liquid state.

At low degrees of epoxy conversion, CCMF of OP-10 does not change (see Fig. 2.7, curves 1–3). Only at 30% conversion does CCMF begin to decrease (curve 4). With increase of the degree of conversion, one observes the decrease of the system surface tension at the surfactant concentration that corresponds to CCMF. The kink on the surface tension isotherm at low surfactant concentrations can be related to the formation of the surfactant "pre-associations" in the system.

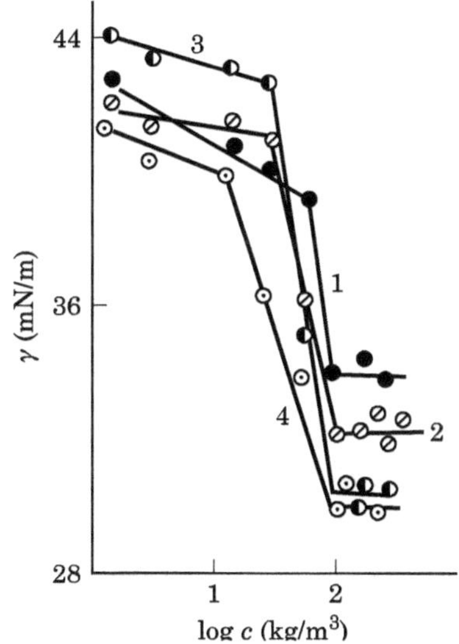

Figure 2.7 Isotherms of the surface tension of OP-10 solutions in DEG-1 at various conversion levels of the epoxy groups: (1) 5%, (2) 10%, (3) 20%, and (4) 30%.

Figure 2.7 shows that with increase in the degree of epoxy conversion, "pre-associations" occur at lower surfactant concentrations. The surfactant concentration at which "pre-micelles" and micelles are formed is related to the decline of surfactant solubility in the oligomer during its polymerization.

The size of micelles is determined by energetic factors and concentration factors. The decrease of CCMF in a viscous system must result in decrease of the size of the micelles. For increase of the surfactant concentration, the micelle size increases, and spherical micelles transform into elliptic ones and then into plate-shaped ones as increase in the surfactant concentration in common solvents is accompanied by a free energy advantage. It can be assumed that such a process occurs with the increase of the degree of conversion up to 30%. In this case CCMF shifts into the area of lower surfactant concentrations without any change in the surface tension (see Fig. 2.7).

Consider the change in the system surface tension with the extent of epoxy group conversion at different surfactant contents (Fig. 2.8). The transition of the surfactant molecules into the associated state with increase of the degree of epoxy groups conversion (with increase of the oligomer molecular weight) results in depletion of the surface layer of surfactant molecules by diffusion to lower-concentration regions, which causes the increase of the system surface tension (see Fig. 2.8, curves 2 and 3). Consequently, the formation of the

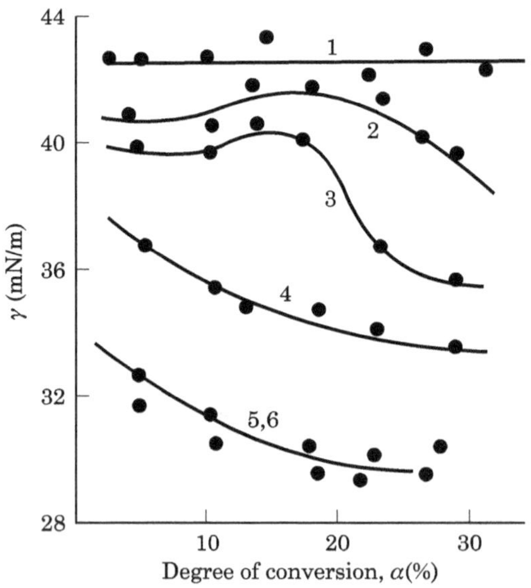

Figure 2.8 Dependence of the surface tension on the conversion level of epoxy groups for OP-10 solutions of various concentrations: (1) 0, (2) 1.2, (3) 20.0, (4) 40.0, (5) 60.0, and (6) 117 g/l.

"pre-associations" produces only a small advantage of the free energy that does not compensate the increase of the system surface tension due to the desorption of surfactant molecules. The formation of micelles at epoxy groups conversion level above 20–30% or at surfactant concentration over 40 kg/m^3 produces a large free energy advantage that causes the decrease of the surface tension along with the increase of the conversion level (see Fig. 2.8).

Figure 2.9 presents the results of a study of the surface tension of the POPT-based system. The polymerization level of the oligomer was controlled by the amount of toluene diisocyanate (TDI) added to the system, the reaction of which with the polyester resulted in the formation of the polyurethane. Completion of the reaction and the system conversion level were monitored by IR spectroscopy using the change of the adsorption band in the region of 2280 cm^{-1} that corresponds to isocyanate groups. The surface tension was measured at 90°C.

The surface tension of the system increases with increase of the concentration of urethane groups; DDN is not surface active with respect to POPT. The surface activity develops with the increase of the concentration of urethane groups (see Fig. 2.9). It seems to be caused by the decline of the solubility of DDN in the oligomer. With the increase of urethane group content, CCMF decreases, and this decrease is most abrupt in the region of considerable degrees of conversion of the system. A similar phenomenon was observed for the epoxy resin: the closer the system is to the gel-formation point, the worse is the surfactant solubility in it. It is probable that at the observed high levels of system conversion

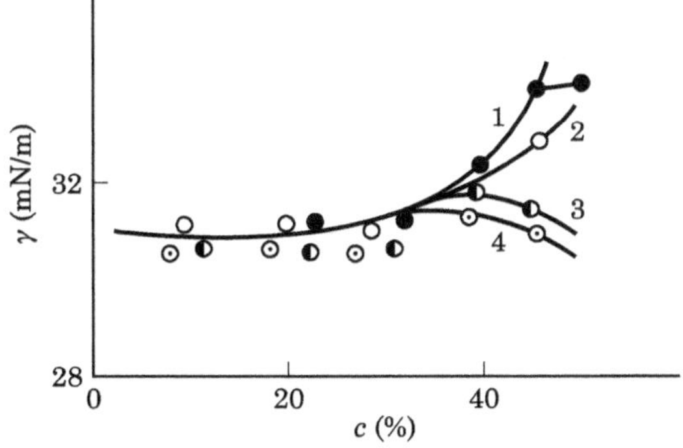

Figure 2.9 Dependence of the surface tension of DDN solutions on concentration of urethane groups: (1) 0, (2) 20, (3) 50, and (4) 100 g/l.

there appear the creation of the three-dimensional network, the compatibility of which with the low-molecular weight substances is sharply decreased.

The increase of aggregation of the surfactant molecules and the transformation of micelles of one type into another one was confirmed by a study of the light scattering of the surfactant solutions of various concentrations with various levels of system conversion. Note that the oligomers investigated involve colored products that hinder the detection of the small-sized colloid particles. The light scattering was determined in the course of reaction at 20°C. Since the system is in the liquid state and the polymerization rate is low, one can assume that the process of building up the micellar formations is close to the equilibrium.

Figure 2.10 shows the optical density of polymerizing systems containing various quantities of surfactant. There is a clear abrupt increase of the system optical density at certain degrees of conversion. With increased surfactant concentration the increase in optical density is observed at lower degrees of conversion. These data indicate that with increase of the surfactant concentration or of the system conversion level there is an increase in the quantity and size of the colloid particles formed in the system. At 15% surfactant in the system and at 10% conversion, there is a decrease of the optical density (curve

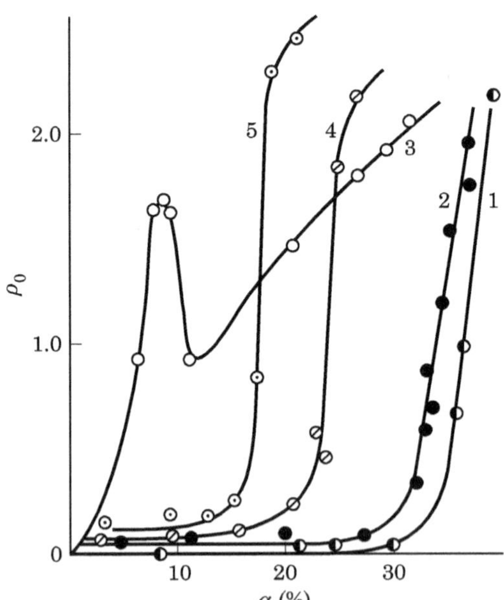

Figure 2.10 Change of the optical density ρ_0 in the course of polymerization of DEG-1 at various surfactant concentrations: (1) 6.5%, (2) 7.0%, (3) 15.0%, (4) 8.0%, and (5) 100% of DDN.

3). This is related to the formation of a great number of large colloid particles at such a high surfactant concentration. Since at 10% conversion the system viscosity is still low, the colloid particles coalesce and the system separates into two phases (polymer and surfactant). The separation of surfactant from the oligomeric solvent results in the decrease of the system's optical density.

Thus, there is a critical surfactant concentration at which incompatibility with the polymer occurs. This concentration depends on the system's degree of conversion. Similar processes can obviously be observed at degrees of system conversion higher than the gel formation point, with compatibility being evident at lower concentrations of surfactant than for systems in the liquid state. It is to be expected that the solid state of the system governs the specificity of the surfactant aggregation process.

At a certain degree of conversion of the system, the size and concentrations of the OP-10 and DDN particles increase in a jump, occurring at a lower conversion level for DDN than for OP-10 (see Fig. 2.11). This seems to be caused not only by differences in the nature of these two surfactants but also by the chemical alterations that arise under epoxy resin polymerization. Actually the formation of the hydroxyl groups in the resin increases its affinity with OP-10, which also contains hydroxyl groups. It shifts the onset of incompatibility in the direction of higher conversions.

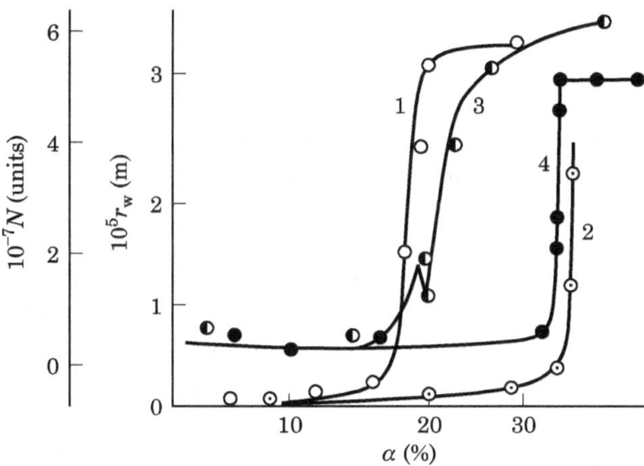

Figure 2.11 Change of size r_w (1, 2) and of number density N of colloid particles (3, 4) in the course of reaction of a 10% solution of DDN (1, 3) and a 14% solution of OP-10 (2, 4) in DEG-1.

The system splits into phases at the maximum possible size of the DDN colloid particles (see Fig. 2.11, curve 1). After the particles acquire this size, only growth is observed, with the particle concentration (at which the system separates into phases) increasing with conversion level and consequently with increase of the system viscosity. In the case of OP-10, the system incompatibility occurs at higher conversion levels, which is why the system separation into phases is not observed visually up to 40% of the epoxy group conversion.

The processes that occur in the polymerizing system have various effects on its ability to separate into macrophases. Increase of the molecular weight of the oligomer results in increase of the surfactant molar fraction in the system, which facilitates formation of the surfactant aggregates. As seen in Section 2.2, surfactant aggregation must provide for the compatibility of the colloid particles with the oligomer.

Upon achieving the limiting size of the colloid particles and with increasing polymerization of the oligomer, the concentration of particles per unit volume of the solvent increases, causing loss of stability and producing aggregation with formation of a separate phase. Separation of surfactant from the polymer results in the abrupt increase of its concentration in the surface layer; when the system becomes incompatible a decrease in its surface tension is observed.

The regularities of the surfactant behavior that have been described in the curing of epoxy oligomers are also seen in the study of other systems. Below a certain conversion level in the system of POPT cured by reaction with TDI and DDN, there is growth of the colloid particles, caused by the decline of the solubility of DDN in the oligomer. It is interesting that the increase of particle size is accompanied by a decrease of their concentration. It seems that from the energetic point of view the existence of particle of a certain comparatively large size is most advantageous and the amount of surfactant separated from the system is insufficient to compensate the decrease of the particle concentration due to the increase of their volume.

The information presented in this section demonstrates that the conditions of wetting of the substrate by the adhesive (which is the first and most important stage of formation of an adhesive-bonded joint), determined by the adhesive's surface tension, depend not only on the type of surfactant and its concentration but also on the change in the surfactant structure in the course of the adhesive polymerization.

The findings presented in Sections 2.3 and 2.4 allow us to characterize differences between IS and RS substances, the majority of which show a trend opposite to the change of solubility in the course of oligomer polymerization: the solubility of IS substances decreases and that of RS substances increases. The separation of surfactant from the polymerizing adhesive with added surfactant can cause abrupt decreases of the adhesion strength because the reaction between the adhesive and the substrate in this case will be carried out via a layer of low-molecular weight substances that has low cohesion strength. For RS substances there is no threat of formation of the weak boundary layers because the polymerization of systems with RS substances does not result in separation into macrophases.

The information presented was obtained under equilibrium conditions, whereas an adhesive-bonded joint is formed under nonequilibrium conditions. Taking account of the fact that in the course of adhesive polymerization there are a number of processes running at different rates, the kinetics of the cementing process must exert a fundamental effect on the properties of adhesive-bonded joints.

2.5 Effect of Surface-Active Substances on the Thermodynamic and Physical-Chemical Properties of Solid Polymers

The surface tension of an oligomer depends on the lifetime of the surface relative to the fact that relaxation and diffusion processes in bulk and at the boundary run comparatively slowly. The viscosity of the system increases due to the oligomer polymerization, which increases the time of establishing equilibrium at the boundary surface. Thus, the surface tension of the polymer depends on the relationship between the rate of polymerization and that of establishing the equilibrium density of molecules at the boundary between the oligomer and the air. That is why studies of changes in the system surface tension in formation of a solid polymer from a liquid one permit one not only to estimate the surface tension of the former but also to determine the means and nature of this change. Let us designate the unbalanced value of the system surface tension as σ_S to differentiate it from the equilibrium value γ. Consider the character of the change of the surface tension during polymerization of systems that do not contain surfactant (ED-20 and DEG-1 epoxy resins cured by PEPA; POPT with molecular weight 750 cured by TDI) as well as during formation of the non-chemical network (gel formation) of concentrated gelatin solutions. Investigations were carried out using the sessile drop method. The system density was measured in the course of polymerization by the dilatometric method.

Figure 2.12 presents the change of the surface tension of the DEG–PEPA system in the course of the reaction at various temperatures. At temperatures of 30, 40, and 60°C for 24 h, the system comes to thestate in which the surface tension no longer changes with time, i.e., it is close to equilibrium. This system at 20°C and the DEG-1

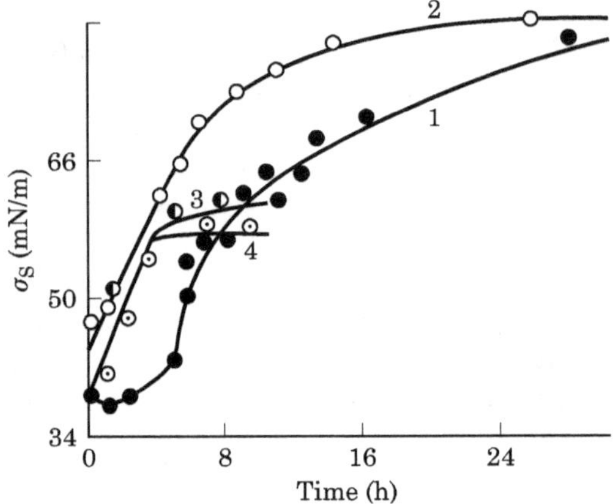

Figure 2.12 Change of the surface tension of the DEG–PEPA system in the course of reaction at various temperatures: (1) 20, (2) 30, (3) 40, and (4) 60°C.

system at 50°C do not achieve equilibrium in the same period of time. This is related to the low rate of the curing reaction in these systems.

The equilibrium value of the polymer surface tension decreases with increase in temperature (see Fig. 2.12), which follows from basic thermodynamic relationships. Thus, the increase of the adhesion strength for cementing at higher temperatures is caused, apart from other factors, by the improvement of wetting of the substrate by adhesive due to the decrease of the adhesive surface tension.

The regularities observed in the course of gel-formation of concentrated gelatin solutions are largely similar to those seen in studies of oligomer polymerization. The surface tension of the reactive blend changes while the reaction runs. Once the reaction is completed the system comes eventually to a state that corresponds to equilibrium. It must be borne in mind that actual (thermodynamic) equilibrium cannot be achieved in some simulated systems, which is why by "equilibrium systems" we mean systems (as in this particular case) in which the surface tension changes very slowly.

Let us consider the general regularities of the change in surface tension as the system polymerizes. At low temperatures and in early stages of curing one observes some drop of the system surface tension. This can be related to the formation of the macromolecular branching, to the increase of diphilicity when the oligomer macromolecule reacts with a molecule of the curing agent, and to the process of orientation of the macromolecules in the surface layer. With an increase of the degree of oligomer polymerization, its surface tension is expected to increase. Note that at 20°C the surface tension of the DEG–PEPA system is practically unchanged for almost 3 h from the time the polymerization begins.

The times of emergence of the surface tension isotherms and the kinetic curves onto the plateau practically coincide. Evidently the conditions are such that the process of building the equilibrium density on the boundary surface, the kinetics of which is determined by the mobility of molecules (segments), catches up with the process of surface change. This assumption was checked by investigating the rate of increase of the equilibrium surface tension of epoxy systems of various viscosities at various temperatures. The system viscosity was varied from 5 to 500 Pa·s by adding different quantities of PEPA to the resin. For all system conversion levels studied, the change of the surface tension was in step with temperature variation, which increased or decreased at the rate of 0.1°C/min. At high conversion levels the rate of building to equilibrium decreases, although the rate of polymerization in the last stages of the reaction is rather low, so it can be assumed that in this case the observed surface tension will not differ noticeably from its equilibrium value. Thus, there is opportunity for processes of completion of system polymerization and achieving the equilibrium surface tension to coincide.

The change of the system surface tension when POPT reacts with TDI is characterized by a number of features. In the initial stages of the reaction the surface tension of the system increases (see Fig. 2.9), which is related to the formation of more polar urethane groups. The increase in the urethane content results in the increase of the surface tension up to a certain value, only after which does it fall.

It is not permissible to adopt data acquired from changes of the liquid resin's surface tension for solid state polymer. Zisman [72] does it supposing that the reversible adhesion work of the solid polymer must be close to that estimated for the liquid state. The conclusion follows from the assumption that the forces that act on the phase separation boundary spread out to a depth that does not exceed the size of some molecules. As a result, the interaction on the boundary cannot depend on the change of state of the substance. One must accept this because the determined values of the surface tension of solid polymers significantly exceed those for the liquid oligomers. If we deal with undercured products the values usually exceed those for the polymer surface tension acquired by wetting agent critical surface tension methods.

Let us consider the surface tension of a polymerizing oligomeric system containing surfactant. The surface tension kinetics of systems based on DEG-1 resin cured by PEPA and containing different quantities of OP-10 is presented in Fig. 2.13. It can be assumed that in this case the rate of building up the equilibrium surface tension will be

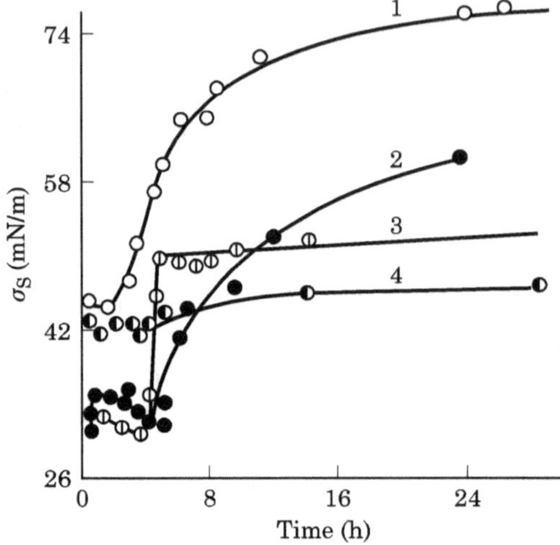

Figure 2.13 Change of the surface tension in the course of polymerization of the DEG–PEPA system (20°C) at various concentrations of OP-10: (1) 0%, (2) 2%, (3) 4%, and (4) 5%.

decreased compared with systems that do not contain surfactant, because the rate of diffusion of the comparatively large surfactant molecules to the boundary must be low.

At low degrees of conversion the trend of the isotherms coincides with the change of the equilibrium surface tension when surfactant is present (see Figs. 2.6, 2.8). The equilibrium density of the surfactant molecules appears to build up in the surface layer at low temperature (the reaction was performed at 20°C, at which temperature the rate of interaction is low). With increase of the degree of conversion systems containing surfactant, the surface tension increases similarly to that of systems without surfactant. The increase of the surface tension is especially noticeable in the region of transition of the system from the liquid state to the solid.

The primary peculiarity of the system under consideration is the fact that increase of the quantity of surfactant does not always result in decrease of surface tension. Thus, the surface tension of polymer containing 1% of L-19 is higher than that of a polymer containing 0.05% or 0.5% of L-19. Additionally, increase of the surfactant content of the liquid oligomer always results in the decrease of the surface tension (naturally when the surfactant concentration in the system is lower than CCMF).

The increase of surface tension in the DEG–PEPA system is related to the change of surfactant solubility. The rate of aggregation of the surfactant molecules depends on the viscosity and the chemical composition of the polymerizing oligomer. Increase in system viscosity means that under the dynamic conditions one cannot always realize the most thermodynamically advantageous surfactant structure, which depends on the relationship of the rates of system polymerization, and on physical-chemical and chemical (in case of RS substances) surfactant conversions. Thus, under dynamic conditions, the change in surface tension of a system that contains surfactant can be different.

The second reason for the anomalous change of surface tension of a solid polymer containing surfactant lies in the structural and conformational conversions of the polymer itself under the influence of the surfactant. Such factors as the increase of the polymer surface tension when surfactant is added cannot be explained by the surfactant adsorption on the polymer surface only (see Fig. 2.13). Later we will consider this in detail. As was noted above, if the rate of aggregation of the surfactant molecules is higher than or equal to the rate of polymerization, the system surface tension alters during polymerization in the same way as in the course of the equilibrium process. At a high rate of polymerization, the formation of micelles of the maximum possible size can be hindered by the rapid increase of the system viscosity. In this case, when an IS substance is applied the split into two phases is not observed and the system appears to be more oversaturated by surfactant than in the first case.

The change in the surface tension of systems that contain RS substances is determined to a great extent by the relationship of the rates of the system polymerization reaction and of the reaction of RS substance with the oligomer. It is clear that if the rate of the first reaction is much higher than that of the second, the influence of RS substances upon the system properties does not differ from that of IS substances.

The relationship between the rates of the polymerization reaction and of surfactant aggregation must affect the trajectory of the isotherm of the system surface tension because these processes shift the value of the surface tension in opposite directions.

The above processes are also influenced by the change in the system temperature. Raising the temperature results in increase of the polymerization reaction rate and at the same time in decrease of the tendency of surfactant molecules to aggregate because of the increase of thermal motion that follows from the Debye equation:

$$W = -3kT \ln c_{\mathrm{k}} \tag{2.16}$$

where W is the adsorption work and k is the Boltzmann constant. Thus, by changing the system temperature one can alter the relationship between the rates of oligomer polymerization and of surfactant aggregation. Experimentally these processes were studied on the DEG–PEPA system with added OP-10, DDN, and L-19. The system was polymerized at 20, 40, and 60°C. The temperature change noticeably affects the character of the surface tension isotherms. The surface tension of systems polymerized at 20°C changes gradually (see Fig. 2.13), while an increase in temperature results in the appearance of maxima on the isotherms (see Fig. 2.14). This difference in the path of the curves is explained by the fact that at low temperature, when the rate of polymerization is low, the state of aggregation of surfactant molecules falls into line with the state of the polymerizing solvent, which is demonstrated by the coincidence of the values of the surface tension determined under the equilibrium and dynamic conditions in the course of system polymerization. Under the influence of these two factors, the surface tension changes smoothly. The ascending branch of the isotherm (see Fig. 2.14) characterizes the increase in surface tension due to the oligomer polymerization at 60°C. In this case the surfactant has no time to fall into line with the oligomer state. At a certain conversion level, surfactant separates from the system. Such a process is characterized by abrupt decrease of the surface tension.

The above considerations are confirmed by turbidity data from a cured DEG–PEPA system containing different quantities of OP-10, DDN, and L-10.

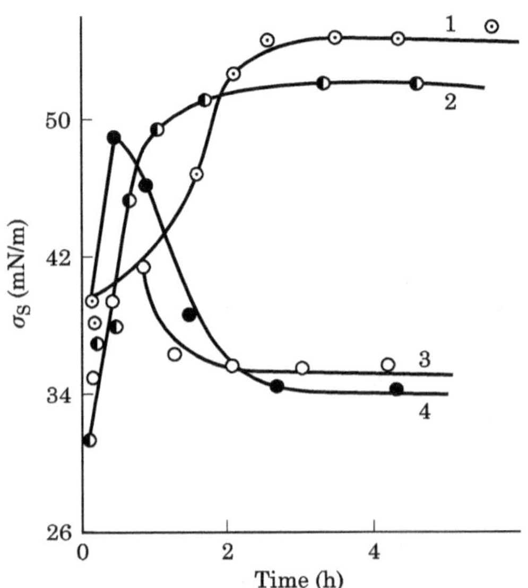

Figure 2.14 Change of the surface tension in the course of polymerization of the DEG–PEPA system (60°C) at various concentrations of OP-10: (1) 0%, (2) 0.05%, (3) 0.50%, and (4) 1%.

Table 2.5 shows that increase of polymerization temperature at low surfactant concentrations results in the decrease of the optical density of the solid polymer and illustrates that temperature increase causes an abrupt decrease of concentration and size of the colloid particles in the polymer. Increase in the surfactant concentration results in the formation of colloid particles at conversion levels of the system for which its viscosity is still comparatively low. At the point of a sharp increase in the viscosity of the reactive system, enough colloid particles appear to be formed, which is why the rate of the viscosity increase practically does not influence the number and the size of the particles. As the table shows, the optical density of polymers including 14% OP-10 or 8% DDN hardly changes as the polymerization temperature is increased from 20 to 60°C.

Let us consider the features of the polymerization of systems that incorporate surfactant. Depending on the reaction time, the character of the alteration of the surface tension is similar for DEG–PEPA systems that contain OP-10 or L-19. For the explanation of this, refer to the information presented in Section 2.3. At the start of the reaction, when the concentration of L-19 that has reacted with DEG-1 is still low, L-19 behaves similarly to an IS substance, i.e., OP-10. In the course of system polymerization, the compatibility of RS substances with the oligomer increases (see Table 2.5), which is why the decrease of the surface tension cannot be caused by the separation of surfactant from the system.

TABLE 2.5 Dependence of Optical Density of a Cured DEG–PEPA System on Temperature of Curing and Concentration of Surfactant (given in parentheses)

	Optical density		
Concentration (%)	20°C	40°C	60°C
		OP-10	
5.0	0.64 (0)	0.1 (0)	–
8.2	2.0 (0)	1.0 (0)	0.3 (0)
14.0	2.7 (0)	2.6 (0)	2.65 (0)
		DDN	
4.0	0.22 (0)	–	–
6.0	3.0 (0)	1.9 (0)	–
8.0	3.0 (0)	2.7 (0)	2.7 (0)
		L-19	
3.0	0.04 (0.03)	–	–
6.0	0.13 (1.03)	0.06 (1.03)	–
8.4	0.21 (1.2)	0.13 (1.2)	0.07 (1.2)

The products of the reaction between DEG-1 and L-19 have higher solubility in the oligomer, and the extent of their aggregation in the course of micelle formation is much lower than that of the initial surfactant. In this respect, when the surfactant concentration is higher than CCMF the micelles are the source of surface-active molecules—i.e., the probability of saturated sorption on a surface increases. It is to be expected that such accumulation of surfactant in the system depends on the rate of the reaction between surfactant and the oligomer, which is much lower than the rate of the reaction between the oligomer and the curing agent. Temperature elevation essentially does not change the relationship between the rates, which is why the processes described will result in a noticeable decrease of the system surface tension only some time after the reaction begins.

The products of interaction of OAA with epoxy resins have an extent of aggregation no greater than 2, which is why they do not form large micelles. In addition, during accumulation of PR in the system there is an increase of solubility of OAA themselves in the system, so that the system is enriched in molecules that incorporate residues of both oligoepoxide molecules and OAA. The increase in OAA solubility causes the CCMF to increase and the surfactant micelles to disintegrate. In so doing, the rise in OAA concentration in the system results in more intensive formation of reaction products and an increase of the micelle failure rate. Actually, the measurements of the optical density of the DEG–PEPA system cured with L-19 indicate that increase in the surfactant concentration does not result in noticeable increase of the optical density of the cured system. Even at very high concentrations of surfactant, the solid polymer turbidity is 5–15 times less than that of the initial blend and much less than the turbidity of systems that contain IS substances, i.e., OP-10 or DDN.

Thus, despite the similarity of the surface tension changes of systems that incorporate RS and IS substances, the observed phenomena are based on different mechanisms. The principal feature of the presence of RS substances in polymerizing systems is the trend toward lower aggregation level in the course of system polymerization whereas IS substances exhibit the opposite behavior.

Increasing surfactant concentration does not always result in decrease of the solid polymer surface tension. As Fig. 2.15 shows, the surface tension of the DEG–PEPA and ED–PEPA systems passes through a maximum with surfactant growth.

The surfactant can block the polar groups of the polymer, disturbing the intra- and intermolecular bonds [73–75]. This results in alteration of the molecular chain conformation and of the intermole-

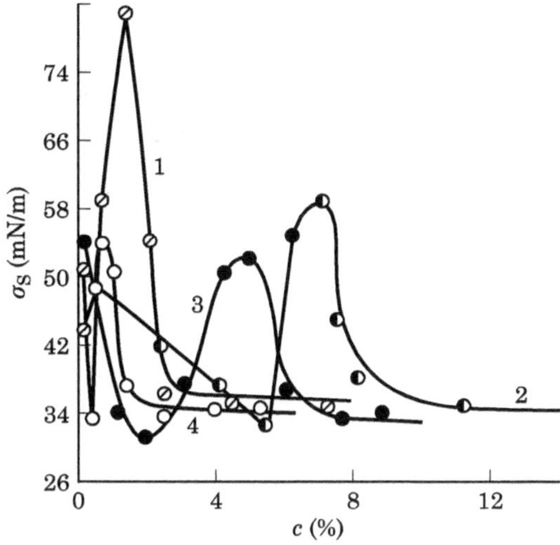

Figure 2.15 Dependence of the surface tension of cured systems on the surfactant concentration: (1) ED–PEPA (L-19); (2) ED–PEPA (OP-10); (3) DEG–PEPA (OP-10); (4) DEG-PEPA (L-19).

cular interaction that depends on the linear, branched or globular shape of the molecule. In turn, the energy and the entropy changes involved in the transition of the macromolecules to the surface differ depending on the shape of the macromolecules. In this respect the structure of the surface layer produced in the course of the dynamic process also changes [76, 77]. Surfactant blocking of the polymer polar groups must result in decrease of the intermolecular interaction, although the alteration of the macromolecular conformation due to such a blocking can cause increased interaction in connection with the macromolecules stretching and the formation of new contacts. Similar processes seem to be observed in the systems we have studied. As Fig. 2.15 shows, the surface tension of polymers with higher content of surfactant initially decreases then passes through a maximum, and finally becomes independent of the surfactant concentration. This behavior can be explained on the basis of the above considerations.

For low surfactant concentrations the shape of the macromolecules does not change noticeably. Adsorption of surfactant on the interface and the decrease of interaction between the macromolecules of the polymer when its polar groups are blocked by surfactant cause a decrease of the polymer surface tension. At higher concentrations of surfactant there is an abrupt increase of the polymer surface tension. Subsequent increase of the surfactant concentration results in its filling the surface layer, which is indicated by a drop in the polymer surface tension and its independence of the surfactant concentration.

The passage of the surface tension through a maximum is not detected for oligomers in the initial state because they have comparatively low molecular weight for which the effect of the macromolecular shape alteration is not manifest or is only slight. The structurizing effect of surfactant begins to be observed at the stage of the oligomer polymerization at which macromolecules of sufficient flexibility are formed.

These considerations are confirmed by studies of viscosity of the systems described. Measurements were made using the REOTEST viscometer. Figure 2.16 shows practically no change in the viscosity of the initial ED-20 oligomer with up to 14% OP-10 or L-19. With increase of the oligomer chain length by reaction with 1% PEPA (the oligomer viscosity increased in this case from 20 to 180 Pa·s), the surfactant structurization begins to be seen. The dependence of resin viscosity on the PEPA content displays a maximum and a minimum that correlate with the similar dependence of the polymer surface tension (see Fig. 2.16). Stretching of the macromolecules and formation of new contacts results in increase of both the system surface tension and the viscosity, while decrease of the intermolecular interaction results in their decrease.

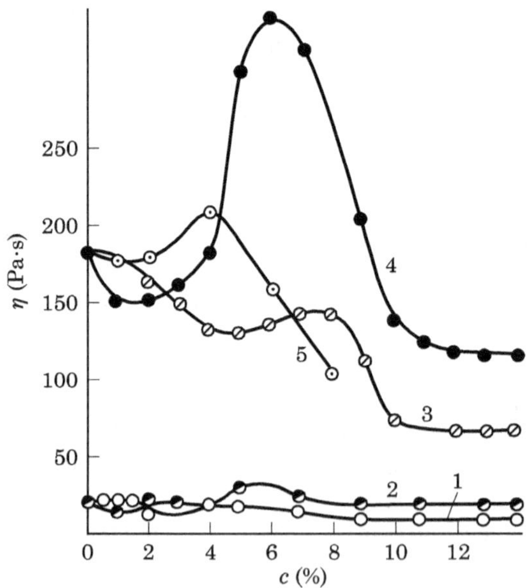

Figure 2.16 Dependence of viscosity of ED-20 resin on the surfactant concentration: (1, 3) OP-10; (2, 4) L-19; (3) PEPA fraction; (1, 2) initial resin; (3–5) product of reaction of the resin with 1% PEPA.

Let us consider to what extent the structure of the surfactant manifests itself in the above effects. We studied the alteration of the system viscosity by an amine that is an initial component in the production of L-19 and constitutes a hydrophilic part of the surfactant molecule. This amine is a PEPA fraction obtained by distillation at 100–150°C (10^3 Pa) [78]. Thus, the influence of two substances on viscosity are compared. One of these substances has a hydrocarbon "tail," i.e., the structure of a typical surfactant, and the other differs only by the absence of such a "tail." Adding 4% PEPA fraction to the system gives a 15% increase of the viscosity, less than the 75% achieved when surfactant is applied (see Fig. 2.16). The system viscosity was determined during the period when chemical reaction between the resin and the PEPA fraction had not been completed. Consequently, it is the molecular diphilicity of the added substance that in this case is the primary factor increasing the system viscosity.

Polymers with the surfactant additives that provide for achieving the maximum surface tension are characterized by a number of features, namely, the absence of the region of high elasticity and the considerable (20–50°C) elevation of temperature of the maximum deformation on the initiation of destructive flow of the polymer. These effects can be explained by the increase of the number of contacts of the macromolecules when the polymer is formed from the oligomer with the addition of surfactant. With a substantial addition of surfactant a plasticizing effect is produced on the polymer that results in decrease of the temperature of the maximum deformation of the polymer and the appearance of the region of high elasticity.

These findings are confirmed by study of the thermodynamic parameters of mixing of the cured epoxy resin with OP-20. At 6–7% content of surfactant, corresponding to the maximum surface tension of the polymer, a kink and an area of decrease of the Flory–Huggins parameter are observed in the dependence of $\chi_{2,3}$ on the surfactant concentration. This anomalous dependence can be explained in terms of the rearrangement of the intermolecular bonds in the polymer–surfactant system. With ED-20 initial resin, there are no extrema on the curve. Alteration of the macromolecular conformation affects the supermolecular structure of the polymer. Adding surfactant to ED-20 resin changes the form and causes a noticeable decrease of the size of the polymeric supermolecular formations.

Consider the surface tension change in the course of polymerization of the POPT–TDI polyurethane system with added surfactant. As mentioned, compared with epoxy systems this one has a number of peculiarities evidently due to alteration of the structure of the polyurethane formed. Increase in the concentration of an IS substance such as DDN results in the decrease of the polymer surface tension that (as above) is related to the separation of the low-molecular weight surfactant into the surface layer. Comparing the equilibrium value of the surface tension with that obtained under the dynamic conditions of the polymerizing system, the surface tension of the POPT–TDI system with added DDN is higher for the polymerizing system at similar conversion levels. This is related to the lag of the rate of the surfactant physical-chemical conversions behind the rate of system polymerization, since the equilibrium and dynamic values of the surface tension of the POPT–TDI system coincide when there is no surfactant added. With increase of the surfactant concentration, the incompatibility is displayed at lower conversion levels and the maximum value of the system surface tension shifts into the region of lower conversion.

Consider the alteration of the surface tension of the polymerizing POPT–TDI system with the addition of an RS substance such as L-19; the use of L-19 displays a number of features. Interaction of the TDI isocyanate groups with the L-19 amine groups occurs at a higher rate than reaction with the POPT hydroxyl groups. The amine groups of L-19 and of the product of the reaction with TDI are capable of catalyzing the reaction between TDI and POPT.

Increase of the amount of surfactant in the system does not always result in decrease of the polymer surface tension. Thus, surface tension decreases to the maximum extent at an L-19 content of 0.92%. Increasing the amount of surfactant to 1.9% results in a 2-fold increase of the polymer surface tension. A similar phenomenon is observed in studies of epoxy systems (see Fig. 2.15). In this case the surface tension increase also seems to be related to the surfactant structurizing effect.

Let us compare the effects on the POPT–TDI system of L-19 and of the PEPA fraction that differs from L-19 by the absence of the hydrocarbon "tail." Table 2.6 shows that L-19 causes a notably greater

TABLE 2.6 Surface Tension (mN/m) of Polyurethane with Added Surfactant

Concentration of L-19 or of PEPA fraction calculated for L-19 (%)	PEPA fraction	L-19
0	35.0	35.0
0.49	27.2	20.4
1.37	40.5	53.6
2.70	31.6	36.0
3.50	22.0	36.0

increase of the polymer surface tension than does the PEPA fraction. This indicates that in this case the structurizing effect of the surfactant is caused by its diphilicity as well.

Consider now the effect of surfactant on the properties of the initial and the cured polyester resins. Such systems show a number of particular features. The structure and properties of the cured polyester resins are determined to a great extent by the compatibility of the copolymerizing components of the system. There is almost no information on the possibility of controlling this compatibility by means of surfactant, which is why it is of great interest to study the influence of surfactant upon the thermodynamics of compatibility of the components of the cured polyester resin, as well as the effect of this compatibility on the molecular mobility of the polymeric chains and on the physical-mechanical properties of the polymer.

Studies were performed with PN 609-21M resin, which is a 60% solution of maleate phthalate polydiethylene glycol (MP PDEG) in dimethylacrylate triethylene glycol (TGM-3) (curing was achieved with the methyl ethyl ketone peroxide–cobalt naphthenate redox system). The surfactant effect was provided by OP-10 and by the product of interaction of 1 mol of adipate polydiethylene glycol with 2 mol of toluene diisocyanate (MDI), which is surface-active [80]. MDI differs from OP-10 in the reactive isocyanate groups capable of reacting with the carboxyl and hydroxyl groups of the polymer and of being included into the polymer. The interaction of MDI with the initial oligomers is not observed in practice because its rate is much less than that of polymer formation. The polymer was investigated after the complete consumption of the isocyanate groups.

The molecular mobility in the polymers was determined by dielectric studies. Polymer films 300–500 μm thick cast between glass plates at room temperature were used. Dielectric measurements were made using the E8-4 ac bridge within a broad temperature range. The superimposed frequency of the external electric field was 1 kHz. The dependence of the coefficient of transmission of the cured polyester resin on the OP-10 content was determined using the FEC-56M photoelectric calorimeter.

Figure 2.17 presents the dependence of the tan δ maximum for the cured PN 609-21M resin containing 2–10% of OP-10. It is evident that the tan δ maximum for the resin without OP-10 appears in the region of 215 K. For the polyester resin containing 5% of OP-10 the maximum shifts toward lower temperatures (218–227 K). For the resin containing over 5% of OP-10 the maximum of tan δ appears at 227–216 K, i.e., the maxima are shifted toward higher temperatures. The appearance of the tan δ maxima of all the studied systems in the low-temperature range indicates dipole-grouped processes of relaxation. Plasticization is

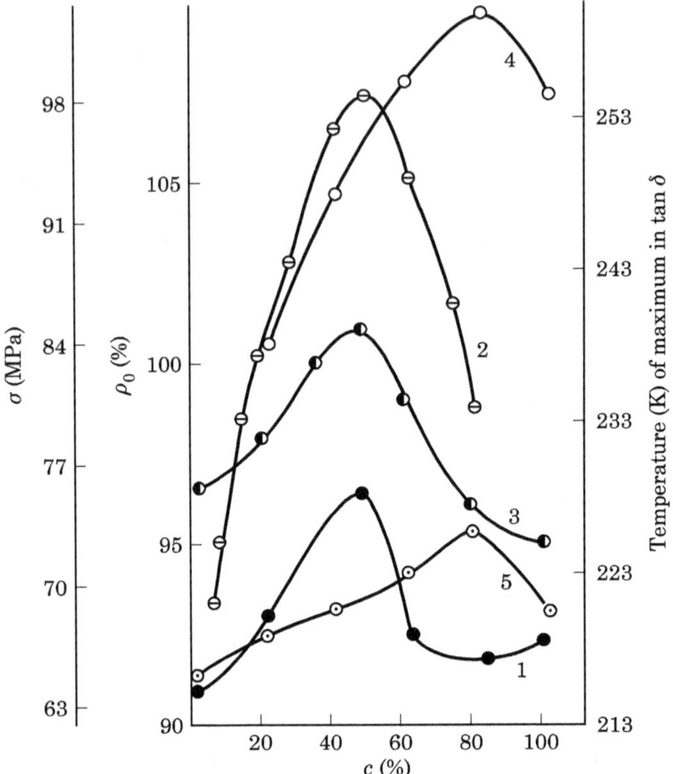

Figure 2.17 Effect of OP-10 (1–3) and MDI (4, 5) content in cured PN 609-21M resin on temperature of the maximum tan δ for the dipole-grouped relaxation process at failure stress (1, 5), on the failure stress σ at static bending of the resin (2, 4), and on the optical density ρ_0 of the resin (3).

known to shift the process toward lower temperatures. The shift of the tan δ maxima toward higher temperatures when up to 5% of surfactant is added, as well as the extremal character of curve 1 (Fig. 2.17), reveals the anomalous effect of plasticization of this system. The observed effect is related to the change of the thermodynamic compatibility of the system components under the influence of surfactant.

Figure 2.18 (curve 1) presents the dependence of the $\chi_{2,3}$ parameter of the thermodynamic interaction between the components on the composition of a mixture of MP PDGE–TGM-3 oligomers at 330 K. This mixture is thermodynamically compatible throughout almost the whole range of concentrations studied because, according to theory, the components are thermodynamically compatible: $\chi^*_{2,3} < (\chi_{2,3})^*_{\text{crit}}$ In this case the latter is equal to 0.442. It

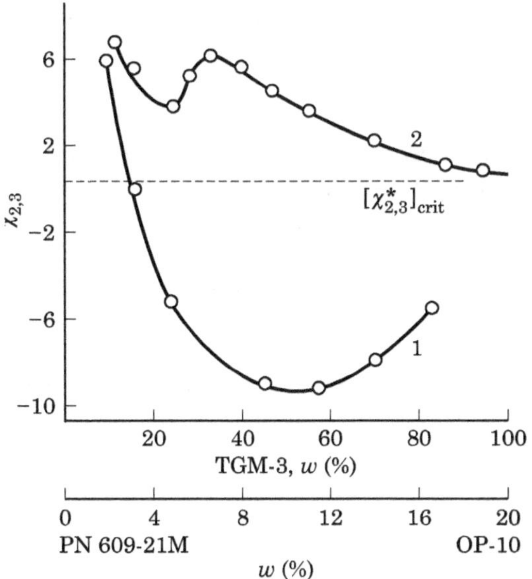

Figure 2.18 Concentration dependence of the parameter $\chi^*_{2,3}$ for the MP PDEG–TGM-3 system at 333 K (1) and for PN 609-21M [cured by MEKP(O)] and NC–OP-10 systems at 383 K (2). The dashed horizontal line represents the limit of thermodynamic compatibility.

is evident that the 40–60% range of the TGM-3 concentration that provides optimal physical-mechanical characteristics of the cured polyester resins corresponds to the maximum thermodynamic stability of the system.

In the curve of compatibility of the cured resin with surfactant (Fig. 2.18, curve 2), a kink is observed around 5% surfactant content and a region of decrease of the Flory–Huggins parameter. Varying the thermodynamic compatibility of the components of the polymeric system must result in alteration of the shape of the macromolecules, and the character and strength of their interaction. The improvement in compatibility will cancel the stretching of the growing macromolecules that increases the number of contacts between them. The anomalous concentration dependence of $\chi^*_{2,3}$ seems to be explained by the rearrangement of the intermolecular links that exist in a polymeric system due to the change in the polymer macromolecular conformations, which must result in alteration of the relaxation processes of the elements of the polymeric network and in alteration of the polymer physical-mechanical properties.

As Fig. 2.17 (curve 2) shows, the strength of the cured polyester resin containing OP-10 has a maximum within the area of 5% of surfactant that corresponds to the minimum value of $\chi^*_{2,3}$ (Fig. 2.18, curve 2) and to the highest temperature at which the maximum of tan δ appears (Fig. 2.17, curve 1). This is confirmed by studies of the turbidity of the cured polyester resin containing OP-10. The transmission coefficient of the polymer has a maximum for 5% content of OP-10 (see Fig. 2.17, curve 3), indicating the minimal level of the phase lamination of the components of the copolymerizing system. Consider the dipole-group relaxation process of the cured PN 609–21M polyester resin with MDI. As MDI content increases up to 8% the relaxation process shifts toward higher temperatures compared with the temperature at which the tan δ maximum of the initial system appears (see Fig. 2.17). The polymer has maximum strength at the surfactant content at which one observes the extremal value of the temperature maximum (see Fig. 2.17, curve 5).

The similar character of the dependence of polymer molecular mobility on the concentration of both IS and RS substances confirms the assumption that the effect of surfactant on the system properties is also related to the surfactant's influence on the process of polymer formation, which occurs under better conditions of compatibility of the copolymerizing components.

2.6 Oligomer–Metal Interphase Tension

Earlier we considered the regularities of the effect of surfactants on the oligomer surface tension. The properties of adhesive-bonded joints are also determined by the character and the magnitude of the interaction at the adhesive–substrate interface, which to a greater extent is characterized by the magnitude of the interphase tension [82]. Study of the interphase tension is of great theoretical and practical interest, but at present there are no direct methods for its determination.

The adsorption of surfactant can be successfully studied by electrochemical methods. Thus, the polarographic method of the analytical determination of surfactant is based on the surfactant's capacity to suppress the maximum. Methods involving measuring the capacity of the electric double layer are also widely used for studying the adsorption. Nevertheless, data from polarographic studies and of measurements of the capacity of the electric double layer alone cannot unequivocally characterize the adsorption level of an organic compound. Polarographic studies are usually carried out at a potential at which some compounds adsorb to the maximum extent and others to a lesser extent. Capacity measurements have no such shortcoming and permit determination of both the potential of the maximum adsorption and the range of potentials within which the adsorption occurs; though cases are known where adsorption of an aromatic compound does not practically affect the capacity of the double electric layer [83]. Consequently, fuller information about the adsorption of organic compounds is provided by a combination of these. Using these methods one can estimate the surface activity of both low-molecular weight organic compounds and oligomers and polymers relative to one another and to the known surfactant [84].

Polarographic measurements were carried out using the LP-60 instrument with the EZ-2 autographic recorder and with the dropping mercury electrode. For aqueous solutions a silver chloride electrode served as anode; in the case of nonaqueous solutions the bottom mercury was the anode. The anode potential did not change significantly in the aqueous and nonaqueous solutions with surfactants; cathode polarization was checked by means of the three-electrode circuit relative to the normal calomel electrode.

The level of suppression of the persulfate ion polarographic maximum was used to assess the surface activity of compounds under study in aqueous solutions with an accurately determined amount of surfactant against background of 100 mol/m^3 NaCl. Electrochemical reduction of divalent copper was used in the study of the surface-active properties in nonaqueous media. It is known that spontaneous reduction of Cu^{2+} ion against background of 100 mol/m^3 of $N(C_4H_9)_4I$ gives a clear maximum at -0.2 V. Careful removal of oxygen is necessary to obtain regenerated waves because oxygen distorts the shape of the wave of copper reduction at -0.3 V. Within the concentration range 0.26–0.43 kg/m^3, the height of the Cu^{2+} maximum is proportional to the concentration of the copper ions. Under these conditions, surfactant adsorption on the electrode has noticeable effects on the height of the maximum.

It is first necessary to select the range of copper ion concentrations in which the maximum of the wave is most sensitive to the additions of the compounds under study. For this purpose an equal quantity of solution of the compounds under study is added to solutions of various copper concentration and the difference in the maximum heights is determined before and after surfactant is added. The difference in magnitude increases initially with increase of the copper concentration, then, having reached a certain limit, remains practically unchanged. With these considerations, the copper concentration for the studies was taken as 0.26–0.43 kg/m^3 because, in this case, linear dependence is observed between the maximum height and the copper concentration.

The procedure is as follows. A solution containing 0.26 kg/m^3 copper chloride in DMFA is polarographed against the background of 100 mol/m^3 of $N(C_4H_9)_4I$; then a precisely defined quantity of surfactant is added and the solution is polarographed again. The relative surface activity of the compounds under study is judged by the maximum height difference of the copper before and after surfactant is added. The circuit comprising polarization with the amalgamated silver microcathode and the bottom mercury as anode, and with the prescribed current, was used for the oscillopolarographic measurements. This method allows estimation of the adsorption of organic compounds. The alteration of the electric double layer capacity and of the dE/dt polarization rate, well known in the course of the surfactant adsorption, is observed by means of dE/dt oscillograms.

Since the alteration of the polarization capacity (c) and rate, and the capacity itself, depend on the potential, a better picture of the adsorption can be obtained by using the relative values of $\Delta c/c_1 = \Delta V/V_2$, where V_2 is the rate of the polarization with surfactant present. From the $\Delta V/V_2$–E relationship one can estimate the adsorption of surfactant under study within a potential range limited only by the depolarization process.

Table 2.7 presents the results of polarographic study of the adsorption of some organic compounds in water and in DMFA. The relative adsorption values as percentages were determined by the difference in heights (H) of the maxima with and without surfactant. The surface activity was determined relative to that of Twin-60, assumed as 100 units. As Table 2.7 shows, the surface activity of OP-10 is less than that of Twin-60 by 17 units in aqueous solution, while that of the solvents under study is lower by 65–68 units. Judging from these data, the Twin-60 is much more surface active in aqueous solution than in DMFA.

TABLE 2.7 Polarographic Study of Surfactant Adsorption in Water and in DMFA

Surfactant	Concentration of surfactant (kg/m^3)	$\frac{H_0 - H_1}{H_0} \times 100\,(\%)$	Surface activity by correlation to Twin-60
		In water	
Twin-60	0.30	62.5	100
OP-10	0.30	52.0	83.3
Chloroform	0.31	22.0	35.0
DMFA	0.30	20.0	32.0
Methanol	0.30	20.0	32.0
		In DMFA	
Twin-60	0.30	20.0	100
OP-10	0.30	4.9	24.6
TMGF-11	0.32	17.0	85
MGF-9	0.30	15.0	75
MDF-2	0.31	11.4	57

The oligoester acrylate in DMFA adsorbs to great extent on the surface of the mercury at $-0.2\,\text{V}$ (Fig. 2.19). The maximum surface activity under the above conditions is that of TMGF-11 (85 units relative to the Twin-60), while the relative surface activity of the higher-molecular weight oligoester MDF-2 is 57 units and that of OP-10 is only 24.6 units relative to Twin-60 (Table 2.7). At the same time, even these low adsorptions cause considerable alteration of the double electric layer capacity under the given conditions (Fig. 2.19a, b). The adsorption maximum of all the compounds studied in aqueous solution lies around the point of zero electrode charge.

The pictures of adsorption of each compound presented in the figures have both common and differing features. Thus, Twin-60 begins to be adsorbed at the potential $E = -0.2\,\text{V}$, and at $-0.6\,\text{V}$ its adsorption is maximal. Twin-60 adsorbs completely at 1.8 V. With increase of the Twin-60 concentration the adsorption potential E and the potential that characterizes complete desorption of this compound shift toward more positive values; at the same time the range of the adsorption potentials remains unchanged and is $-1.6\,\text{V}$. The maximal value of $\Delta V/V_2$ is 35 units. The area of the OP-10 adsorption potential is somewhat less (1.5 V). The maximum of this surfactant adsorption shifts toward positive values of the potentials along with increase of concentration. The maximal value of $\Delta V/V_2$ is 32 units.

After the zero-charge point the adsorption of chloroform is maximal, and within the range of potentials from -0.6 to $-1.6\,\text{V}$ is constant,

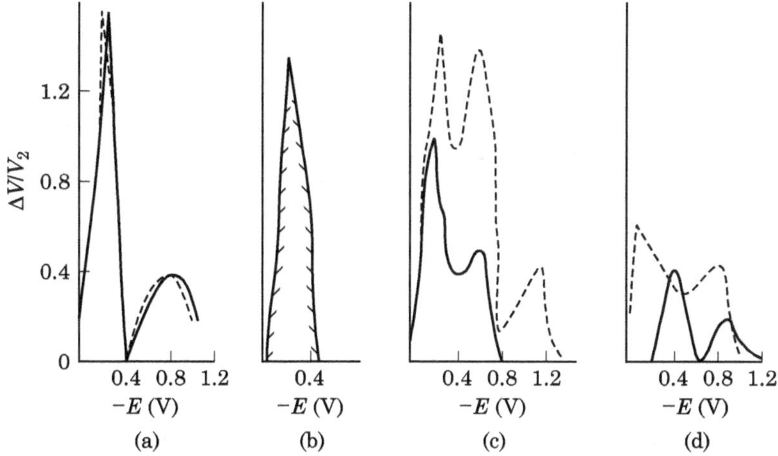

Figure 2.19 Dependence of $\Delta V/V_2$ vs. E with surfactant in DMFA against $N(C_4H_9)_4I$ solution (1 mol/l) at surfactant concentration of 0.6 g/l (solid line) and 0.3 g/l (broken line): (a) Twin-60; (b) TMGF-11; (c) MGF-9; (d) OP-10.

judging from the capacity change. DMFA produces an adsorption peak at the zero-charge point which then drops to 0 at $-1.2\,V$. With increasing DMFA concentration, the picture of the adsorption suddenly changes. There are two adsorption peaks observed in the $\Delta V/V_2$–E relationship: the first is at the zero-charge point, the second is at $E = -1.5\,V$, with the desorption within the potential range from -0.9 to $-1.1\,V$. Taking into account that in this case the maximal value of $\Delta V/V_2$ is less than at the initial concentration of DMFA, it can be assumed that the DMFA forms associations capable of adsorption. The appearance of the second adsorption maximum with the considerable shift to negative values can be explained by disintegration of the associations inside the thin layer at the electrode surface.

In DMFA with $100\,mol/m^3$ of $N(C_4H_9)_4I$ present, the picture of surfactant adsorption is somewhat different because both DMFA and $N(C_4H_9)_4I$ are surface-active. Thus, Twin-60 produces one sharp peak of adsorption at a potential of $-0.4\,V$ and a second one, which is less intense, at $-0.6\,V$. The adsorption potential range is $1.3\,V$. OP-10 adsorption begins at zero and the maximum is reached at $-0.3\,V$.

Adsorption of the oligoester acrylate is characterized by a number of peaks. As Fig. 2.19 shows, TMGF-11 produces two peaks and MGF-9 (at $0.3\,g/l^3$ concentration) produces three peaks of adsorption. In addition, the change of the electric double layer capacity with concentration is less. This can be explained by the fact that at low concentration the molecules of the oligoester acrylate are stretched out in the electrode's field and oriented parallel to its surface. With increase of concentration they are arranged perpendicular to the electrode surface. In line with the investigation of these compounds' adsorption in water and in DMFA, correlation of the polarographic and oscillopolarographic data allows them to be arranged in series according to decreasing surface activity: in water, Twin-60 > OP-10 > chloroform > mixture of DMFA with methanol; in DMFA, Twin-60 > MGF-9 > TMGF-11 > OP-10. These findings show that the surface activity of OP-10 in DMFA differs greatly from that in aqueous medium.

On the whole the polarographic and oscillopolarographic data coincide, apart from TMGF-11, which suppresses the polarographic maximum with the same intensity as Twin-60 and at the same time has no very noticeable effect upon the capacity of the electric double layer. Evidently we observe the case of an organic molecule with a system of conjugated double bonds involved in the formation of the electric double layer as though increasing its thickness with small effect upon capacity.

Thus, this method allows evaluation of the surface activity of a number of compounds in terms of two characteristics quickly and fairly reliably. The short time needed for the experiment enables us to determine the change of adsorption in time, which is especially important in the study of mixtures of substances that have different surface activities and where surface activity changes with time, for example, in the course of chemical reaction in solution.

For the study of the interphase tension at the boundary between the oligomer and a metal, a model system was used in which liquid mercury was used for the metal. The mercury was thoroughly purified and distilled prior to use. The interphase tension was determined by the sessile drop method [230]. The error was estimated as 3 mN/m. The oligomers and adhesives based on them were ED-20 epoxy resin-based adhesive with PEPA curing agent (12%) and PN 609-21M polyester resin-based adhesive with MEKP(O).

Figure 2.20 presents isotherms of the interphase tension of surfactant in the PN-1, ED-20 and PN 609-21M oligomers. Comparing them with the data described in Sections 2.1 and 2.2, one can conclude that there is no correlation between the surfactant activity on the boundary of mercury and air for the cases under consideration. Thus, OP-10 and L-19 are active at the boundary between the oligomer and air and are not active at the boundary with mercury. In contrast, the newly developed product ATG is active on the boundary with mercury only. As Fig. 2.20 shows, the interphase tension of the ED-20 resin increases with added OP-10. The increase, earlier described, of the surface tension at the boundary between the oligomer and the air with addition of surfactant to the oligomer was related to the availability of the surface-active impurities in the oligomer. In fact, commercial industrial

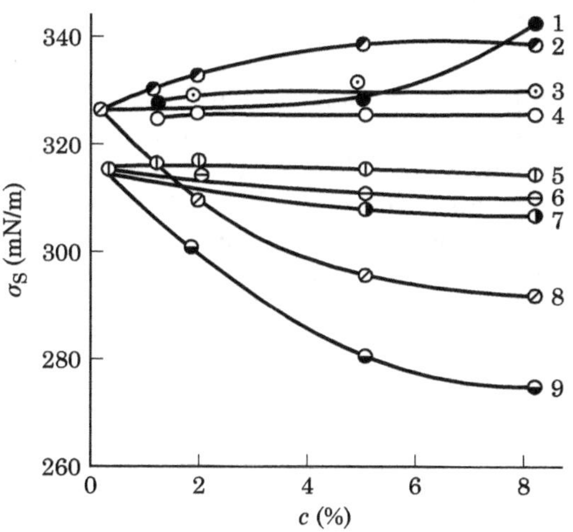

Figure 2.20 Dependence of surfactant content of surface tension of oligomeric systems at mercury interface: (1) ED-20–OP-10; (2) ED-20–decane; (3) ED-20–octane; (4) ED-20–L-19; (5) PN 609-21M–OP-10; (6) PN 609-21M–dibutyl phthalate (DBP); (7) PN 609-21M–dimethyl phthalate; (8) ED-20–ethanol; (9) PN 609-21M–ATG.

oligomers always contain substances whose molecules differ from the oligomer molecules in chemical composition and structure. These substances represent oxidized oligomer molecules with altered functionality and branching, and the remains of catalysts, solvents, etc. Similar impurities can coincide poorly with the main bulk of the oligomer and can adsorb at the boundary between the oligomer and mercury. Adding surfactant to the oligomer was shown to result in increase of the solubility of the impurities and in their desorption from the boundary, which increases system surface tension. The increase of the interphase tension of ED-20 resin with added OP-10 seems to occur for the same reasons.

The equilibrium interphase tension for the studied oligomers is achieved in 10–100 min. The low rate of reaching equilibrium even for the noncuring oligomers is evidently explained by both the increased viscosity of the oligomer in the boundary layer and the diffusion processes on the boundary.

The principal peculiarity of the kinetics of the interphase changes at the boundary between the adhesive and the mercury in comparison with those of the surface tension at the boundary with air is the approach of the kinetic curves to the equilibrium value long before gel-formation of the system, while the surface tension undergoes maximum changes close to the gel formation point. The change in the interphase tension at the point of transition of the system from the liquid state to solid is not observed for systems that do not contain isocyanate groups. This seems to be related to the fact that polymerization at the interphase boundary in such systems occurs at a lower rate than in the bulk, which can be caused by the selective sorption of some components of the system by the high-energy surface, disturbing its equimolarity, by the decrease of the macromolecular mobility in the surface layer, and by blocking of the system functional groups by the surface.

One reason for the stability of the interphase tension during transition of the system into the gel state might also be adsorption of the low-molecular weight impurities at the boundary with the mercury. In this case there is a layer of chemically unreactive liquid between the cured oligomer and the mercury surface, while the observed value of the interphase tension does not naturally reflect the state of the system. Figure 2.20 shows that some low-molecular weight compounds such as dibutyl phthalate (DBP) and dimethyl phthalate can diffuse toward the interphase boundary and adsorb onto it, decreasing the interphase tension. For the systems containing isocyanate groups, the kinetic curve has a minimum. Such extrema seem to be due to the fact that the surface can increase the probability of the reaction between isocyanate, urethane, and other polar groups adsorbed on the surface, resulting in the formation of branched macromolecules [29]. The formation of branched macromolecules increases the interphase tension as a result of their less dense packing on the surface compared with the linear molecules.

2.7 Control of Polymer Adhesion Strength by Means of Surfactant

The fracture stress of adhesive-bonded joints (adhesion strength) is a consequence of processes that occur in the course of their formation, but all attempts to formulate a fundamental relationship between formation and failure of adhesive-bonded joints have so far been unsuccessful. This is mainly due to the lack of methods for measuring adhesion that would permit determination of the failure equilibrium work. Accordingly, the relationship between the experimentally determined value of the adhesion strength and the thermodynamic characteristics can be one of correlation only.

In our view, the failure to determine correlation between the thermodynamic characteristics of adhesion and the strength of adhesive-bonded joints lies in the fact that what was being studied was the relationship of the initial compound properties (in the liquid state) with the strength of joints in which the adhesive was in the cured state. In addition to ignoring the essential differences of the cured adhesive from the initial liquid, there was no account for irregularity of the process of formation of adhesive-bonded joints itself, which means that the most favorable adhesive structure (from the thermodynamic point of view) is not realized for kinetic reasons. The necessity for accounting for these factors has been established in previous sections.

Let us consider the effects upon the adhesion strength of substances that are active at the boundary between the adhesive and the air. The comparison of experimental data presented in Sections 2.2 and 2.5 shows that the dependence of the surface tension of the initial liquid oligomers on the surfactant concentration differs noticeably from that of the cured adhesive. For oligomers this dependence is characterized by a monotonic decrease of its value with increasing surfactant concentration up to a certain limit, while for polymers the dependence passes through a maximum. Given that these characteristics reflect both the dynamics of the adhesive-bonded joint formation process and the essential difference between the cured polymer surface properties and those of the initial liquid mixture, it is of considerable interest to study the relationship of these characteristics with the adhesion strength of the joint. In determining this relationship one should remember that the strength of the adhesive-bonded joint is affected by internal stresses, which decrease its value. Thus, the relationship determined is a qualitative one, although it indicates possible ways to control the adhesion strength.

Let us consider the dependence of the adhesion strength of adhesive-bonded joints on the content of surfactant (active at the boundary between the polymer and the metal) in the initial oligomer. As Fig. 2.21 indicates, with the increase of the quantity of IS substance, the adhesion strength passes through a minimum, after which the strength of the adhesive-bonded joint increases sharply, decreasing again with further increase of surfactant content. Such a dependence is observed for both epoxy and polyurethane systems. It should be noted that analogous dependences have been obtained for polyethylene and other polymers.

A somewhat different dependence is observed for RS substances (see Fig. 2.21, curve 1). For small quantities of surfactant this dependence for RS substances coincides with the analogous dependence for IS substances; at greater quantities of surfactant this dependence does not indicate any drop in the adhesion strength in that it becomes independent of the surfactant content and the change in the adhesion strength shows the same dependence as the polymer surface tension. This correlation is not random: insofar as the adhesive spreads well on the steel surface, the thermodynamic work of adhesion of the cured adhesive to a solid surface is essentially twice the value of the surface tension, and consequently the dependence of the adhesion strength on the surfactant content must coincide with the change in the surface

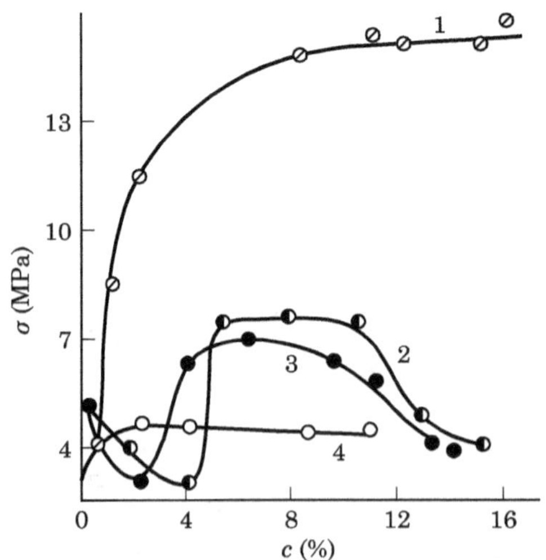

Figure 2.21 Dependence on surfactant content of the adhesion strength of the ED–PEPA [(1) L-19, (2) OP-10)] and DEG–PEPA [(3) OP-10, (4) L-19)] systems.

tension that is observed experimentally. Additionally, the increase of the polymer surface tension under the influence of surfactant is caused, as shown earlier, by the change in the polymer structure, which affects the cohesion strength and naturally results in an increase of the adhesion strength.

Comparison of Figs. 2.15 and 2.21 shows that the availability of this correlation permits easy interpretation of the dependence of the adhesion strength on the quantity of surfactant, taking account of the processes that occur in bulk and on the surface of the polymeric phase.

Let us consider the mechanism of the effect of surfactant on the adhesion strength and the effects of RS and IS substances in accounting for the observed correlation dependence. At a low surfactant content the surface tension of a solid adhesive decreases, which decreases the adhesion tension. Further increase of surfactant content, and the related growth of the thermodynamic work of adhesion, increases the adhesion strength as sharply as the growth of the surface tension. But at these quantities of surfactant, as Figs. 2.15 and 2.21 show, the increase of the adhesion strength occurs over a greater range of concentrations than that of the surface tension. This seems to be related to the fact that internal stresses decrease with the increase of the surface tension.

Increase of the surfactant content above a certain value results in a drop in the thermodynamic work of adhesion due to the decrease of the surface tension. It can be assumed that the drop in adhesion strength for IS substances is related not only to the change of the thermodynamic work of adhesion but also to the formation of micelles that form a weak zone at the boundary between the adhesive and the substrate. The accumulation of such micelles occurs gradually with increasing surfactant concentration, as a consequence of which a comparatively smooth decrease of the adhesion strength is observed.

In the case of RS substances things are somewhat different. The surfactant solubility increase in the course of the reaction and chemical interactions with the polymer mean that such surfactants do not form micelles. No decrease of the adhesion strength was observed even with large quantities of surfactant.

Thus, differences in the effects of surfactant of both types on the adhesion strength are exhibited only at high surfactant content and consist of an opposite trend of the compatibility in the course of polymerization to that found in studies of surface tension in the course of oligomer polymerization. When cementing low-energy surfaces or in liquid media, the factors under consideration will be superimposed on the change of wetting of the substrate by the adhesive due to the effect of surfactant.

Now let us consider the effect on the adhesion strength of surface-active substances at the boundary between the adhesive and the metal. Figure 2.22 presents the dependence of the adhesive-to-steel adhesion strength on the difference between the interphase tension of the solid adhesive–mercury system and the adhesive surface tension. Because it is impossible to measure the adhesion strength at the boundary with mercury, it was determined at the boundary between the adhesive and steel. The interphase tension will surely differ here from that at the boundary with mercury, although this is only a quantitative difference and the qualitative relationship will be maintained.

As Fig. 2.22 shows, the adhesion strength increases as $\sigma_{\mathrm{adh\,Hg}} - \sigma_{\mathrm{adh\,air}}$ decreases. RS substances were used to control the interphase tension. The thermodynamic work of adhesion is related to the surface tension at the boundaries between adhesive and air, between adhesive and solid, and between solid and air by the relationship

$$W_{\mathrm{adh}} = \sigma_{\mathrm{adh\,air}} + \sigma_{\mathrm{adh\,air}} + \sigma_{\mathrm{adh\,solid}} \tag{2.17}$$

Given that the metal surface tension was the same in all cases, the thermodynamic work of adhesion depends on the interphase tension and on the adhesive surface tension. In this case (see Fig. 2.22) a correlation is seen between the thermodynamic work of adhesion and the adhesion strength. This correlation was determined earlier by Lipatov and Myshko by applying the modified equation of Dupré–Young [102]. Epoxy, polyester, polyurethane, and polyacrylate adhesives were used: the type of the adhesive did not have a significant effect on the correlation dependence.

Such a dependence is surprising at first sight. Actually, as was shown in Chapter 1, the primary reason for disintegration of adhesive-bonded joints is the formation of a weak zone at the bound-

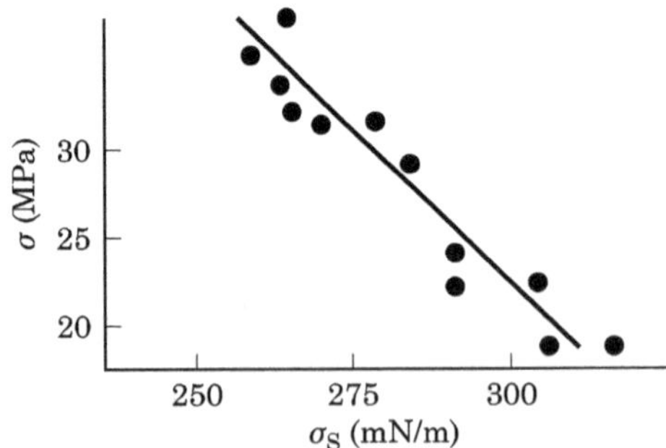

Figure 2.22 Dependence of the adhesion strength of adhesive-bonded joints on the difference of the interphase and surface tensions of adhesive.

ary between the adhesive and the substrate, which can be caused by layers of gases and vapors adsorbed from the air, by low-molecular weight impurities, or by certain ingredients of the adhesive on the surface of the substrate. In addition, the substrate can induce defects of the polymer boundary layers. In the course of cementing surfaces with adsorption layers, the adhesive can force them apart or spread all over their surface. Most realistic are intermediate cases where the thickness of the adsorption layers decreases with the application of the adhesive onto the substrate. Clearly, the lower the level of the interphase tension produced by RS substances, the higher the probability of desorption of substances previously adsorbed from the substrate surface and the lower the probability of the selective sorption of some ingredients of the adhesive by the substrate.

Let us consider the effect of surfactant on the properties of the polymer boundary layers. Octyltrimethylammonium bromide (OTMAB) integrated into the compound was used as surfactant. Figure 2.23 shows that surfactant that has little effect on the cure rate of the epoxy compound in the bulk does increase the rate and the degree of conversion in the boundary layer on the surface of KRS-5. The adsorption of surfactant on the interface seems to remove the selective sorption of the epoxy resin, which excludes the opportunity for formation of the undercured boundary layer and results in increase of the adhesion strength (Table 2.8).

Addition of surfactant makes it possible to monitor the glass-transition temperature in polymer films only 0.01 μm thick obtained at room temperature (Fig. 2.24). Heating the polymer raises the glass-transition temperature in the boundary layers to higher values than in the bulk, as was noted earlier for polymers cured on a low-energy solid surface (see Fig. 1.3). Beginning with a surfactant content of 10%, the adhesion strength decreases due to

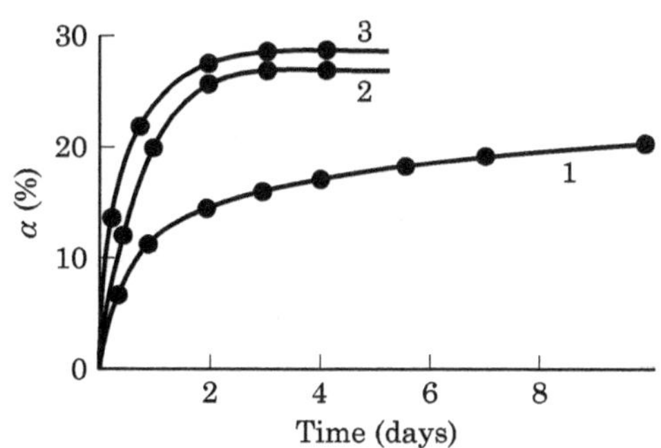

Figure 2.23 Time dependence of the conversion level of epoxy groups: (1) ED-20 (PEPA)–OTMAB (boundary layer); (2) ED-20 (PEPA); (3) ED-20 (PEPA)–OTMAB.

TABLE 2.8 Effect of OTMAB on Adhesion Strength and Glass-Transition Temperature of Epoxy Adhesive

OTMAB concentration (%)	Adhesion strength (MPa)	T_g from	
		Reversed-phase gas chromatography	Calorimetry
0	11	370	382
1	13	367	–
5	24	365	–
8	30	363	374
10	28	360	371
20	20	355	366

formation of polymolecular boundary layers of surfactant. At this content of surfactant, the thermodynamic compatibility of the surfactant with the polymer increases, which confirms the formation of micelles in the system.

Thus, the inclusion of surfactant in the adhesive may result in the presence of defects in the polymer boundary layers. Let us consider the effect of OTMAB on the physical-mechanical properties of epoxy polymer boundary layers. A nonmonotonic dependence of the modulus of elasticity of the coatings on both substrates on the surfactant concentration was observed for fine coatings in the systems investigated

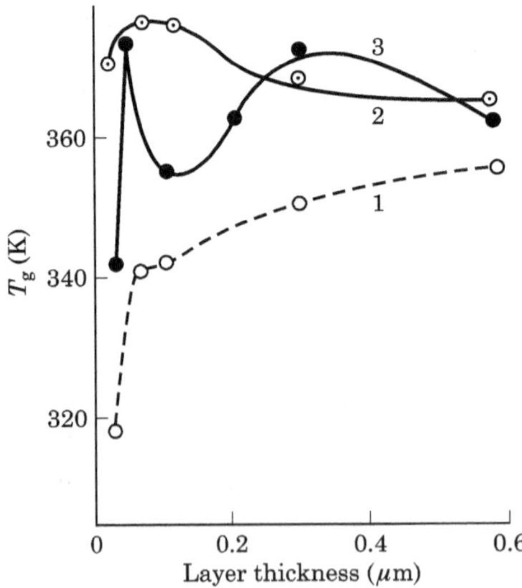

Figure 2.24 Dependence of T_g on the layer thickness for epoxy polymers on iron: (1) ED-20–PEPA–OTMAB; (2) ED-20–PEPA–OTMAB with heating for 1h at 423 K; (3) ED-20–PEPA with heating for 3 h at 423 K.

(Table 2.9). Addition of 5% of surfactant increases modulus, although increase of the surfactant content up to 10% in the reactive mixture results in a noticeable decrease of the coating's elasticity, which is more important in the area of fine coatings on aluminum high-energy surfaces. When the coatings are 5 μm and more thick, this dependence is weaker. The extreme character of the dependence of fine polymer coatings on the substrates can be explained by the fact that integration of surfactant into a compound may cause formation of an adsorbed monolayer on the surface of a solid even at low concentration. This monolayer has higher strength and forms a new interface with the epoxy coating with its own adhesion interaction. It is also bound to hinder the effect of the selective sorption of molecules of the epoxy oligomer, which is one reason for undercure of the epoxy compound on a high-energy surface. This produces boundary layers with higher cohesion properties, which seems to result in increase of the modulus of elasticity of fine coatings and the increase of the adhesion strength of adhesive-bonded joints that is observed with addition of 5% surfactant. On further increase of the surfactant content in the reactive mixture up to 10%, the observed decrease of the elasticity of the coatings is related to plasticization by the low-molecular weight surfactant whose contribution to the cohesion of the boundary layer becomes dominant.

TABLE 2.9 Modulus of Elasticity E_n (10^3 MPa) of Epoxy Coatings on Substrates at Various Surfactant Contents in the Reactive Blend

	Surfactant content (%)		
Coating thickness (μm)	0	5	10
	Aluminum		
1	27.0	34.0	16.0
2	10.0	14.5	6.8
5	5.0	5.5	3.4
10	4.3	4.7	2.4
20	3.2	2.9	2.2
	Polyethylene terephthalate		
1	19.5	25.0	15.7
2	11.2	17.2	12.0
5	6.7	6.8	5.2
10	4.6	4.8	4.0
20	3.4	3.4	3.0

The information presented convincingly illustrates the possibility of analyzing the processes that occur during formation of an adhesive interlayer by thermodynamic approaches and by colloid-chemical concepts. The regularities described make possible the prediction of adhesive properties and the creation of adhesives with predetermined properties.

2.8 Influence of Surfactants on the Structure of Polymers

2.8.1 Influence of surfactants on the structure of polyurethanes

The properties of adhesives depend largely on the structure of the polymers; accordingly, research on the ability to control this structure with the help of surfactants is of considerable scientific and practical interest. For this reason, we have undertaken the investigation of the structure-forming action of the surface-active substance KEP-2 in the formation of polyurethanes and polyepoxides, dimethylsiloxanes and alkylene glycol block copolymers. KEP-2 (a dimethylsiloxane–alkylene glycol block copolymer) is an effective surfactant that exhibits surface activity even at very low concentrations.

The reaction kinetics has been studied in the system of 2,4-toluene diisocyanate (TDI), polyoxypropylene glycol of $\bar{M}_n = 1000$, and trimethyl propane (TMP) with different contents of KEP-2 (0.02–0.2%) at 353 K [224]. The observed reaction rate constants of urethane formation on the first and the second sections of the kinetic curves (Table 2.10) were calculated from the calorimetric data according to the second-order rate equation. As is evident, the presence of KEP-2 accelerates the reaction. Two concentrations of surfactant (0.03% and 0.15%) are exceptions. The observed acceleration becomes more notice-

TABLE 2.10 Values of Reaction Rate Constants for Formation of Space-Network Polyurethanes of PPG-1000, TDI, TMP in the Presence of Different Amounts of KEP-2 at 80°C

Surfactant concentration $C_{\mathrm{KEP-2}}$ (%)	$10^2K_1\left(\frac{1}{\mathrm{mol\cdot NCO\cdot min}}\right)$	$10^2K_2\left(\frac{1}{\mathrm{mol\cdot NCO\cdot min}}\right)$
0	1.45	1.0
0.02	1.80	1.40
0.03	1.52	1.06
0.05	1.80	1.41
0.11	2.04	2.00
0.15	1.70	1.10
0.2	2.06	1.45

able at high degrees of the completion of reaction (above 60% transformation isocyanate groups). Since surfactants are active structure-forming agents, it is supposed that the observed acceleration by surfactant is connected with this capacity. For this reason, further study deals with the effect of KEP-2 on the structural parameters of the polyurethane reaction mixture in the estimation of viscosity and light diffusion during hardening. The process of gel formation begins later for higher amounts of surfactant. This nature of the effect of KEP-2 in hardening systems seems to be connected both with the details of surfactant behavior at the interfaces and with the mechanism of formation of network structures by the polycondensation mechanism. As is shown for the one-stage method of polyurethane production [225], microparticles of the crosslinked polymer appear in the reaction medium when the reaction is 55% complete. As soon as the other phase appears, molecules of surfactant can sorb onto the already formed surface of microheterogeneities just as on the interface. The adsorption layer of surfactant prevents hydrogen bonding and other forms of physical interaction of these particles with the rest of the medium, and also promotes chemical interaction of the microgels formed. These effects are manifest in the decrease of viscosity and in the delay of the sharp increase of viscosity, respectively.

A light diffusion method was used to determine the amount and character of changes in the microheterogeneities during hardening of the polyurethane reaction mixture. A general view of the dependence curves up to 85% transformation shows no differences from other processes of formation of spatially crosslinked polymers studied by this method [226, 227]. The observed behavior, due to the viscosity change in the course of the reaction, may be interpreted as follows. Before the considerable growth of viscosity in the presence of surfactant, larger scattering particles are formed than without surfactant. The particle size increases with the amount of KEP-2 in the system. Hence, it can be supposed that this method estimates the number of "scattering nuclei," which are polymer regions of different density from the rest of the mass and which carry on their surface a thicker layer of surfactant the higher the amount added to the reaction mixture. As a result, the reaction rate increases and more extended regions of altered density are formed in the reaction mass. The beginning of the sharp growth in viscosity corresponds to the regions approaching the maximum and following the decrease of the size of formations. Thus, the redistribution of surfactant between the surface and the volume of the reaction mass takes place under the network polyurethane formation, which must result in the impoverishment of the surface layer of surfactant at the reaction mass–air interface. Estimation of the surface tension during the hardening of polyurethane with 0.1% KEP-2 has shown a 35% growth of the surface tension in the gel formation region.

Thus, the hardening of the toluene diisocyanate-based polyurethane system in the presence of microquantities of the surface-active agent KEP-2 occurs at all stages of the process in conditions different from the ordinary hardening mixture of TMP–TDI–oligoglycol. These differences become apparent in kinetic and structural studies of the hardening process. The final product is a polymer with very weak adhesion compared with the initial specimen (hardening temperature 353 K).

The increase of urethane formation reaction rate constants in the presence of KEP-2 (Table 2.10) may be connected with the decrease of the reaction mass viscosity due to the weakening of intermolecular interactions and the increase of the mobility of the reacting molecules. In general, the mechanism of the effect of surfactant is not clearly understood. The surfactant is active at borders with low surface energy and inactive at those with high surface energy. It may be supposed that essential structural rearrangements occur in the system under these concentrations of surfactant. It thus proves necessary to define thermodynamic functions for the surface layer. A rigorous approach to this definition is possible in the simple systems—binary systems, for example.

Of all the reagents of the reaction mixture under study, polyoxypropylene glycol possesses the highest chemical affinity for KEP-2 and defines the distribution of the latter in the reaction system. The model system of oligoglycol–KEP-2 has been studied to elucidate the structure-forming effect of surfactant. The concentration dependence of the surface tension γ for the concentration range of surfactant from 0 to 0.3% has been analyzed and isotherms typical of surfactant solutions were obtained (Fig. 2.25). The equation for binary solutions of surfactant was used:

$$\gamma = (\gamma_1 - \gamma_2)e^{-bc} + \gamma_2 \qquad (2.18)$$

where γ = surface tension of the solution containing surfactant; γ_1 = surface tension of the solvent; γ_2 = surface tension of surfactant; c = surfactant concentration; and b = a coefficient of proportionality with dimensions of inverse concentration ($1/b$ = the concentration of such a solution for which the adsorption value for the Gibbs equation reaches a maximum).

Comparison of experimental and estimated curves (Fig. 2.25) shows that Equation (2.18) satisfactorily describes the experimental data only when KEP-2 concentrations exceed 0.03%. A sharp change of the surfactant surface activity at this concentration is the cause of divergence between the estimated and experimental data. This can

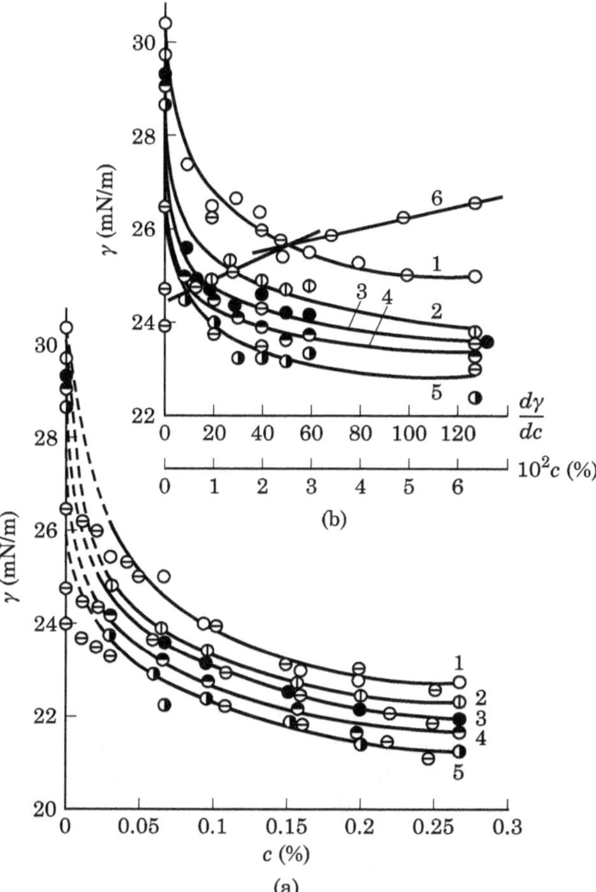

Figure 2.25 Dependence of surface tension of the binary system PPC-1000–KEP-2 on concentration at different temperatures: (1) 60°C, (2) 65°C, (3) 70°C, (4) 75°C, and (5) 80°C. Curve (6) shows the dependence of $d\gamma/dc$ on the surface tension in the system [estimated values of γ (curves 1, 3, 5)].

be supposed to occur as a result of reorientation in the surface layer of surfactant molecules rolled up in the form of plane disks into a helix-like shape [229]. This possibility was considered in the derivation of the equation and thus the estimated values of the surface tension are lower than the experimental ones (Fig. 2.25). Reorientation of surfactant molecules in the adsorption layer can draw oligoether molecules into this process. This is indirectly confirmed by the sharp change of morphology of the binary mixture surface layer and may be confirmed by nuclear magnetic resonance (NMR) data (Fig. 2.26) [230]. The chemical shifts of proton signals do not change with

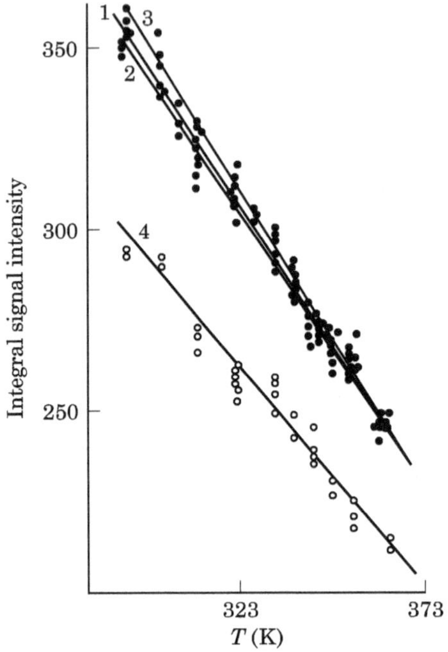

Figure 2.26 Dependence of integral intensity of proton signals on temperature in the system PPC-1000–KEP-2 with various KEP-2 contents: (1) 0%, (2) 0.1%, (3) 0.15%, and (4) 0.03%.

amounts of KEP-2, while the integral intensity decreases regularly with increase in measurement temperature. As is evident from Fig. 2.26, the considerable decrease of integral intensity (~20%) in the whole temperature range under study is observed only for one surfactant concentration, namely 0.03%. Such a decrease of integral intensity may indicate the limitation of mobility [231] of the oligoether molecules to such an extent that a proportion of the protons do not contribute to the NMR signal. The experimental result indicates the change of volume properties in the binary mixture under the effect of surfactant.

It is of interest to estimate the thermodynamic functions of surface layers of different composition. Since a binary system is considered, it is possible to make thermodynamic calculations for mixtures of different concentrations at different temperatures (Fig. 2.27). This was done using the equation that connects the surface pressure with the change in entropy and enthalpy of the surface layers [232]:

$$\Delta H_{\mathrm{S}} = \pi + T\frac{d\pi}{dT} \tag{2.19}$$

where ΔH_{S} = change of the surface layer enthalpy; π = surface pressure; and $d\pi/dT$ = change of the surface layer entropy.

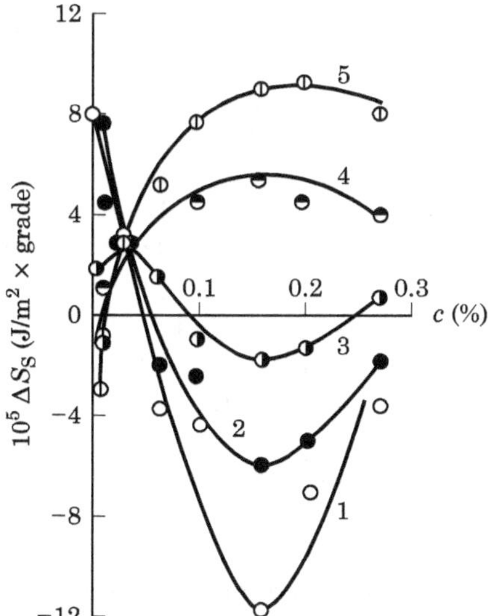

Figure 2.27 Change of surface entropy with concentration of KEP-2 at different temperatures in the system PPC-1000–KEP-2: (1) 335 K, (2) 338 K, (3) 343 K, (4) 348 K, and (5) 351 K.

As seen from Fig. 2.27, two concentrations of surfactant (0.3% and 0.15%) have particular characteristics. The independence of ΔS_S of temperature is characteristic of the first concentration (Fig. 2.27), indicating the existence of structure that is not affected by the thermal factor, according to the NMR data, within a rather wide temperature range. Thus, the behavior of a 0.03% solution of KEP-2 in oligoglycol may (with a good degree of probability) be interpreted as reorientation of surfactant molecules in the surface layer, the glycol molecules being drawn in, resulting in the formation of a stable, ordered combined structure.

As can be seen from the isotherms of surface tension (Fig. 2.25), beginning from a KEP-2 concentration of 0.25%, increase of surfactant content does not lower the surface tension of the system, i.e., this concentration corresponds to the critical concentration. The extremum in the change of thermodynamic functions of the surfactant surface layer entropy (Fig. 2.27)—in particular—is observed at 0.15% KEP-2, below the CMC. This testifies to structural changes in the adsorption layer [229] and may be explained in terms of current ideas of the process of micelle formation. Micelle formation, like adsorption, is entropic as well as energetic in nature [233], as is shown well in thermodynamic studies of water solutions of proteins [234]. Also, micelle formation is characterized by the concentration region but is not a critical point [235]. From the calculation of parameter b in Equation (2.18) one can conclude that the positions of extrema in the concentration dependences of ΔS_S and ΔH_S (at 0.15%) correspond approximately to the concentration of KEP-2 for which the Gibbs adsorption is maximum [228]. Thus, it may be considered that the process resulting in micelle formation begins even at a KEP-2 concentration of 0.15% in this system. Depending on the system temperature, it is characterized either by the maximum energy and the highest degree of randomness or by the highest packing density and minimum energy (Fig. 2.27). Under these conditions, change of sign of thermodynamic functions may be connected with change in the packing density [229] and conformation of molecules taking part in the formation of a micelle [233].

Thus from analysis of the KEP-2–olyoxypropylene glycol mixture it is established that there are two specific concentrations of surfactant. The state of the surface layer at 0.03% is characterized by the highest stability and reflects the behavior of separate molecules. The extremal values of thermodynamic functions describing the onset of micelle formation in the adsorption layer—i.e., the cooperative behavior of KEP-2 molecules—are characteristic of the second specific concentration. These results lead to the conclusion that as the role of the interface is increased in the polyurethane formation reaction, one should expect changes in the kinetic parameters. Because KEP-2 has high surface activity, the effect of the surface (adsorption) layer in the systems investigated extends a considerable distance from the interface. Using data from [235]; interface thickness in the range of the surfactant concentrations studied is calculated as $\sim 1.3 \times 10^{-6}$ m.

Comparing the results of kinetic studies with these conclusions, we see that the surfactant availability in the reaction system does not affect the formal kinetics of polyurethane formation precisely at the transition state of the surfactant adsorption layer. The acceleration of the reaction is achieved when KEP-2 is introduced into the system at concentrations other than 0.03% and 0.15%. It is shown that there is an important relation between the effect of surfactant on the formation kinetics of polyurethane networks and the structure of surfactant layers in the oligoglycol in the formation of these polymers.

Hence, one can suppose that for nonionogenic surfactants there must exist two concentration regions with the described effect on polymer formation in systems with specific interaction of surfactant with reagents. These concentration regions correspond to the arrangement of lyophobic groups in the interface plane and perpendicularly to it.

2.8.2 Influence of surfactants on the structure of polyepoxides

Proceeding to study of the role of KEP-2 adsorption layers in polyepoxide formation, we apply the conclusions derived for the important role of thermodynamic functions in characterizing surface layers of surfactant and their relationship with the processes of formation of the chemical and physical structure of polyurethanes. This is why we begin from the thermodynamic study of the binary model system of phenylglycidyl ether (PhGE)–KEP-2. We have studied the effect of temperature (283–313 K) and concentration of surfactant (0.001–0.1%, within the limits up to CMC) on the surface tension of these surfactant solutions in PhGE, and thus, on the surface pressure (π) calculated from [232]. The dependences of extrema in π on $c_{\text{surfactant}}$ at the temperatures show that the greatest changes in surface pressure are observed at lower temperatures and for surfactant concentrations up to 0.01%. Further increase in concentration of added surface-active compounds affects the surface pressure insignificantly. Information on the value of surface pressure at each temperature studied for each concentration allowed π to be calculated from the values of surface entropy (ΔS) and enthalpy (ΔH) characterizing the thermodynamics of formation of surfactant adsorption layers under those conditions [228]. The concentration dependences of ΔH_S and ΔS_S exhibit the extremal character of changes in thermodynamic functions in the temperature range 288–313 K (Fig. 2.28, curves 5–12) and their approximation to the minimum constant value at 283–288 K (Fig. 2.28, curves 1–4). Lowering the temperature leads to broadening of the minimum and its shift towards higher concentration (from 0.01 to 0.075%). The decrease of both thermodynamic functions to minimum values indicates formation of the most highly ordered structure of the adsorption layer. In accordance with the ideas for surfactant based on polydimethylsiloxane, when the monolayer is compressed its structure rearranges, from molecules rolled up in the form of plane disks [229], through compaction of these two-dimensional formations, to the subsequent change of molecular conformation to a helix-like shape. The movement of ΔS_S into negative values proves this. Using the analogy between compaction of the siloxane surfactant layer under compression stress [229] and the change in its structure with increasing number of surfactant molecules, let us consider the results. From the change in thermodynamic functions (Fig. 2.28), one can conclude that ΔH and ΔS in the region of negative values describe the process of packing of coiled molecules to the maximally ordered state (the region of the extremum). Gradual disordering of the adsorption layer occurs with further increase of surfactant concentration [229],

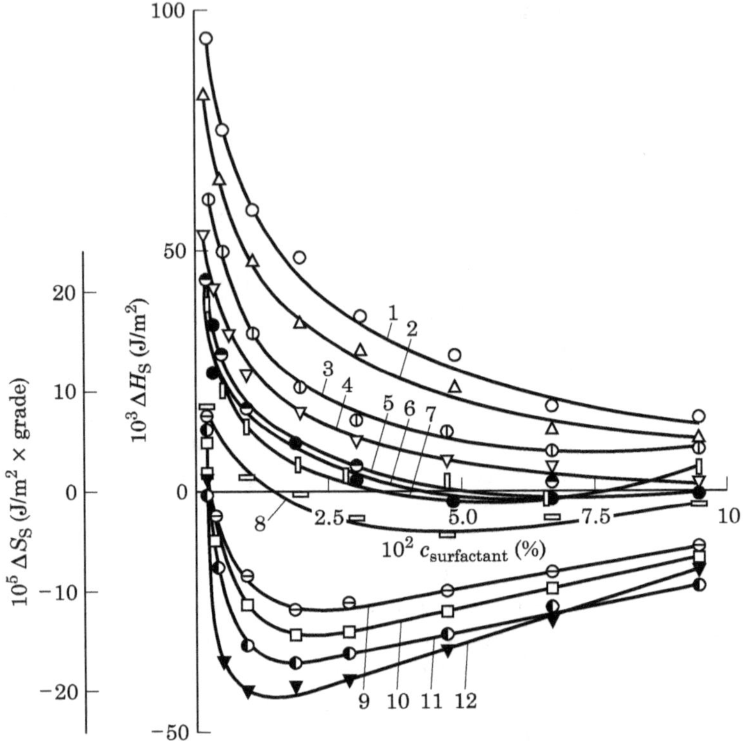

Figure 2.28 Dependence on KEP-2 concentration of thermodynamic functions ΔS_S (4, 5, 6, 8, 9, 11) and ΔH_S (1–3, 7, 10, 12) of adsorption layers of surfactant solutions in PhGE at 283 K (1, 9), 288 K (2, 5), 293 K (3, 4, 6), 298 K (7, 8), 308 K (9, 10), and 311 K (11, 12).

and this is accompanied by the increase of ΔH_S and ΔS_S values (Fig. 2.28, curves 5–12).

It thus follows that from the concentration dependences of thermodynamic functions we can determine the $c_{surfactant}$ values that promote formation of the surface layer with the closest packing of molecules. These concentrations are in the region of values significantly below the critical values and depend on temperature. (The reported results are for the model system of phenylglycidyl ether–surfactant.)

The result based on the thermodynamic characteristics of the surfactant surface layers in PhGE may be referred to the binary system of ED-20 resin–surfactant. The dependence of viscosity on the amount of added surface-active compound is characterized by a decrease of viscosity to minimum value at 0.05% that is in good agreement with the position of the extremum in the concentration dependences of ΔS_S and ΔH_S (Fig. 2.28, curves 7 and 8). It follows that when surfactant is introduced into ED-2, the resin viscosity decreases and is lower the closer the packing of surfactant molecules in the adsorption layer. Volume viscosity is characterized here, so we will deal with the effect of two interfaces: solution–air and solution–surface, determined by the microheterogeneous structure of the epoxide oligomer itself [236].

The process of gel formation with formation of aminoepoxide begins very quickly and to investigate changes in the properties of polyepoxide hardened in the presence of different amounts of surfactant one must observe the extrema in the dependences of the thermodynamic functions of surface layers against concentration obtained from the analysis. The position of maximum strength and closeness of polyepoxide crosslinking corresponds to a surfactant concentration of 0.05%. Therefore, the existence of closely packed surfactant layers in the hardening system promotes the formation of a polyepoxide network with more closely crosslinked structure and higher adhesion characteristics. For concentrations in the region of 0.003% of surfactant, the presence of surface-active additives results in formation of a polymer with a defective structure and low strength.

No monotonic dependences of adhesion strength of polyepoxides on surfactant concentration have previously been reported [223]. The decrease of adhesion strength at low concentrations of surfactant was interpreted in terms of the decrease of the surface tension of solid glue and the thermodynamic work of adhesion. However, the reason for such a decrease of characteristics at low concentrations of surfactant remains obscure. In this connection we studied the morphological structure of the surface of polyepoxide formed in presence of surfactants. Photomicrographs show that network polymers without additives are characterized by a globular type of structure, with sizes ranging from 50 to 900 nm. Separate supermolecular formations are grouped into folded structures from 50 to 1000 nm in width. These studies showed that the introduction of surfactant promotes a decrease of the average size of the globular formations in polyepoxides. At surfactant concentrations of 0.05%, the structure of a polymer is characterized by strictly bounded formations with diameter ~50–80 nm, which are grouped into structures of irregular shape with maximum extent of ~250 nm. The introduction of 0.003% of surfactant into the polymer promotes the formation of microstructure with characteristic globular formations of irregular shape and low packing density that correlates with the changes found in the polymer properties. Hence, it can be concluded that the structure of polyepoxide formed in presence of the surfactants studied depends both on the concentration of surfactant molecules in the adsorption layers (this determines conformation) and on their packing density.

Thus, for block copolymers based on dimethylsiloxane with alkylene glycols as surface-active additives there is a relation between the structure of the epoxide system in the liquid state and that in the hardened state. This allows regulation of the properties of polyepoxides using the thermodynamic characteristic of these surfactant solutions in monofunctional epoxide compounds. In addition, it is found that the introduction of surfactants of the epoxide series to polyepoxides affects the parameters of the polymer structure rather differently, depending on the arrangement (parallel or perpendicular) of lyophobic groups of KEP-2 in a monolayer with respect to the interface.

The use of quaternary ammonium salts [237] as surfactants for polyepoxides is of interest. As shown in [223, p. 301], these additives increase the rate and extent of transformation of the epoxide groups in the boundary layer. The typical (extremal) concentration dependence of polyepoxide properties with unreactive surfactant is observed. Quaternary ammonium salts should be considered inactive surfactants in systems containing epoxide compounds. Salts of the type

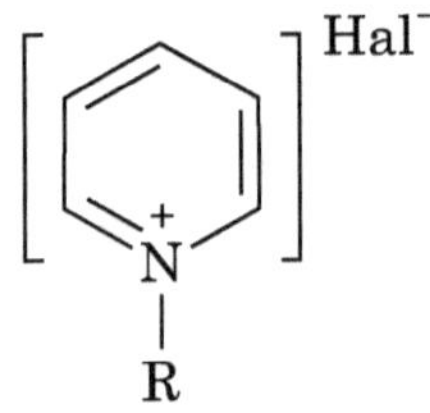

where R is ethyl, octyl (in octylpyridinium bromide, OcPyBr), heptyl (in heptylpyridinium chloride, HepPyCl) or benzene, and Hal is Cl or Br, when dissolved in the epoxide compounds, show color with a characteristic band at 404 nm and consumption of epoxide groups is observed. To more completely characterize the behavior of a number of halogenides of pyridinium *N*-arylates and *N*-alkylates, we have studied their interaction with PhGE as the model compound and made comparisons with the binary system of pyridine–PhGE (Fig. 2.29) [238]. As is evident from the electronic spectrum, the same band appears in the pyridine–PhGE system as in the case of OcPyBr in epoxy compounds. From experimental results and given the ionic nature of the mechanism of polymerization of epoxide compounds under the influence of tertiary amines, the appearance of a complex band with maximum at 404 nm may be attributed to the formation of intermediate complexes of epoxy compounds with introduced additives, which then leads to the opening of the epoxide cycle or the most active center of ionic polymerization.

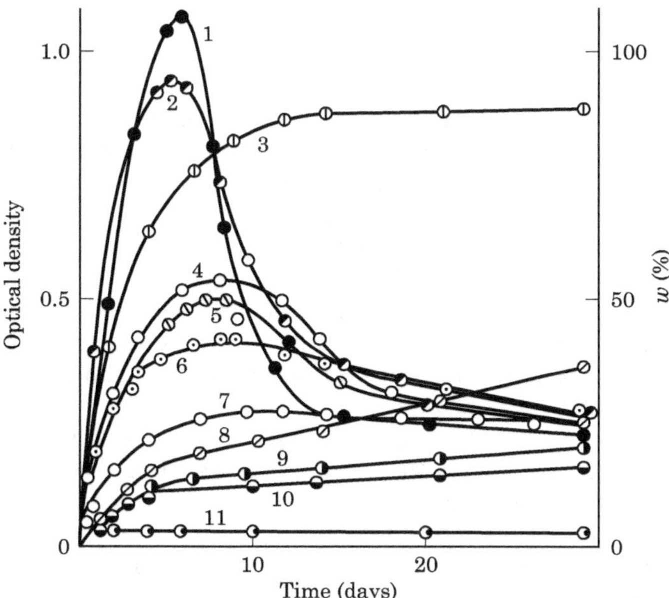

Figure 2.29 Kinetics of change of intensity for the optical band with maximum at 404 nm (1–7, 11) and content of epoxide groups (8–10) of solutions in PhGE [1% pyridine (5, 10), with addition of absolute ethanol (10-fold surplus of pyridinium salt) (2) and (11), respectively]; 10% EthPyBr or OcPyBr (3); 10% HepPyCl or BenPyCl (6–9) [(7, 9) using 90% solution of salt in water]. Spectra are taken for alcoholic solutions of concentrations of 6×10^{-2} (1–5, 11) and 6×10^{-3} mol/l (6, 7) PhGE.

The kinetics of accumulation of the compound characterized by the band at 404 nm (Fig. 2.29, curves 4 and 6) indicates a dependence between the amount of pyridine halogenide introduced into the epoxy compound and the concentration of the product formed: the maximum intensity of the absorption band in the visible region increases with the amount of additive. The character of the curves for accumulation and consumption of spectrally active particles differs for bromides and chlorides (curves 3 and 6). It follows from the data in Fig. 2.29 that the changes in the band with maximum at 404 nm (curves 4 and 6) are connected with consumption of epoxide groups (curves 8 and 10).

For example, the reaction rate constants for systems containing 10% and 1% of HepPyCl are 9.2×10^{-6} and $4.2 \times 10^{-6}\,\mathrm{L\,mol^{-1}\,min^{-1}}$, respectively. The interaction of growth of viscosity and intensity of the observed band is found in the reaction with participation of ED-20.

As was mentioned earlier, the introduction of proton-donor additives in the case of tertiary amines considerably accelerates their interaction with epoxides [22–24], which is also demonstrated by the data on Fig. 2.29 (curve 1). As is evident from Fig. 2.29, a nucleophilic reagent accelerates pyridine quaternary salts used for formation of spectrally active particles. At the same time, the presence of a proton donor in solutions of pyridine halogenides can completely inhibit the appearance of the particles (Fig. 2.29, curve 11) or produce considerable deceleration (curve 7). The latter is accompanied by a twofold decrease of the rate constant of consumption of epoxy groups: for curve 9, $K_2 = 4.7 \times 10^{-6}\,\mathrm{L\,mol^{-1}\,min^{-1}}$.

Hence, it can be supposed that the opening of the epoxide cycle by pyridine and pyridine halogenides proceeds by different routes with these substances. As a result, different end products—crosslinked or linear or branched structures—are formed.

Ultraviolet microscopy using the band characterizing the aromatic ring (250–260 nm) was also used to analyze the reaction between epoxy compounds and pyridine or its derivatives. Both aromatic and aliphatic monoepoxides were used. Changes in the intensity of the band during the reaction may be interpreted unambiguously when butylglycydyl ether is used. The number of aromatic molecules decreases in the course of interaction of pyridine or HepPyCl with the epoxy compound. Because the other aromatic structure is excluded in these systems, the observed decrease of the band within the 250–260 nm region must be related to the consumption of pyridine heterocycles or the quaternary salt.

As the reaction proceeds a band appears in the 300–310 nm region. However, the use of a heterocyclic amine with epoxy compounds does not always result in such changes in the UV spectrum. For instance, with 2,6-dimethylpyridine the consumption of aromatic molecules on interaction with PhGE is not observed. The reaction of 2,6-dimethylpyridine with PhGE is characterized by the appearance and accumulation of substances producing two bands in the visible spectrum (at 408 and 505 nm), and by a decrease in the number of epoxide groups.

Thus, from comparison of the particular changes in the maximum at 255 nm during reaction with unsubstituted pyridines and pyridines substituted in both α-positions, it can be concluded that the reaction proceeds with participation of the ring CH-groups near the heteroatom. IR and NMR spectroscopic studies have investigated the mechanism of reactions proceeding through interaction of epoxy compounds with heterocyclic tertiary amines and their derivatives.

In ^{1}H NMR study of the reaction between epoxy- and pyridine-containing compounds, the areas of signals in the ranges $\delta = 8.57$–6.88 ppm, 9.70–8.50 ppm, and 9.70–8.80 ppm, which characterize the protons of the aromatic ring of pyridine and HepPyCl, decrease ~1.5 times more rapidly than the reduction mobility due to increase in viscosity. The measure of proton consumption from the reactive groups is the ratio of integral intensity (II) of proton signals from both heterocycles to the II of protons of the aliphatic radical HepPyCl or OCH_2 in the glycidyl group, decrease in the signal being connected only with loss of mobility of the corresponding protons. Such processing of the NMR spectra permits one to observe the proportional consumption of the pyridine and epoxide rings in the reactions of pyridine and its salt with epoxy compounds. There is a time limit for the method because of rapid and significant rise of viscosity in the course of the reaction. To broaden the scope of the study, we used the ability of proton-donor reagents to inhibit reactions of quaternary chlorides with epoxy compounds. Use of a 90% solution of heptylpyridinium chloride in D_2O permitted us, first, to obtain the spectrum of the salt–phenylglycidyl ether system immediately after mixing, and, second, to inhibit the viscosity increase in the system during the reaction. Thus, it was found that dissolution of HepPyCl in the epoxy compound produces an instantaneous interaction. All the signals of protons of quaternary action shift toward strong field: the 0.27, 0.20, and 0.24 ppm shifts of the ring protons in the α-, β-, and γ-positions, respectively, as well as the 0.25, 0.2, 0.10, and 0.0 ppm shifts of protons of the heptyl radical bound to the nitrogen atom as it moves away from the latter. These results show that, like the case with pyridine [239], chemical interaction of reagents is preceded by complexation or association due to electrostatic forces. Association is likely as the large dipole moments of ionic pairs, as in quarternary salts, induce strong interaction with polar molecules [240]. It should be noted here that complexation of a tertiary amine with an epoxy compound is effective for the following oligomerization only in the presence of a proton donor [239]. In our experiments the appearance of color in the pyridine–epoxide system occurs 3–5 hours later, depending on the ratio of reagents.

Different changes of II of signals for α-, β- and γ-positions of aromatic heterocycles are seen as a result of interaction inhibition; in NMR spectra of the heptylpyridinium cation there is a considerable decrease of the signal at 9.43 ppm characterizing the protons of the aromatic heterocycle [241]. Hence, the total decrease in II of a signal from pyridine ring protons is mainly determined by the consumption of α-protons; and the observed decrease in the number of aromatic molecules is connected not with the rupture of heterocycles but with the change of their structure. Also, in a study of the interaction of BytGE with pyridine by NMR it was found that in the region of $\delta = 6.2$–5.7 ppm there appears a weak signal whose position is characteristic of β-protons and α-pyridones [242]. This is confirmed by the observed NMR spectrum of N-methyl-α-pyridone. The results of studies of the reaction of epoxide compounds with pyridine or its quaternary salt by electron and NMR spectroscopy permit us to represent the scheme of their interaction as follows:

Such a reaction mechanism is in good agreement with IR spectroscopic studies, which have shown that a simple ester bond of symmetric structure is not detectably formed for 20 days in the PbGE–pyridine and PhGE–HepPyCl systems. One observes the exhaustion of epoxide groups and aromatic heterocycles and the appearance and growth of bands at 1640, 1660, and 1700 cm^{-1} that characterize the absence of conjugated double bonds and carbonyl groups, respectively [241]. The molecular mass of the reaction products grows in the early stages of the reaction as a result of the growth of oligomer molecules (product IV) due to the opening of the double bonds of 3,5-dienes (product II). Thus, the bands that appear in the IR spectrum may be related to the valency changes of no conjugated double bond in the oligomer and of the carbonyl group of pyridone (product II). The band characteristics of the carbonyl group [241] of pyridones also appears in the electron spectrum at 300–310 nm. After 20 days of interaction a small band appears at 1100 cm^{-1}; the likelihood of reaction leading to formation of this band is probably determined by the decrease of mobility of the alloxyanion formed because of the intensive increase of the viscosity of the reaction mixture. With 2,6-dimethylpyridine and PhGE as reagents, no new bands appear either in the 300 nm region or in the IR spectrum in the 1600–1700 cm^{-1} region, although the appearance and accumulation of C—O—C bonds are observed; i.e., only the epoxy compound oligomers are formed in this case. Such behavior is in good agreement with the results of the reaction of epoxide compounds with 2-oxyethylpyridine [243].

This study thus indicates that at room temperature pyridine and its quaternary haloid salt form oligomers when reacting with epoxides by several mechanisms. In the initial stages of the reaction (15–20 days) one mainly observes the formation of heterocyclic 3,5-dienes, their oligomerization resulting in formation of a chain of heterocycles containing one nonconjugated double bond. As demonstrated, this reaction is a result of the attack by alkoxyanion on α-hydrogens of aromatic heterocycles; *N*-substituted pyridone is formed as a by-product. Parallel homooligomerization of the epoxide compounds by the anion mechanism typical of epoxide–tertiary amine systems proceeds in the following stages.

Under what conditions might such a scheme be realized of interaction of quaternary ammonium salt with the epoxide compound in the presence of primary or secondary amines? Reaction proceeding by this mechanism is possible when the amine proton-donor reagents are exhausted and when there is essentially no water. Thus, depending on the state in which the quaternary salt is introduced to the reaction system (as crystalline salt or as aqueous solution), it can act as a reactive or an inert surface-active additive.

2.8.3 Influence of surfactants on curing processes and structure of unsaturated polyesters

Polymer modifiers that influence the polymer structure in some way are widely used for regulation of the properties of cured polymers and to produce composite materials with specified chracteristics. For mixtures of unsaturated polyesters, the use of small quantities of surfactant additives is the universal method for influencing their structure and properties from the point of view of raw materials and operational capabilities. This section explores the possibility of regulating the properties of cured systems based on polyester resins by the introduction of nonreactive surfactants, since it is known that regulation of the level of intermolecular interaction can alter the molecular structure and physical-chemical properties of polymer networks [236]. It should be noted that as long as the surfactants used in polyester compositions are nonreactive, they exert their influence in a definite concentration range corresponding to the formation of the absorption layer. With large surfactant contents at the substrate boundary, multilayers are formed, leading to reduction of adhesion strength.

2.8.3.1 The influence of surfactants on the formation of microdiscontinuities. The structure and properties of the cured polyester resin are determined to a great extent by the compatibility of the system's copolymerizing components. Let us therefore consider the possibility of regulating their compatibility with the help of surfactants.

We have studied the influence of indifferent surfactants (alkyl phenyl ester of an oligoethylene glycol (OP-10) and alkyl ester of an oligoethylene glycol (DC-10)) on the optical transmission coefficient of the cured styrene-free polyester resin PN-609-21M, a solution of polydiethyleneglycol–maleate–phthalate in triethyleneglycol–dimethacrylate. The redox system of methyl ethyl ketone peroxide (PMEK) and cobalt naphthenate (NC-1) was used for curing. Results are shown in Fig. 2.30, which shows that the optical transmission coefficient versus time for different surfactant contents has extrema in the area of 4–7% concentration, with maximum at 6% by weight. The data demonstrate the various influences of OP-10 and DC-10 surfactants on the curing process. Accumulation of microgel formations [244] in reaction mixtures containing OP-10 occurs at earlier stages under the influence of surfactant in the system. DC-10 does not influence the rate of gel formation. Introducton of specific OP-10 concentrations promotes better compatibility of the cured system components and the system is characterized by a minimum degree of phase breaking of the copolymerizing system components in the region of the maximum of the transmission coefficient. This is obviously the result of the redistribution of surfactant in the system [236]. Introduction of small quantities of DC-10 (2–6 wt%) has no influence on the compatibility of the system components. With increasing DC-10 content, up to 10 wt%, the compatibility of the system components declines (Fig. 2.30, curve 2).

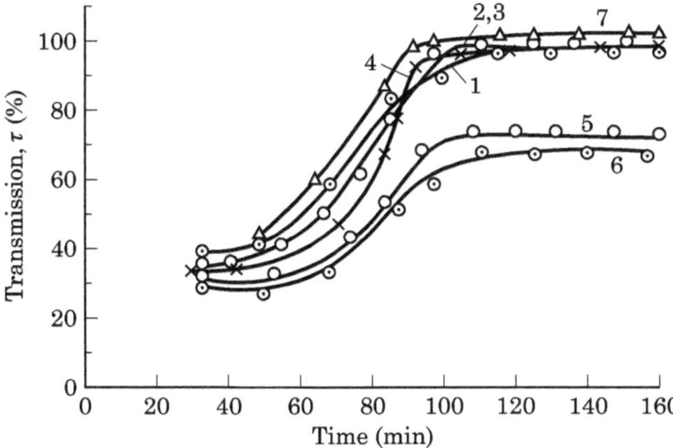

Figure 2.30 Dependence of optical transmission coefficient of PN-609-21M polyester resin on surfactant content in the curing process. DC-10 wt%: (1) 0%, (2) 2%, (3) 4%, (4) 6%, (5) 8%, and (6) 10%. OP-10 wt%: (7) 6%.

Investigations of turbidity spectra of PN-609-21M polyester resin in the presence of surfactant have permitted us to determine the concentration interval in which a surfactant significantly affects the system's condition and to define the character of this influence on copolymerizing systems. Investigation of the influence of these surfactants on the rheological behavior of the system was done with the aim of confirming the regularities and assumptions described above.

2.8.3.2 The rheological behavior of polyesters in the presence of surfactants. Methods of measuring the rheological characteristics were chosen to suit the features of the systems investigated. Unsaturated polyesters in the form of monomers in solution are Newtonian liquids since flow does not cause changes in their structure and their viscosity does not depend on voltage shift.

Results from investigations of the viscosity of polyester resins with addition of OP-10 and DC-10 surfactants are presented in Fig. 2.31. The vicosity of the initial resin is virtually unchanged by introduction of surfactants, probably due to the comparatively small molecular weight of the initial resin, for which the effect of change of macromolecular shape due to blocking of polar groups by surfactant is not yet evidenced [244]. A different pattern is observed with introduction of 4 wt% of the PMEK–NC-1 redox system (Fig. 2.31, curves 3 and 4). During curing in the presence of small quantities of OP-10 (1.2 wt%) the system viscosity increases only slightly, but on increase of surfactant content to 6 wt% a sharp increase of the viscosity of the cured

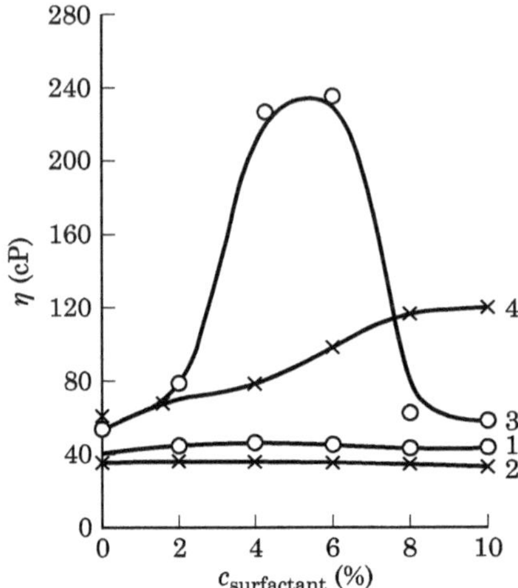

Figure 2.31 Dependence of PN-609-21M polyester resin viscosity on concentration of surfactant: (1) OP-10; (2) DC-10; (3) OP-10 in the process of curing; (4) DC-10 in the process of curing.

system occurs. Increase of OP-10 up to 8 wt% leads to reduction of the viscosity of the polyester resin. The viscosity does not change with further increase of surfactant concentration. The introduction of various quantities of DC-10 into the cured polyester resin produces only a gradual viscosity increase throughout the range of surfactant concentrations (Fig. 2.31, curve 4). Thus, it is concluded that during curing of polyester resins the structure-forming action of surfactants begins to become evident with the increase of the molecular mass of the resin [244].

2.8.3.3 Kinetics of the curing of unsaturated polyesters in the presence of surfactants. Figure 2.32 depicts viscosity data for the cured PN-609-21M resin in the presence of OP-10. The viscosity of the polymerized system increases sharply after a certain induction period. As seen, the introduction of surfactants to the cured system exerts a considerable influence on the dependence. Thus, for systems cured in the presence of OP-10 the curves of viscosity increase are located to the left of those for the systems without surfactants, and the curves for OP-10 content of 4–6 wt% (curves 3 and 4) are distinguished by the most rapid rise and greatest leftward shift of the vertical branches. It follows that in reaction mixtures containing surfactants the gel formation process begins at earlier stages of interaction. Introducton of increasing concentrations of DC-10 somewhat accelerates the beginning of polyester resin gel formation over the entire concentration interval.

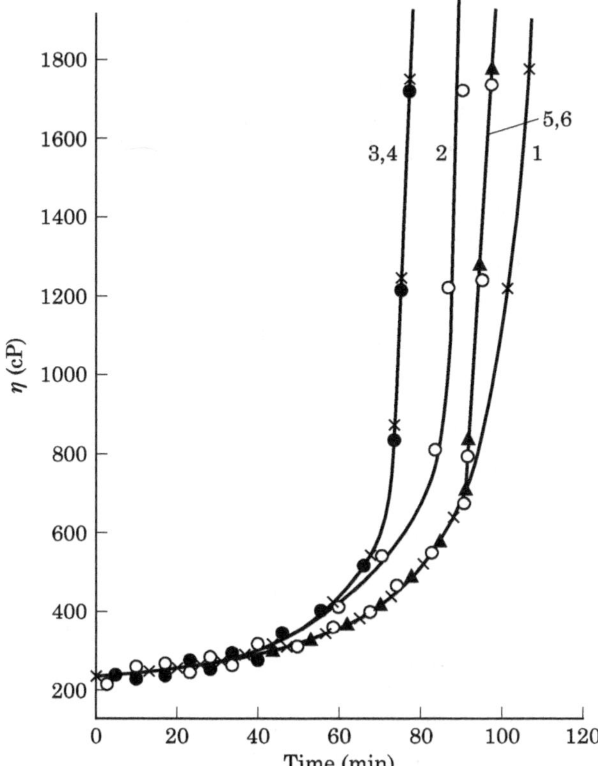

Figure 2.32 Dependence of PN-609-21M polyester resin viscosity in the process of curing on concentration (wt%) of surfactant OP-10: (1) 0%, (2) 2%, (3) 4%, (4) 6%, (5) 8%, and (6) 10%.

The influence of OP-10 is of extreme character and the largest acceleration of gel formation occurs with an OP-10 content of 4–6 wt%. The acceleration of the initiation of gel formation in the system containing DC-10 depends directly on the concentration.

The data correlate well with turbidity data (Fig. 2.31, curves 2–4) and show that acceleration of the start of gel formation with 4–6 wt% OP-10 occurs because of accumulation of microgel formations under the influence of the surfactant at an earlier stage. Acceleration of the start of gel formation wih DC-10 is insignificant and remains directly dependent on DC-10 concentration, corresponding to the turbidity data that indicate that this surfactant does not influence the compatibility of the system components and speed of gel formation.

2.8.3.4 The influence of surfactants on surface tension in reaction systems based on unsaturated polyesters. Changes in the structure and chemical composition of the polymerized oligomer and in the colloid-chemical properties of the surfactant influence the surface tension of polymerized systems. The oligomer surface tension depends on the time of existence of the surface because relaxation and diffusion processes in the bulk and at the boundary run comparatively slowly. The system viscosity increases due to oligomer polymerization, thus extending the time of establishment of equilibrium at the boundary. Consequently, the polymer surface tension depends on the correlation of the rates of the polymerization reaction and the reaching of molecular equilibrium density at the oligomer–air boundary [236].

Let us consider the changes in surface tension during polymerization of polyester resin with surfactant additives. Experimental results are presented in Fig. 2.33, which shows that the dependence of the solid polymers' surface tension on the surfactant concentration has a minimum at low surfactant concentrations (1–2 wt% of OP-10 and 1–5 wt% of DC-10). With increase in surfactant concentration, the surface tension also increases. In the case of OP-10, the maximum surface tension is reached at a content of 4 wt%; for DC-10 the corresponding content is 7 wt%. Further increase in the concentration of both surfactants leads to a sharp reduction of the surface tension of the polyester resin and ultimately to its independence of surfactant concentrations. As shown in [236], the surfactants can block polar groups of the polymers, breaking intra- and intermolecular bonds. This can result in change of the conformation of molecular chains and of intermolecular interactions that depend on whether the molecule is linear, unfolded, or globular in shape. Blocking of the polymer polar groups by the surfactants should result in reduction of intermolecular interactions, but the change in macromolecular conformation due to such blocking can

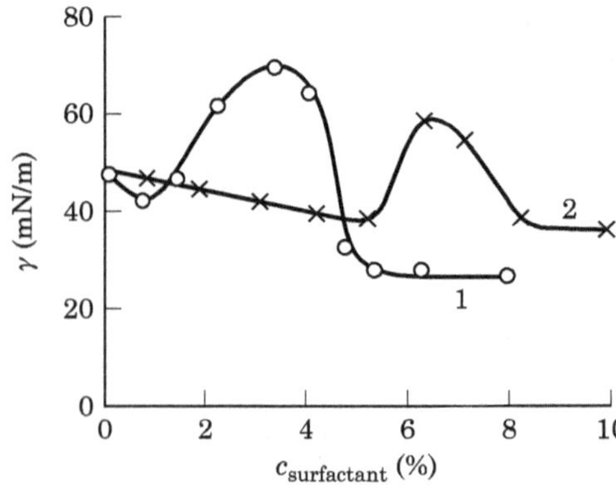

Figure 2.33 Dependence of PN-609-21M polyester resin surface tension on surfactant concentration in the curing process: (1) OP-10; (2) DC-10.

increase interactions as a result of macromolecules unfolding and forming new contacts.

In the systems considered, similar processes evidently take place. Having adsorbed at the boundary, surfactants block the polar groups, leading to reduction of the interaction between polymer macromolecules and worsening the surface tension. With increase of surfactant content in the system, polymer surface tension begins to increase sharply and reaches a maximum value that can be related to macromolecular unfolding. At larger surfactant concentrations, the surface tension decreases until it reaches a constant value that does not depend on surfactant concentration and is apparently connected with filling of the surface layer with surfactant [236].

Thus, if the oligomer surface tension in the presence of surfactant decreases monotonically to a certain limit, the solid polymers' surface tension is first reduced as the surfactant content increases, then it goes through maximum, and finally becomes independent of the surfactant concentration.

2.8.3.5 Cohesion and adhesion characteristics of polymers based on unsaturated polyesters with surfactant additives. As was shown earlier, the dependences of the surface tension of the initial liquid oligomers and of the cured polymers on surfactant concentration are notably different. For this reason it was of interest to study these relationships with changes in the physical and mechanical properties of cured polyesters. Results from these investigations are given below. The breaking stresses $\sigma_{str.}$, $\sigma_{comp.}$, and $\sigma_{bend.}$ increase with OP-10 concentration in the system, and only for OP-10 concentrations of 10 wt% does a certain reduction of breaking stress of stretching, compression, and bending take place. With introducton of DC-10 into the system, the breaking stresses of stretching, compression, and bending decrease monotonically throughout the range of surfactant concentrations.

The dependence of the modulus of elasticity under compression on OP-10 and DC-10 concentration has extremal character, with a maximum at 5–6 wt%; the DC-10 curve is located somewhat lower than the similar curve for OP-10. $E_{bend.}$ does not depend on OP-10 concentration, and only the introduction of 10 wt% of OP-10 leads to a decrease of elasticity modulus under bending. $E_{bend.}$ decreases monotonically throughout the range of DC-10 concentrations.

Introduction of OP-10 leads to an increase of cohesion strength at all concentrations considered, and only a content of 10 wt% results in decrease. The influence of DC-10 on cohesion properties is the inverse of this: cohesion strength becomes worse at all concentrations in the range.

Adhesion strength increases with time as a function of OP-10 concentration over 10 days; it then reduces slightly, but further increasing the OP-10 concentration results in reduction of adhesion strength. Similar behavior is seen with DC-10, with the only difference being that introduction of 6 wt% of DC-10 leads to a noticeable increase in adhesion strength. Introducton of 8–10 wt% of DC-10 leads to a definite reduction of adhesion characteristics during 10 days after adhesion, but adhesion strength then increases with time, and after 30 days it exceeds the adhesion strength of polyester resin adhesive joints without surfactants.

Comparing the results obtained, the various influences of OP-10 and DC-10 on polyester resin adhesion can be assessed. Whereas introduction of OP-10 leads to reduction of adhesion strength over the entire concentration range, with DC-10 the introduction of small concentrations (2 wt%) causes a reduction of adhesion strength over time. Increasing DC-10 concentration in the system increases the adhesion strength, and maximum adhesion strength increase is achieved with introduction of 6 wt% of that surfactant.

The dependence of polyester resin adhesion strength on the surfactant concentration shows extrema. Adhesion strength with increasing surfactant concentration first goes through a minimum; the adhesive joint strength then increases, and with further increase of surfactant concentration decreases again. To explain the increase in cohesion and adhesion strength, let us consider the data on change of turbidity coefficient, viscosity, and surface tension of the system under the influence of surfactant. From the turbidity data, OP-10 surfactant promotes better compatibility of copolymerized system components, while DC-10 has no influence and at higher DC-10 concentrations the compatibility of system components becomes worse. Hence, it can be supposed that with OP-10 the change of parameters characterizing behavior of the cured system in bulk is greater; the second case relates to the change of surface properties of the cured polyester resin. The turbidity data are well correlated with the results of the investigations of the cohesion and adhesion properties of polyesters. Naturally, the introduction of concentrations of OP-10 that do not actually worsen the adhesion characteristics leads to improvement of the cohesion properties of unsaturated polyesters, and with certain DC-10 concentrations the adhesion strength of the system increases.

As shown earlier, the dependence of the surface tension of solid polymers on the surfactant concentration exhibits a minimum and maximum, and the increase in surface tension under the influence of surfactant is rather significant (Fig. 2.33). This is apparently due to the demonstrated structure-forming action of surfactant, which is possible only at stages of reaction when sufficiently flexible molecules are formed [236]. These results are supported by the fact that the viscosity of the initial systems is practically unchanged with introduction of surfactant (Fig. 2.31, curves 1, 2). In the course of curing in the presence of surfactant, the system viscosity increases sharply as the molecular mass grows, i.e., the surfactant structure-forming action begins. The dependence of the viscosity of unsaturated polyester resin during the curing process on surfactant concentration is well correlated with the same dependence for the surface tension (Figs. 2.31 and 2.33).

Thus, dependence of adhesion strength on the surfactant concentration correlates with the dependence of the system surface tension. The increase of surface tension under the influence of surfactant is caused by changes in the polymer structure that determine the cohesion strength and should therefore lead to increase of the adhesion strength.

From the investigations reported, it may be concluded that the concentration dependences of the physical and mechanical properties of unsaturated polyesters formed in the presence of surfactants are of extremal character. This extremal character corresponds essentially to the same dependences of surface tension, turbidity, and viscosity of system with surfactant present.

Thus, surfactant influences the processes of polyester formation, improving the physical and mechanical characteristics only when there is improvement of the compatibility of the components of the polymerized system.

Chapter

3

Properties of Adhesives Based on Polymeric Mixtures

3.1 General

In the course of blending polymers, the following systems can be formed: one-phase systems, two-phase (colloid) systems, or systems in a metastable state of transition from a one-phase into a two-phase system. The properties of polymer mixtures are determined to a great extent by the phase equilibrium in the system formed and their properties can be changed by controlling the processes of phase separation, which occur by two mechanisms: by nucleation and growth or by the spinodal mechanism.

Two-phase mixtures of polymers differ from classical colloid systems mainly by a transition layer that exists between the system components and is of special significance. The formation of such a layer in mixtures of linear polymers is governed by the kinetic factors of the retarded process of phase separation, by the collid-chemical mechanism of formation, or by the adsorption interaction as well as by the segmental solubility [103, 104]. In mixtures of crosslinked polymers its formation is governed also by the conditions of synthesis. Note also that in the thermodynamically nonequilibrium mixtures of polymers in the two-phase systems, the processes of segmental solubility usually have time to reach completion while the macromolecules do not move inside the high-viscosity medium, which ensures the stability of the structure and its mechanical properties [103].

The most promising method of obtaining polymeric materials is by the use of so-called hybrid binders. These constitute polymeric composite materials in which crosslinking indicates the heterogeneity of the distribution of the system components determined by thermodynamic incompatibility and thus hinders further phase separation. Areas of microheterogeneity in colloid sizes are formed in the polymerizing system and are considered as virtual phase particles or phases. The system reaches a metastable state (which is not thermodynamically at equilibrium). For such systems there is a great increase of the mechanical properties in a narrow range of compositions [19].

Surface physical-chemical phenomena are involved in all heterogeneous polymeric systems with a developed interphase surface. These include wetting and all kinds of adhesion (friction, fracture, dispersion, homogenization, structuring). The development of the idea of the interrelation of the dispersion state and surface phenomena, due to P.A. Rebinder, leads to the necessity for the addition of surfactant in small amounts. The mechanism of the action of these additives on different interphase boundaries is considered in detail in [105].

Integration of surfactant in small quantities facilitates the dispersion of the structural elements of dispersions, solutions, and melts of the polymers. In the case of surfactant with groups that can be involved in the formation of hydrogen bonds between the structural elements, the effect of particle dispersion increases with increase of the length of the surfactant hydrocarbon radical and with increase of the polymer polarity. To improve the properties of the polymer mixture, a third component is sometimes added. A rather promising method of controlling the properties is to combine polymer mixtures with their block polymers. The latter in small quantities tend to concentrate on the interphase boundary, thus improving the interaction of the mixture components; they seem also to serve as emulsifiers.

The properties of polymer mixtures depend on the method by which they are obtained and are determined by many factors: by sizes of particles of the dispersed phase, by their shape and number in bulk, and by the thermodynamic affinity of the components for one another [19]. Linear polymers blend either in the course of their mutual dissolution or, in two-phase systems, under conditions of thermodynamic incompatibility of the components, when the dispersion is forced. The mixtures formed can be compatible (forming true solutions of one polymer in another), incompatible (representing a typical colloid system), quasicompatible (characterized by microscopic homogeneity at a level above heterogeneity on the molecular level), or pseudocompatible (with a strong adhesion interaction on the boundary) [106].

In the majority of cases the compatibility of the polymers is characterized by the glass-transition temperature T_g, determined by methods such as dilatometry, differential scanning calorimetry (DSC), reversed-phase gas chromatography (RGC), radiation thermal luminescence (RTL), dynamic mechanical spectroscopy (DMS), nuclear magnetic resonance (NMR), or dielectric loss. The existence of two values of T_g, corresponding to T_g of the individual components in a two-component system, is a certain indication of the incompatibility of the components. When there is a single T_g that depends on the change of composition, a mixture of two polymers can be considered compatible. The sensitivity of the method used must be taken into account. For example, DMS detects a much smaller quantity of the second phase than does DSC, while NMR uniquely reveals small quantities of the second phase ("clusters") that are not detected by other methods.

The compatibility of polymer–polymer mixtures can also be determined by nonradiative energy transfer (NRET), by measuring the intensity of light scattering, and by electron and neutron diffraction studies. Electron photomicrographs of two-phase systems suggest a colloid system of a solid in a solid; for compatible systems the second phase does not show up. But electron microscopy does not unequivocally resolve the question of phase composition, in that 10–15 nm particles can be either actual particles of the second phase or heterogeneities that do not have the properties of the second phase. A macromolecule coiled a certain number of times will have a size of this order, so that microscopic observation of heterogeneity does not necessarily indicate that a second phase is present; additional information from the change of T_g is necessary.

It is more reliable to estimate the compatibility of polymers in terms of thermodynamics. The thermodynamic compatibility or mutual distribution of molecules of one substance within another, which extends up to the state of true solution and does not depend on the mixing procedure, is always accompanied by Gibbs energy decrease [107]:

$$\Delta G_{\text{mix}} = \Delta H_{\text{mix}} - T\,\Delta S_{\text{mix}} \tag{3.1}$$

where ΔH_{mix} and ΔS_{mix} are the molar enthalpy and entropy of the mixing, which determine the compatibility of the components. The compatibility of polymers is favored as a rule by negative values of ΔH_{mix} [39], but these do not mean thermodynamic compatibility in all cases: the role of ΔS_{mix} is also important. To ensure a negative value of ΔG_{mix} (i.e., to produce a homogeneous solution) it is necessary to create favorable energetic and geometric conditions [40].

The concept of the solubility parameter δ, which is frequently used in studies of polymer compatibility, is not always applicable. The solubility parameter is equal to the square root of the substance's cohesion energy density ($\delta = \sqrt{\text{DEC}}$) and reflects only intermolecular interaction without accounting for the entropy factor. This parameter is an integral characteristic of only intermolecular interaction, and the components' miscibility is determined by the presence of functional groups capable of specific interactions in the polymer molecules. This is why one can predict that polymers with poorly matched solubility parameters will not blend, although the opposite—good miscibility for polymers with close values of solubility parameters—does not always hold [108]. Thermodynamic compatibility is estimated using phase diagrams of the three-component systems of solvent (1)–polymer (2)–polymer (3). In addition, in similar systems the region of the ply separation increases with increase of the molecular weight of the components during blending. The ply separation of the polymers in a solvent depends not only on the compatibility of the polymers but also on their change of compatibility in the solvent. Thus, mutual solubility of the polymers is influenced both by their mutual interaction and by the nature of the solvent.

It is accepted that the most reliable information on thermodynamic compatibility of polymers is obtained from phase diagrams of binary systems, which were introduced by G. Allen using the method of turbidity spots [109]. Such diagrams can also be plotted from results of optical interferometry, RGC, and other methods. To plot phase diagrams it is necessary to determine the Gibbs free energy of mixing ΔG_{mix} and from its concentration and temperature dependences locate the point that corresponds to $\partial^2 G/\partial\varphi_1\partial\varphi_2 = 0$; this represents a spinodal on the Gibbs energy of mixing diagram and is determined by a double graphical differentiation of the $\Delta G_{\text{mix}} = f(\varphi)$ curves at various temperatures and by extrapolation of $\partial^2 G/\partial^2\varphi$ to zero at various concentrations.

A.A. Tager [110] determined the change of ΔG_{mix} by producing thermodynamic cycles using static sorption of solvent vapors on the polymeric specimens. However, in solution the polymer compatibility is influenced by the difference of the thermodynamic interaction parameters of the common solvent with each indvidual polymer. Additionally, compatibility in solution does not always correspond to that in the solid state (without a solvent). These shortcomings limit the applicability of the ΔG_{mix} estimation method that Tager developed.

In the procedure proposed by Farooque and Deshpande [111], ΔG_{mix} is found by the method of dynamic sorption of ROC, which removes the uncertainty in determining this parameter. These authors improved the procedure in such a way that the molecular weight and the polydispersity of the polymer stationary phase influence the experimentally determined value—the sorbate retention volume V_{q}.

Quantitative estimation of the compatibility of polymers and oligomers can be made from the concentration dependence of the Flory–Huggins thermodynamic interaction parameter $\chi_{2,3}$. The shape of the phase diagram is judged to a certain extent by the type of this dependence in that there is a relation $\Delta G_{\text{mix}} = RT\chi_{2,3}\varphi_2\varphi_3$ between the parameters $\chi_{2,3}$ and ΔG_{mix} [111]. The critical miscibility temperature is determined from the concentration dependence of $\chi_{2,3}$ at various temperatures.

Of interest are the studies of polymer compatibility depending on the excess volume of the blended components ΔV [112]. As these authors point out, $\Delta V < 0$ for thermodynamically compatible systems in the majority of cases. This was the basis for concluding that in the course of blending the mobility of the macromolecular chains decreases to values comparable with those of individual components.

Patterson [113] reports that compatibility of polymers can be considered in terms of the dependence of the difference of their free volumes τ_n on the interaction parameter $\chi_{2,3}$ and on molecular weight. In particular, he showed that only polymers with small value of τ_n at $\chi_{2,3} > 0$ are compatible. However, if $\chi_{2,3} < 0$ (when there are specific interactions between the components), compatibility is possible at greater difference of their free volumes.

These considerations indicate that structural (X-ray analysis, and optical and electron microscopy), relaxation (mechanical and dielectric relaxation, NMR, and RTL), and thermodynamic (phase diagrams, thermodynamic cycles, RGC) methods of investigation are currently used to determine the limits of the mutual solubility of polymers. At the same time it must be noted that evaluation of polymer compatibility is complicated by kinetic factors and because thermodynamically unstable systems are formed in the course of mixing of polymer. The effects of the nature of the polymers and of foreign impurities on compatibility do not yield to a complete explanation and this problem is far from being solved, despite its importance in the fact that the problem of modifying the properties of polymeric material is related to evaluation of the compatibility of polymers.

3.2 Adhesives Based on Interpenetrating Polymer Networks

An interpenetrating network (IPN) can be defined as a mixture of two crosslinked polymers when at least one of them is synthesized and (or) crosslinked with another [114]. The components that make up an IPN are thermodynamically incompatible and a transition region of two phases is formed in such a system. The whole complex of IPN properties is determined by the availability and features of this region.

Let us consider the properties of an adhesive based on the following mixture of oligomers: an unsaturated polyester resin and a prepolymer with end isocyanate groups (or macrodiisocyanate) based on polydiethylene glycol adipate of molecular weight 800 and on TDI (a mixture of the 2,4- and 2,6-isomers in ratio of 65:35), with the surfactant ATG added. The adhesive is cured by the effect of a redox system, for which MEKP(O) and CN are commonly used.

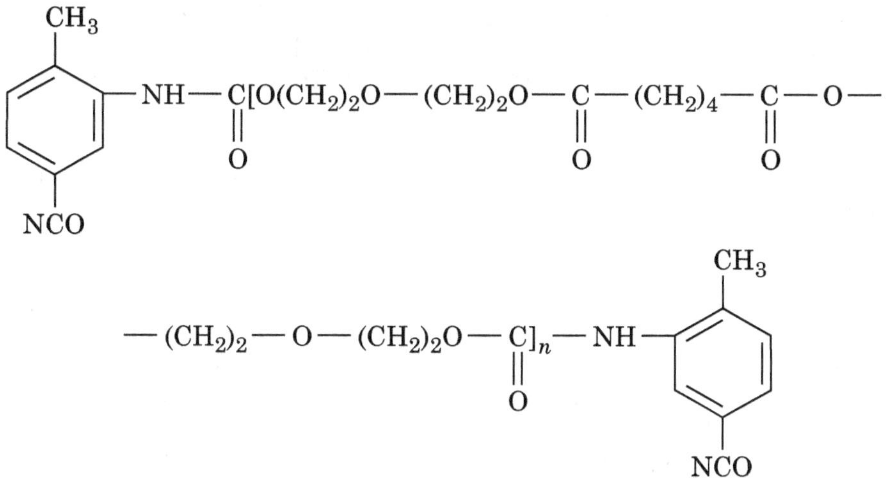

$$CH_3C_6H_3(NCO)\text{—}NH\text{—}\underset{\displaystyle O}{\overset{\|}{C}}[O(CH_2)_2O\text{—}(CH_2)_2O\text{—}\underset{\displaystyle O}{\overset{\|}{C}}\text{—}(CH_2)_4\text{—}\underset{\displaystyle O}{\overset{\|}{C}}\text{—}O\text{—}$$

$$\text{—}(CH_2)_2\text{—}O\text{—}(CH_2)_2O\text{—}\underset{\displaystyle O}{\overset{\|}{C}}]_n\text{—}NH\text{—}C_6H_3(CH_3)NCO$$

The “Sprut-5M” adhesive consists of unsaturated polyester resin 100 parts by weight, ATG-M 80 parts (a mixture of 50 parts of macrodiisocyanate and 30 parts of ATG), polymerization initiator (MEKP(O)) 2 parts. Usually 1 wt% of CN is integrated additionally into the adhesive base. Sprut-5M can be produced using various polyester resins, but the most applicable one is the adhesive based on PM-I resin owing to its availability and low cost. For this reason we will consider the properties of the adhesive based mainly on this resin although it has adhesion strength 20–40% lower than that of adhesives based on PN-11, NPS 609-21M, and other resins, especially when cementing in water.

MP PDEG in the composition of the polyester resin has end hydroxyl and carboxyl groups capable of interacting with the isocyanate groups. To prevent the chemical interaction of the individual IPN networks, the MP PDEG end groups were pre-blocked with a monofunctional isocyanate (phenyl isocyanate); later this reaction was excluded with almost no effect upon the adhesive properties.

The adhesive cures for 20–40 min, during which time the isocyanate groups in fact have no time to interact noticeably with the hydroxyl groups and especially with less active carboxyl groups of MP PDEG. Immediately after the components were mixed and in the two weeks after the cure (the adhesive was in the shape of a free film), the intensity of the 3500–3600 cm^{-1} absorption band, which is referred to valence oscillations of polyester hydroxyl groups, did not change, indicating the low conversion of these groups in the course of adhesive cure. At 20°C, after PN-1 and ATG-M are mixed, a clear change of viscosity is observed at 3–4 h, which is evidence that there is no chemical reaction between them up to that time.

It has been reported that under the influence of radical polymerization initiators styrene can copolymerize with TDI with the formation of the substitute polyamide, identified by the IR absorption band of the carbonate group at ~1730 cm^{-1}. In the system under consideration no formation of the substituted polyamide is observed. Thus, the polyester and polyurethane components essentially do not react one with another in the course of the adhesive cure.

With time, the free isocyanate groups of the solid adhesive react with the moisture that diffuses through the adhesive, as confirmed by the decrease of the IR absorption of the isocyanate groups at ~2260 cm^{-1}. This process results in polymerization of the macrodiisocyanate.

Interaction of the adhesive isocyanate groups with water alters a number of adhesive properties such as the modulus of elasticity, the adhesion strength, electric properties, etc. Thus, the modulus of elasticity of adhesives determined at 3 days after curing for specimens cemented and kept in air, in water, and in oil was 40, 110, 160 MPa, respectively; at 100 days these values increased to 400, 120, 320 MPa.

Figure 3.1 presents the maximum stress surfaces when St3 steel is cemented with the Sprut-5M adhesive and the polyester and poly-

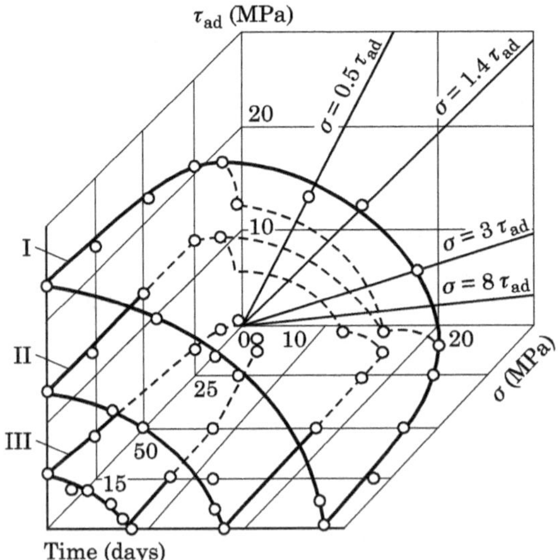

Figure 3.1 Surfaces of maximum stress of steel specimens cemented with Sprut-5M (I), polyester (II), and polyurethane (III) adhesives.

urethane individual components of this adhesive. It is clear that the adhesion strength of the Sprut-5M considerably exceeds not only those observed using the indvidual components but also their sum. To clarify this phenomenon, the interphase tension of the polyester, polyurethane and of their mixture was measured:

Polyurethane content in the mixture (%)	0	2	5	10	20	100
Interphase tension (MPa)	331	311	300	290	296	370

As can be seen, the interphase tension of the polyester and the polyurethane is much higher than that of their mixture. One can assume that the formation of the loose transition region in the IPN type mixtures facilitates the adsorption of the polymer's polar groups on the substrate surface and their closer packing.

The decrease of the interphase tension governs the increase of adhesion strength of adhesive-bonded joints and their resistance to liquid media. Note that the adhesion strength of Sprut-5M adhesive is completely determined by the polymer cohesion strength. When other resins that produce stronger polymers replace the PN-1 resin, the adhesive strength of the adhesive also increases. The rupture of the adhesive-bonded joints is of cohesive character in all cases. Their adhesion strength increases when they are held in oil and decreases when they are held in water (by about 25%) then stabilizes at this level. The use of PN-11 and NPS 609-21M polyester resins (which contain triethylene glycol dimethyl acrylate instead of styrene) to replace the PN-1 resin enhances the water resistance of adhesive-bonded joints. With integration of aerosil, which is hydrophobicized by dimethyldichlorosilane, into the adhesive the adhesive strength did not change when the joint was held in water.

Adhesives based on unsaturated polyester resins have not only comparatively low adhesion but also low long-term strength. For example, the limit of PM-1 polyester resin long-term strength is less than 10% of the maximal short-term strength. The application of IPN-based adhesives considerably increases the long-term strength of adhesive-bonded joints. As Fig. 3.2 shows, the strength of Sprut-5M adhesive loaded to 50% of the fracture loading does not decrease but increases in time, with the loaded specimen's strength growing more rapidly than that of the unloaded specimens.

The long-term strength was studied by a method based on the principle of "damage accumulation," i.e., plotting the kinetic curve of the short-term strength dependence on the duration of constant loading. The growth of the long-term strength of adhesive-bonded joints ceases as soon as the isocyanate groups in the adhesive are completely con-

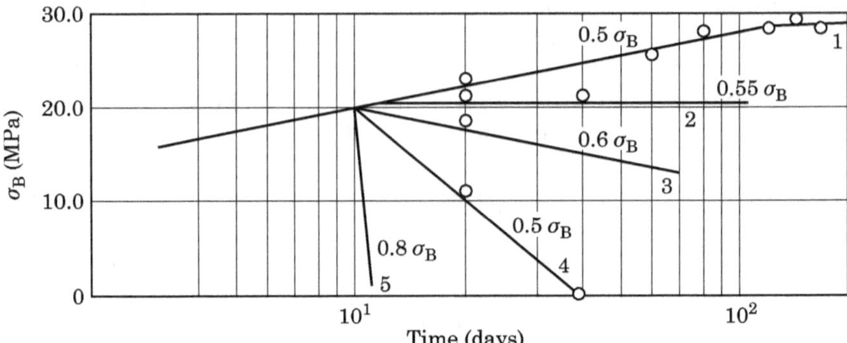

Figure 3.2 Long-term strength σ_B of St3 adhesive-bonded joints at various loadings: (1, 3, 5), Sprut-5M; (4) polyester adhesive.

sumed. Thus, the increase of the long-term strength seems to be related to the presence of the reactive oligomer in the adhesive.

Let us consider a possible mechanism for the increase in the long-term strength of adhesive-bonded joints. Under the effect of loading, the initially elastic adhesive interlayer is deformed. In this case the MDI molecules, which are in the interstructural regions of the polyester network, are oriented with the force field with changes in conformation that can result in increase of the cohesion and adhesion strengths. This assumption is confirmed by the following experiment. If adhesive-bonded joints are held at normal or shear load and then tested at various ratios of normal or tangential stresses, the strengthening of the adhesive interlayer is observed mainly in the direction parallel to the applied load. Adhesive-bonded joints subjected for 100 days to a shear or normal loading of 30% of the fracture load showed increased strength, mainly for shear or normal pull (Fig. 3.3).

There is an increase in the polymer strength when the polymers are formed within a gradient of shear stresses, known to result in the alteration of the macromolecular conformation. For example, polyethylene produced by melt extrusion through capillaries under high pressure possesses unique physical-mechanical properties—its modulus of elasticity is of the same order as that of glass fiber (7×10^{10} Pa) [116].

We investigated the effect on the adhesion strength of shear stresses applied to the adhesive interlayer MDI as a component of Sprut-5M adhesive. An equimolar quantity of butylene glycol was added to MDI to cure it. The shear stresses were created by the torsion of one dolly of standard size relative to another in a special device; the adhesive was placed between the dollies. On reaching the prescribed shear stress (due to the increase of the adhesive viscosity) the torsion of the dolly

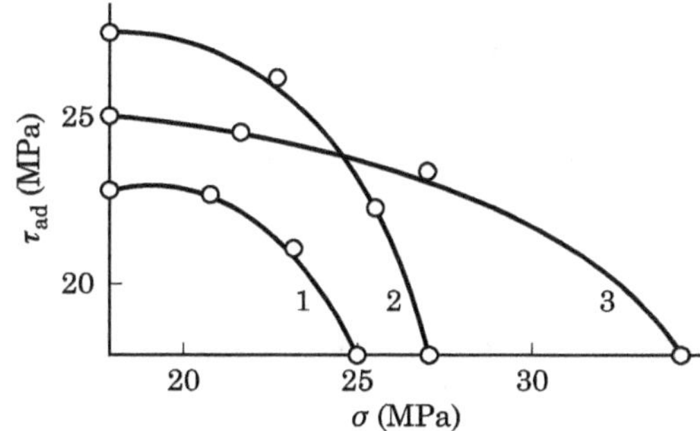

Figure 3.3 Adhesion strength of the adhesive-bonded joints for different relationships of the normal and tangential stresses 100 days after cementing: (1) unloaded specimens; (2) shear load 0.3 GPa; (3) normal load 0.36 GPa.

ceased spontaneously. The adhesive interlayer was post-cured and then the adhesion strength was measured.

It was shown that the polymerization of the adhesive within the gradient of the shear stresses that ensures the stretching and corresponding reorientation of the adhesive macromolecules can increase the adhesion strength more than twofold. If the torsion of the dolly is stopped while the viscosity of the adhesive is still low, the strength does not increase as much, because the adhesive molecules have had time to acquire initial conformation before the polymerization. The cemented dollies were tested for shear.

The increase of strength is observed not only in the direction of action of the load but to some extent perpendicular to it (see Fig. 3.3). Thus, orientation of the marcomolecules alone is not enough to explain the increase of strength, and one must assume other mechanisms.

It was shown earlier that contrary to the well-known case, the strengthening of some resins under single-axial tension occurs not only parallel to the axis of orientation but also perpendicular to it. Some authors [117] assume that a similar effect is produced by the strong forces of transverse compression that result in the increase of contact between the polymer globules. In the case under consideration, a similar mechanism of transverse strengthening of the adhesive cannot operate. As Fig. 3.3 shows, the transverse strengthening of the adhesive is observed under both normal and tangential loads. Although transverse compression of the adhesive may occur in the latter case, it is excluded under normal load because of the adhesion interaction of the adhesive with the undeformed substrates.

Chains of macromolecules are known to fail at even small loads. Electron paramagnetic resonance (EPR) studies showed correlation between the long-term strength of some polymers and the formation and accumulation of free radicals. Depending on the environmental and operational conditions, macroradicals can cause loss of either structure or structurization. In all cases the failure of the polymer under mechanical load depends more or less on the development of these processes. The course of various processes, including chemical ones, is determined to a great extent by the medium the macroradical contacts. The final result of the mechanical processes depends on the direction of the subsequent secondary conversions of free macroradicals, which are multistage, complex, and at the same time typical of the free-radical processes that occur in polymeric systems.

The polymer strength depends on both the probability of destruction of the polymer chains and the recombination of the free macroradicals formed. The potential barrier to restore a terminated chain increases the stress. A liquid with which the free radicals react without secondary chemical reactions further decreases the probability of the recombination of macroradicals. In this case one may consider the medium to be the catalyst of the polymer destruction process.

The alteration of the material strength by different media (alcohol, for example) depends on the length of the molecular chain: the material strength falls with the decrease of the number of carbon atoms in the alcohol molecule. This is related to the fact that with decrease in the size of the alcohol molecules the chance of their recombination and of R—R molecule formation decreases and the probability increases of the stabilization of alcohol radicals due to the separation of the hydrogen atoms from the polymer macromolecules, which results in the formation of R—H molecules.

The macroradicals formed in the course of destruction of the Sprut-5M polyester network or the peroxide radicals resulting from the interaction of the macroradicals with atmospheric oxygen most probably react with liquid MDI. Thus, MDI can serve as a bridge that connects the terminated chains of the polyester. The polyurethane network formed heals the defects of the polyester network. In that the most highly stressed polymer networks are the first to terminate, the formation of such "bridges" must result in more uniform distribution of stresses and in increase of the polymer strength. There was an increase of the fatigue strength of polymethyl methacrylate (E_{MMA}) under the effect of farina, fructose, or glucose dissolved in the water in which the polymer was deformed. These materials result in crosslinking of the terminated macromolecules and in strengthening of the polymer.

Let us consider one more feature of polymer failure and the effect of MDI thereon. The termination of molecular chains causes formation of submicrocracks and cavities in the direction perpendicular to the tension in the polymer. The concentration of the submicrocracks increases the effect of the load over time, which asymptotically approaches a limiting value in the final stages of the failure. Such cavities can be filled with MDI or even with segments of the liquid component.

The formation of cavities in the polymer results in increase in its volume. Thus, the volume increase of a specimen made of PN-1 polyester resin is 0.1% in 1 h after loading, 0.15% in 3 h, and 0.2% in 10 days at 20°C and 5 MPa. The addition of 20–60% of MDI into the resin decreases the volume defect of the specimen twofold to fourfold.

Thus, one can identify at least three factors that increase the long-term strength of Sprut-5M adhesive at the stage of formation of the polyurethane network but which have no further effect after the adhesive cures completely. Actually, in this state of the adhesive there is no indication of the structure that is observed with the orientation of the molecular chains; there is an abrupt reduction of the opportunity for macroradicals to recombine and for the submicrocracks to be filled. At this stage of curing, despite some decrease of the long-term strength under a load of 50% of the fracture load, the Sprut-5M adhesive has much better resistance to the effect of constant loads than does the polyester adhesive (see Fig. 3.2, curve 1 (after 102 days) and curve 4).

After the formation of interpenetrating networks is complete, it is their specificity that results in the increase of the adhesive long-term strength. As we noted before, IPNs are characterized by the availability of the loose transition region. In this case sections of the polymer main chains can shift more freely, and consequently the load between them can be distributed more uniformly, which must enhance the polymer lifetime.

The adhesive layer can be strengthened by the effect of both constant external load and internal stresses. Consequently, the latter can facilitate the strengthening of the adhesive interlayer but not its softening.

3.2.1 Properties of Sprut-5M adhesive-based reinforced coatings

When applying polymeric strengthening coatings, the issue arises of the mutual operation of the metal and the coating in terms of the indifferent moduli of elasticity. Let us consider this issue in detail. Dog-bone-shaped specimens manufactured to GOST 11262-68 standard were used for the investigations. The specimens were made of 3 mm thick St3 sheet hot-rolled steel, of 3 mm thick glass-reinforced plastic based on Sprut-5M adhesive and T-11–GVS9 fiberglass, as well as of biplates representing metal with glass-reinforced plastic molded onto it. The glass-reinforced plastics were molded under contact pressure.

Figure 3.4 presents curves of specimen deformation. As one might expect, the contribution of the polymeric coatings to the total rigidity and carrying capacity of the bispecimens subjected to the loads within the elastic limits of the steel is not great. Calculation indicates that the surplus of rigidity and of the limit of load proportionality amounts to 7.5%.

Let us consider the sections of the tension diagrams that correspond to steel deformation. On section II of the diagram, which corresponds to the steel yield plateau, the polymeric coating takes the whole load gain, which amounts to 7500 N.

The value of the yield limit and the extent of the yield plateau depend on a number of factors related to the steel structure and the test conditions. In our experiments, the yield plateau ended at deformations ε of 1.2%. Under further deformation the steel began to harden (section III of Fig. 3.4). It is evident that at this stage of specimen failure the glass-reinforced plastic coating facilitates the increase of strength.

Thus, Sprut-5M adhesive-based reinforced coatings allow an increase in the strength of the metal, mainly under the conditions in which the load on the specimen exceeds the elastic limit of the steel.

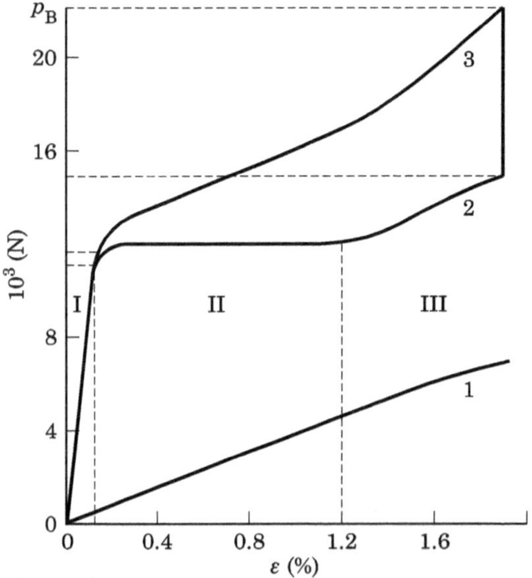

Figure 3.4 Patterns of deformation of specimens made of glass-reinforced plastic (1), steel (2), and metal–glass-reinforced plastic (3).

These are frequently the condtions under which structures operate that are weakened by intensive metal corrosion—pipelines, oil tanks, ships. Sprut-5M adhesive is successfully used for strengthening these.

The serviceability of the coatings in the region of metal elastic deformation can be enhanced in many ways—by using high-modulus reinforcing materials, by altering the temperature at which the coating is applied, or by preloading the metal before applying the coating. The reinforced coating on pipelines is in the stressed condition due to the normal internal stresses that arise in the course of formation of the coating, and enhances the pipeline strength even when the pressure of liquid or gas within is inconsiderable. It should be noted that if the stresses in the coating are too high, its adhesion and cohesion strength will decrease rapidly. Application of Sprut-5M adhesive avoids this danger because the internal stresses for this adhesive do not exceed 5 MPa in any case.

Let us consider some mechanical properties of biplates. These are steel plates onto which glass-reinforced plastics are molded, with Sprut-5M adhesive used as a binder. The types of mechanical load selected were those that were encountered in the course of the coating operation.

For a 2 mm thick coating the average energy of impact resulting in rupture of the binder is 325 J; in the glass-reinforced plastic, the energy of layer separation is 475 J; and in the glass-reinforced plastic fibers, the energy of rupture is 625 J. For a 3 mm thick coating these values increase to 373, 525, and 675 J, respectively.

The relative elongation of the metal on bonding of the biplates to the extent that the fibers of the fiberglass break and wrinkles and folds appear in the external layer of the glass-reinforced plastic amounts to 0.19–0.22%.

Sprut-5M adhesive is widely used to seal blowholes in pipelines. The pressure that breaks a glass-reinforced plastic coating made of 10 layers of the fiberglass through a 15 mm blowhole is 8 MPa. A coating with 15 layers of fiberglass withstands the 8 MPa pressure through a 35 mm blowhole.

When adhesives are applied for the repair of ships, blowholes in pipelines, and other objects, the adhesive interlayers are frequently subjected to impact loads under tension; this occurs, for example, in hydraulic impacts on pipelines, in ships during mooring, and so on. In this context it is interesting to investigate the effect of the modifying additives in Sprut-5M upon the strength of adhesive-bonded joints under tensile impact loads.

The investigations were carried out using the Charpy machine. A special device was used to apply the tensile impact load. The specific impact viscosity a was calculated by the well-known formula $a = A/S$ where A is the work (in kJ) performed to destroy the specimen; S is the specimen cross-sectional area in m^2. Adhesive based on PN-1 polyester resin without modifying additives and Sprut-5M adhesive were the subjects of the investigations. Specimens of 8 mm and 18 mm diameter and made of steel were cemented by butt ends. The investigations were carried out at 20°C.

The impact viscosity of the 8 mm diameter specimens cemented with the PN-1 and Sprut-5M adhesives was 10 and 20 kJ/m^2 at 10 days after cementing, and for 18 mm diameter specimens was 60 and 95 kJ/m^2. It is evident that even in the 10 days after cementing, i.e., long before the Sprut-5M adhesive acquires full strength, its impact viscosity is 1.5–2 times higher than that of the PN-1 based adhesive. The specific impact viscosity increases with the increase of the specimen diameter, which seems to be related to the decrease of the ratio of the effective zone of the concentration factor of the internal stresses inside the edge of the adhesive interlayer to the cementing area.

3.3 Adhesives Based on Thermodynamically Incompatible Polymeric Mixtures

The properties of polymeric mixtures depend on the compatibility of the system components. Mixtures in which the polymeric components are incompatible can be considered by analogy with filled polymers. In such systems one component distributed within the continuous phase of another serves as filler. The principal regularity of the physical-chemical and mechanical behavior of the filled polymers, the boundary layers in particular, is maintained when the filler is also a polymer.

Of special interest are the systems obtained in the course of polymerization of monomers in which another polymer is dissolved but is not compatible with the polymers obtained from these monomers. In such a case, at certain degrees of polymerization the dissolved polymer precipitates in form of the dispersed phase owing to the appearance of incompatibility; the content and size of the particles of the dispersed phase can be controlled by changing the polymers' compatibility, their weight ratio, the system viscosity, and the rate of polymerization.

3.3.1 Adhesives based on acrylic polymer mixtures

Let us consider the effect of PBMA particles precipitated from polymerizing MMA or from its mixture with butyl methacrylate (BMA) on the properties of the polymeric mixture in the adhesive interlayer. PBMA is dissolved in a reactive system.

For the monomer polymerization at room temperature, the adhesive was augmented with a redox system of 3% BP and 0.75% DMA. To study, explain, and predict the development of the elastic failure of the polymer in the adhesive interlayer, an improved method of investigating adhesive layer crack resistance with modeling of the formation and growth of a crack at the adhesive-bonded joint loading was used [119]. Five adhesive-bonded joints with the adhesive mixture compositions shown in Table 3.1 were subjected to static tests for crack resistance at room temperature. The characteristics of the static crack resistance of the adhesive-bonded joint (K_{IC} is the coefficient of the stresses intensity: G_{IC} is the intensity of the elastic energy release; δ_{IC} is the opening in the crack tip) were determined at the moment of onset of the crack in double-cantilever specimens DCB (Fig. 3.5). The specimen cantilevers were made of PMMA of TOCH type.

TABLE 3.1 Compositions of the Adhesive Compounds

	Content (%)		
Composition #	MMA	BMA	PBMA
1	60	—	40
2	60	5	35
3	60	10	30
4	60	15	25
5	60	20	20

The optical transparency of the cemented material allowed visual detection of the moment of crack onset. In this case the load P_1, the shift of the point of its application ϑ_P, and the crack face displacement close to the mid-point of its front were recorded (Fig. 3.5). Determination of the critical load P_C from graphs of $P–\vartheta_P$ and $P–\vartheta_C$ using the secant method gave good agreement with the results obtained by visual detection of the crack onset.

The configuration of the DCB specimen provides for the multiple initiation and arrest of the crack growth within the working range of its length $1.5h < l < W - 1.5h$. The cycle (specimen loading–crack onset–specimen unloading–crack arrest) was repeated on average 15 times on each specimen. All the adhesive-bonded joints under test were subjected to cohesion fracture along the adhesive inerlayer material. The values of the stress intensity critical coefficient K_Q were calculated from the values of the load P_C at the moment of the crack onset and from the equation [120]

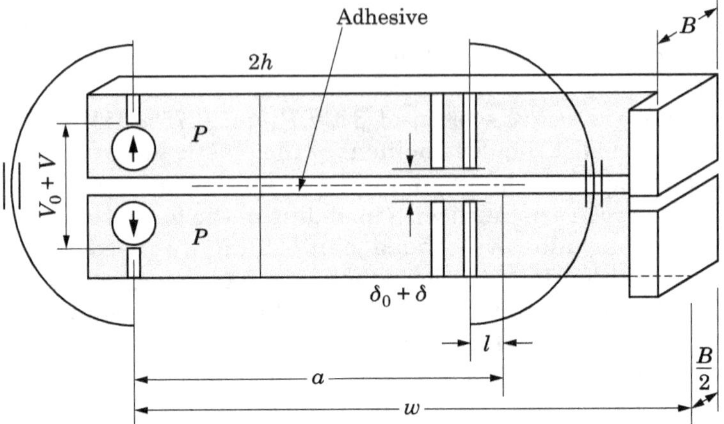

Figure 3.5 Schematic of specimen and diagram measuring the crack displacement and the crack opening displacement measurements.

$$K_Q = \sqrt{\frac{12P^2l^2}{B^2h^3}\left(1 + 1.32\frac{h}{l} + 0.542\frac{h^2}{l^2}\right)} \tag{3.2}$$

where l is the crack length, $2h$ is the height, and B is the specimen thickness.

The values of K_Q for the specimens and the crack sizes met the well-known requirements of applicability of plane deformation conditions that ensure the equality $K_Q = K_{IC}$. The fact that the acquired values of K_Q are equal to the characteristic K_{IC} of the adhesive interlayer indicates their independence of the crack length, the insignificant deviation from linearity of the P–ϑ_P and P–δ_C graphs, the absence of drawing of the crack tip, and the approximation of its front to a straight line.

The possibility of multiple initiation and arrest of the crack in the DCB specimen validates the application of the method of mechanical compliance for determining the characteristic energy G_{IC}. Values of G_{IC} for the adhesive-bonded joints under test were calculated from

$$G_{IC} = \frac{{P_C}^2}{2B}\frac{\partial\lambda}{\partial l} \tag{3.3}$$

in which $\lambda = \vartheta_P/P$ is the elastic compliance measured at various lengths of cracks in the specimen.

Within the framework of failure linear mechanics, the relationship between the cracking resistance characteristics K_{IC} and G_{IC} was written as

$$K_{\mathrm{IC}} = \sqrt{\frac{E_{\mathrm{P}}}{1 - \gamma_{\mathrm{p}}{}^{2}} G_{\mathrm{IC}}} \tag{3.4}$$

where E_{P} is the modulus of elasticity and γ_{p} is the Poisson coefficient.

For adhesive-bonded joints the values of E_{P} and γ_{p} (in Equation 3.4) are customarily related not to the adhesive interlayer but to the materials to be cemented. This commonly accepted assumption is not completely correct due to ignorance of the effect of the adhesive elastic constants upon the failure process. For composition 5 (Table 3.1) the reduced modulus of elasticity E_{nP}, calculated from Equation (3.4) using the independently determined characteristics $K_{\mathrm{IC}} = 3000^{3/2}$ MPa and $G_{\mathrm{IC}} = 0.0033$ N/mm, is equal to 245×10^{3} MPa, noticeably below the modulus of elasticity of PMMA, $E_{\mathrm{P}} = 300 \times 10^{3}$ MPa.

The reduced modulus E_{nP} does not depend on the deformation rate ε of the material in the crack tip. In our tests the rate calculated from

$$\varepsilon = \frac{2(1 + \gamma_{\mathrm{p}})}{E_{\mathrm{P}}} \frac{\sigma_{\mathrm{P}0,2}}{K_{\mathrm{IC}}} K_{\mathrm{IC}} \tag{3.5}$$

(where $\sigma_{\mathrm{P}0,2}$ is the adhesive yield limit under the single-axial tension; K_{IC} is the rate of increase of the stress intensity coefficient) decreased significantly along with the increase of the initial crack length.

The experimentally determined modulus of elasticity E of the adhesive interlayer of composition 5 decreased approximately from 130×10^{3} to 100×10^{3} MPa. Some disparity between the obtained values of the cracking resistance characteristics K_{IC} and G_{IC} is accounted for by the excessive simplification of the common mathematical models of a crack in a thin film of polymer that connects two elastic plates.

With reference to the earlier information, the deformation characteristic δ_{IC} looks more promising for the evaluation of the cracking resistance of adhesive-bonded joints. This characteristic, in contrast to K_{IC} and G_{IC}, can be measured directly on a specimen without any limitation typical of the force and energy analysis of the linear mechanics of failure. The characteristics of an adhesive-bonded joint were determined using the engineering method of measuring the crack displacement close to the mid-point of its front line in the loaded specimen. The displacement δ_{I} in the medium thickness plane of the specimen at various distances from the crack tip for composition 5 (see Table 3.1) are given in Fig. 3.6. Extrapolation for the case when $l = 0$ results in $\delta_{\mathrm{IC}} = 0.021$ mm, which is the same for various crack lengths in the specimen.

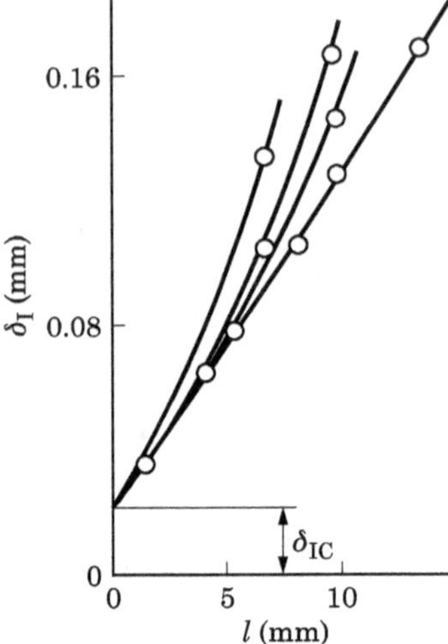

Figure 3.6 Value of the displacement of the crack faces of various length measured at different distances l from the crack tips.

The value of δ_{IC} can be used not only as a comparative but also as an estimated characteristic of the cracking resistance of adhesive-bonded joints. For this purpose it is necessary to find a relationship between the displacements δ_I, load, body geometry, and crack length. The relationship for a crack with yield zone localized within its plane can be written

$$G_{IC} = \alpha_k \delta_{0,2} \delta_I \tag{3.6}$$

where G_{IC} is a known function of the load and of the geometry of the body with a crack and α_k is a coefficient.

For polymers that fail as a result of formation of so-called silver cracks the coefficient α_k is customarily considered as equal to 1. In this case, calculation using Equation (3.6) and using the characteristics G_{IC} and δ_{IC} for composition 5 (see Table 3.1) gives $\sigma_{P0.2} = 1.6 \times 10^3$ MPa. The adhesive interlayer yield limit determined experimentally at deformation rates $\varepsilon = 1 \times 10^3 – 1 \times 10^{-4}\,\text{s}^{-1}$ changes within the range $(2.2–2.6) \times 10^3$ MPa. If the relationship $\alpha_k = \pi/4$ is assumed for the plane stressed state, the estimated value of $\sigma_{P0.2} = 2.0 \times 10^3$ MPa will come close to that found experimentally.

When BMA is added to the system, MMA and BMA copolymerize. The compatibility of the polymer formed with PBMA increases, which results in change of the number and the size of the colloid particles of PBMA formed.

The number of the PBMA particles and their radii in the cured adhesive were determined from the turbidity spectrum. As Fig. 3.7

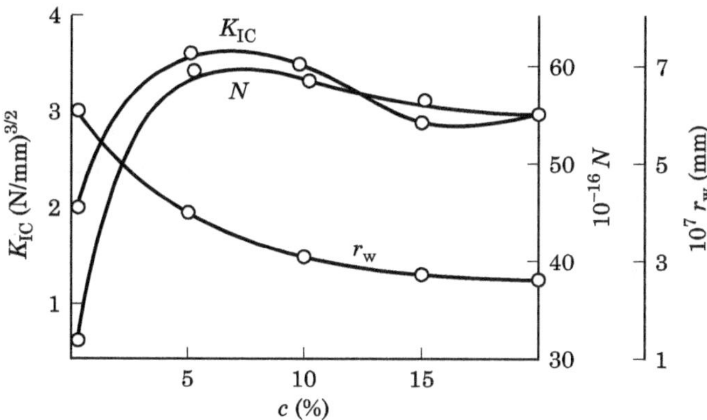

Figure 3.7 Dependence of the critical stress intensity coefficient K_{IC}, particle sizes r_w, and their number N on the BMA content.

shows, the size of the PBMA particles in the polymer decreases with increase of the BMA concentration in the initial mixture. The curve of particle concentration has a maximum around 5% BMA that correlates with the maximum value of the stresses intensity coefficient K_{IC} and with the dependence on the BMA content in the adhesive mixture. The effect of the dispersed phase on the polymer's capacity to resist the growth of cracks seems to consist in energy dissipation and in the decrease of the overstress close to the initial crack tip in the particle of the low-modulus PBMA polymer.

Let us consider the properties of VAK (water-resistant acrylate) adhesives and of Sprut-5M adhesive, both based on the PMMA–PBMA mixture for cementing under water. The VAK adhesive consists of the base (40% solution of PBMA in MMA) 100 parts by weight and ATG 10 parts, BP 3 parts and DMA 1 part. Sprut-5M also has fluorinated alcohol added. In the system under consideration, ATG is surface-active and can decrease the adhesive surface tension at the boundary with the high-energy surface. Adding 10% ATG to VAK adhesive decreases the interphase tension on the boundary between the adhesive and mercury to 240 mN/m. ATG increases the adhesion strength as well as making the adhesive capable of cementing in water and in oil products.

In this system, ATG performs other functions as well. In the course of the adhesive cure, ATG copolymerizes with MMA to form a copolymer that contains isocyanate able to form intermolecular crosslinks consisting of disubstituted urea groups when it reacts with moisture that diffuses into the adhesive interlayer. Such separation in time of the processes of formation of linear polymer and of the intermolecular crosslinks results in decrease of the internal stresses in the adhesive interlayer.

In Sprut-4, the quantity of ATG is increased to 30%. Increase of the number of intermolecular crosslinks and of polar groups enhances the resistance of this adhesive to oil products. The fluorinated alcohol introduced into the adhesive works as reactive surfactant because it can react with the isocyanate groups of ATG. The amount of fluorinated alcohol is selected to bond not more than 30% of the isocyanate groups in the adhesive. The fragment of the macromolecule of the polymer formed in this case has the structure

$$\left[-\underset{\displaystyle CH_2}{\overset{\displaystyle COOCH_2}{\overset{|}{\underset{|}{C}}}} - CH_2 - \right]_n \left[-\overset{\displaystyle CH_2OCONH(C_2H_3)CH_3NCO}{\overset{|}{CH}} - CH_2 \right]_m -$$

$$-\left[-\overset{\displaystyle CH_2OCONH(C_2H_3)CH_3NCO - C{=}O}{\overset{|}{CH}} - CH_2^+ \quad \begin{matrix} | \\ O \\ | \\ CH_2 \\ | \\ (CF_2)_3 \\ | \\ CF_3 \end{matrix} \right]_p -$$

The additional introduction of the fluorinated alcohol into the adhesive enhances its water resistance. Thus, the adhesion strength of adhesive-bonded joints of steel using Sprut-4 adhesive does not change significantly over time even with many years in water.

To demonstrate the roles of the double bond and of the ATG isocyanate group, which make this surfactant reactive, the adhesion strength was determined for adhesive containing, instead of ATG, analogues in which these groups were substituted by others. The adhesion strength of adhesives with substituted ATGs was lower than that with ATG itself (Table 3.2).

Using the ATG product, the minimal possible temperature of the adhesive cure was reduced. MMA containing ATG polymerizes under the effect of the CHP (cumene hydroperoxide)–DMA redox system even at −10°C. Compounds with isocyanate groups generally

TABLE 3.2 Adhesion Strength of Steel Adhesive-Bonded Joints Using Adhesive Based on 40% Solution of PBMA in MMA with Different Additives

Additive	Quantity (%)	Adhesive strength (MPa)
$CH_3C_6H_3(NCO)NHCOOCH_2CH{=}CH_2$	10	40.5
$CH_3C_6H_3(NCO)NHCOOCH_2CH_2CH_3$	5	24.0
$CH_3C_6H_3(NCO)NHCOOCH_2CH_2H_3$	10	22.0
$CH_3C_6H_3(NHCOOCH_3)NHCOOCH_2CH{=}CH_2$	5	25.5
$CH_3C_6H_3(NHCOOCH_3)NHCOOCH_2CH{=}CH_2$	10	18.0

accelerate the polymerization reaction of the unsaturated compounds, but with ATG this acceleration is much stronger. The higher activity of ATG compared with TDI and other isocyanate-containing compounds is assumed to be related to the integration of the isocyanate group into the growing polymer chain.

We also studied the interaction of the ATG isocyanate groups with the MMA double bonds. In this case substituted amide groups must be formed that can act as crosslinks in the branching of the polymer chain $R^1R^2{-}N{-}CO{-}$. A weak band of carbonyl group absorption in the 1730 cm^{-1} region that is typical of substituted amines appears in the polymer IR spectrum, recorded by the differential method. The reaction of the ATG isocyanate group with MMA must result in formation of the crosslinked polymer, but it is not observed with the VAK adhesive—the cured adhesive dissolves completely in acetone. The number of isocyanate groups which initiate the copolymerization reaction seems not to be great. This is confirmed by the band of valence oscillations of the free isocyanate groups with a maximum at 2280 cm^{-1} seen in the IR spectrum of the cured VAK adhesive. The insoluble fraction appears in the cured adhesive when the ATG content is above 15%, while the noticeable accelerating effect of ATG upon the polymerization reaction appears when its concentration in the adhesive is only 1%. Thus the accelerating effect of the ATG isocyanate group upon the MMA polymerization reaction is not explained in this case by chemical reaction of this isocyanate group with MMA.

Let us analyze the strength characteristics of VAK adhesive. The adhesive for investigation was taken from a commercial lot manufactured by the pilot plant of the Institute of Macromolecular Chemistry of the Academy of Sciences of the Ukraine.

The following couples of specimens were used:

Epoxy glass-reinforced plastic–epoxy glass-reinforced plastic

Steel–epoxy glass-reinforced plastic

Steel–steel.

The specimens were tested 10 days after cementing.

It was demonstrated that cohesion failure, i.e., failure of the adhesive, was typical for these specimens. The qualitative similarity of phenomena in the course of specimen failure allowed comparative analysis of the strength characteristics by applying probability-statistical methods. The results of tests of specimens cemented with VAK adhesive for a short-time strength at normal pull-off are given in Table 3.3.

To evaluate the test results for various combinations of specimens, the confidence index θ was calculated and compared with the boundary confidence index θ' (tabulated value). Table 3.3 shows that at a probability of 0.95, θ is less than 1 for all the specimens, i.e., the difference between the specimens can be considered insignificant.

Analysis of the experimental data indicated that the strength index of the adhesive specimens is distributed normally. Comparing the mean values of strength and evaluation of the systematic errors, we proceed from the preconditions that

$$\bar{M}_{\sigma 1} \cong \bar{M}_{\sigma 2} \cong \bar{M}_{\sigma 3} \cong \bar{M}$$

where $\bar{M}$ is the mathematical expectation of the average strength. This relationship was estimated by means of the tabulated values of the distribution of the standard deviations t_α:

TABLE 3.3 Characteristics of Bonded Specimens[a]

Sample number	Combinations of bonded materials	Sample size n	Mean strength, M_δ (MPa)	Variation factor, v_B (%)	Mean square deviation of σ_B	Dispersion of σ_B^2
1	Epoxy glass-reinforced plastic–epoxy glass-reinforced plastic	100	29.2	13.7	40.0	1600
2	Steel–epoxy glass-reinforced plastic	100	30.8	14.2	43.8	1920
3	Steel–steel	100	29.7	16.1	45.9	2110

[a] The specimens were tested 10 days after cementing.

$$t_\alpha = \frac{|\bar{M}_{\sigma 1} - \bar{M}_{\sigma 2}|}{S_\sigma} \qquad \left(S_\sigma = \sqrt{\frac{n_1 S_1{}^2 + n_2 S_2{}^2}{n_1 + n_2 - 2} \left(\frac{1}{n_1} + \frac{1}{n_2} \right)} \right) \tag{3.7}$$

because this value is subject to the distribution with the number of degrees of freedom for two specimens, $k = n_1 + n_2 - 2$. This is why at the prescribed reliability the means can differ not more than the value $t_\alpha S_\sigma$, i.e.,

$$|\bar{M}_{\sigma 1} - \bar{M}_{\sigma 2}| < t_\alpha S_\sigma \tag{3.8}$$

This condition is met for all the combinations of the specimens (see Table 3.4). At $\alpha_B = 0.999$, the results become confident.

Thus, for the combinations of the cemented materials that are considered, one can guarantee a reliable joint under standard (normal) loads that produce tension up to 25 MPa inside the adhesive interlayer. In connection with the retarded formation of the intermolecular crosslinks and with the more complete consumption of the monomer, the VAK and Sprut-4 adhesives acquire the maximum strength in 1–2 months after cementing, although even at 1 day the adhesion strength of the adhesive-bonded joints already exceeds 10 MPa and is sufficient for the majority of applications. Similar character of the change of adhesion strength is observed during cementing in water. When cementing takes place in oil products, the VAK adhesive is elasticized by the oil, as a result of which the adhesion strength decreases with time. Sprut-4 adhesive has no such shortcoming.

TABLE 3.4 Evaluation of the Differences of the Dispersion and the Mean Short-Term Strength of Specimens Cemented with the VAK Adhesive under Various Combinations of Specimens

Combination of specimens	Confidence index of specimen strength mean $\theta = \sigma^2/\sigma_1{}^{2a}$	S_σ	$t_\alpha S_\sigma{}^b$	Difference of combination values of two specimens
1, 2	$\frac{1920}{1600} = 1.2$	5.96	19.6	16
1, 2	$\frac{2110}{1600} = 1.32$	6.12	20.1	8
2, 3	$\frac{2110}{1920} = 1.1$	6.34	20.9	11

[a] The boundary value of θ for all cases is equal to 1.35.
[b] The tabular value of t_α at $\alpha_B = 0.999$ is equal to 3.29.

Let us consider the change of the long-term strength of VAK adhesive in air and in water. The investigations were carried out at 20°C and 60°C in air and at 20°C in water. Analysis of the results by probability-statistical methods with the probability of $\alpha_B = 0.95$ showed that the difference of dispersions and mean values is insignificant. The curves of the long-term strength of steel adhesive-bonded joints under normal pull-off in air and in the water (Fig. 3.8) are approximated by the equations $\log \tau_{ad} = 8.5 - 0.0289\sigma$ (curve 1) and $\log \tau_{ad} = 4.5 - 0.0226\sigma$ (curve 2), where τ_{ad} is the tangential stress. Further holding of the specimens under load in water and in air does not practically affect the adhesive strength properties, and the σ–$\log \tau$ curve kinks and runs parallel to the y-axis.

As Fig. 3.8 shows, the adhesive long-term strength in water is lower than that in air; this could be accounted for by the adsorption and plasticizing effect of water, although at loads <10 MPa the adhesive reliability in water is higher than in the air. Insofar as in the actual full-size adhesive structures the stresses are designed to be below the above limit, VAK adhesive can be characterized as highly water-resistant. In actual operational conditions the adhesive-bonded joint is practically always subjected to the effect of both tangential and normal

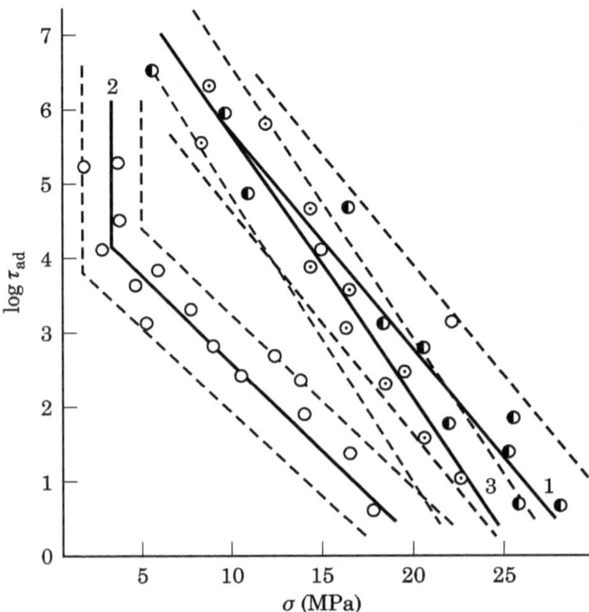

Figure 3.8 Long-term strength of specimens made of St3 steel cemented with VAK adhesive: (1) in air at 20°C and 60°C; (3) in water at 20°C. Time is given in seconds; broken lines indicate the confidence interval.

stresses. In this connection the strength of the adhesive-bonded joints for various combinations is of great interest. A set described in [122] was used for the investigations.

The strength of adhesive-bonded joints under the combined effect of tangential and normal stresses at 20, 40, 60, 90, and 100 days after cementing of St3 steel is displayed in Fig. 3.9 as surface stress plotted in coordinates σ–τ_{ad}–time. Surface I characterizes the extreme stresses of the joints using VAK adhesive; surface II is the same without addition of ATG to the adhesive.

Let us consider the dependence of the shear strength of specimens cemented with VAK adhesive on the stress–strain state of one of the cemented parts. The investigation consisted in determining the strength of the adhesive interlayer for high-frequency (up to 20 kHz) longitudinal-mode oscillations of a specimen in the shape of a cylindrical rod with half-cylinders cemented on. The magnetostriction vibrator to which the specimen was rigidly fixed generated the longitudinal oscillations. The adhesive-bonded joint was loaded and sheared by inertial volume forces, which originate in the specimen under the effect of vibration.

The shear stresses in the adhesive interlayer are

$$\tau_{ad}\nu_k{}^2 A_C \frac{m}{k} \qquad (3.9)$$

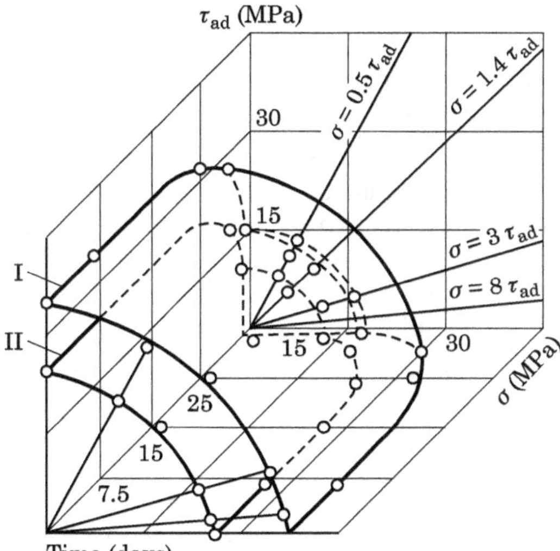

Figure 3.9 Surfaces of the maximum stresses of cemented steel specimens: (I) VAK adhesive; (II) the same without ATG added.

where ν_k is the circular frequency; A_C is the amplitude of the specimen oscillations in the loop; m is the mass of the part; and k is the area of the cemented surface.

The amplitude of the rod's oscillations in the loop zone was measured by means of a microscope, and to determine the amplitude at any point within the section between the node and the loop of oscillations a recalculation was performed using

$$A_x = A_C \sin\frac{4x}{\lambda_B} \tag{3.10}$$

where x is the distance between the node and the point under consideration; and λ_B is the length of the oscillation wave. The relative deformation ε_x in the rod changes in the following way:

$$\varepsilon_x = \varepsilon_y \cos\frac{4x}{\lambda_B} \tag{3.11}$$

Here ε_y is the maximum relative deformation of the rod in the vibrational mode determined in the following way:

$$\varepsilon_y = \nu_k A_C \sqrt{\rho/E_P} \tag{3.12}$$

where ρ is the density and E_P is the modulus of elasticity of the rod material.

The location of labels on any section between the node and the loop of oscillations designates the respective relationship of the shear stress and the deformation of the cemented part.

The above method of determining the adhesion strength for the stress–strain state of one of the cemented parts enables analysis of the reasons for failure of adhesive-bonded joints when, at first glance, there are no loads in the joint, as when using the adhesive for repair of ships' decks or for cementing any part onto a high-pressure bottle. For deformation of 0.01% of the cemented surfaces, the adhesive-bonded joint based on UP 5-177-1 epoxy adhesive is destroyed completely. With application of VAK or Sprut-4 adhesives, the joint strength is decreased by 70% and 54%, respectively, at 0.2% deformation of the substrate. Consequently, there is no failure of the adhesive-bonded joints even at this degree of deformation.

Because of high adhesion strength and deformability of the VAK and Sprut-4 adhesives, low internal stresses in the adhesive-bonded joints ensured their serviceability when used as binders for the formation of reinforced coatings on metal and other surfaces. The thickness of such coatings can reach some centimeters and their strength is comparable with that of metals. Such coatings are suitable to strengthen metal structures, for the repair of pipelines, and for sealing oil tanks and other reservoirs.

Good technological applicability of these adhesives is of great importance in terms of low viscosity (6.0–8.0 Pa·cm) and their capacity for fast cure even at low temperatures and for good impregnation of the glass cloth and other reinforcing materials. Glass-reinforced plastics can be formed using the above adhesives in air, in water, and in oil products. In the last case the glass cloth is preimpregnated with the adhesive in air and the reinforcing material is subsequently used for operations in water and oils. For formation of the glass-reinforced plastics under liquids, e.g., when strengthening pipelines or repairing ships and oil tanks, special mechanisms and appliances have been developed to mechanize the formation process [124, 125].

Let us consider the results of physical-mechanical tests of glass-reinforced plastics based on VAK adhesive and of steel specimens with glass-reinforced plastic molded on to them. The tests were performed to model to some extent the types of loads to which such cemented structures are subjected in actual use.

Table 3.5 displays the data for the strength properties of the glass-reinforced plastic. The reinforced material was T-11-GVS9 glass cloth, which made up 50% of the glass-reinforced plastic. One concludes that the strength of the glass-reinforced plastic, even when molded in water or oil, is in the majority of cases sufficient for mounting various structures or for the repair and strengthening of metal structures in liquid media. It should be noted that the glass-reinforced plastics were contact pressure-molded for tests under conditions simulating those of their operational application.

The glass-reinforced plastics formed using VAK adhesive as binder have good resistance to vibration (Table 3.6). Specimens of 5 mm thick St3 steel with glass-reinforced plastic coating cemented on (bispecimens) were also vibration tested. The coating was applied either to

TABLE 3.5 Mechanical Properties of Glass-Reinforced Plastics Based on VAK Adhesive Binder

Load type	Glass-reinforced plastic failure stress (MPa)		
	In oil	In air	Under water
Tension			
Along warp	320	240	152
Along weft	190	117	96
Compression			
Along warp	290	220	140
Along weft	210	145	98

TABLE 3.6 Vibration Resistance of Glass-Reinforced Plastics Based on VAK Adhesive Binder

Direction of load application	Failure tensile stress (MPa)		
	Initial state σ_B	Under effect of 10^7 cycles, σ_{BC}	$\frac{\sigma_B - \sigma_{BC}}{\sigma_B} \times 100\,(\%)$
Along warp	320	297	7
Along weft	190	173	9

both sides or to one side only. The coating was 1–2 mm thick. The tests were carried out under conditions of alternating cyclic loading at 18–19 Hz frequency of deformation; the flexural vibrations were excited by a vibration electrodynamics test bench. The frequency of the natural free vibrations was practically the same for all bispecimens and was 18–19 Hz. The amplitude of vibration at the ends of the specimens was within the range 25–30 mm. The value of stresses (strains) that act in a marked cross-section of the specimen was determined using wire resistance tensometers. The stresses acting at the boundaries of layers in the marked cross-section were found to be ±40–45 MPa. All the specimens were subjected to 10 cycles of load alternation.

Visual inspection of the specimens and checking with the DIK-1 defectoscope did not register any failure of the glass-reinforced plastic coating or its separation from the metal plate.

Biplates of 800 × 800 mm size with 5 mm thick steel plate were used for impact testing of the glass-reinforced plastic coating, which was 2–3 mm thick and applied on one side. A 14 kg free-falling ball provided the impact onto the plates. Table 3.7 shows the changes that occur in the glass-reinforced plastic coating under the effect of the impact load. These changes are of local character and do not spread beyond the region of application of the force. It is evident that even in the case of considerable impact energy, which results in failure of the glass-reinforced plastic coating, there is no separation of the coating from the metal.

TABLE 3.7 Results of the Biplate Impact Tests

Type of failure	Coating thickness (mm)	Failure engery (J)
Binder crushing	2	425
	3	525
Plastic ply separation	2	625
	3	700
Plastic fiber rupture	2	725
	3	825

For flexural tests, 3 mm, 10 mm, and 14 mm thick metal specimens were coated with VAK-based glass-reinforced plastic on both sides and on one side only. The coating was 1–3 mm thick. The sag of the pushed and pulled fibers was measured using sagmeters.

The relative elongations of the pulled ($\varepsilon_{\text{pull}}$) and pushed ($\varepsilon_{\text{push}}$) fibers of the specimen were determined by the following equations:

$$\varepsilon_{\text{pull}} = \frac{2f}{(a_{\text{h}} - f)^2}$$

$$\varepsilon_{\text{push}} = \frac{2f}{(b_{\text{h}} - f)^2}$$

where f is the sag, a_{h} is the half-base of the sagmeter, and b_{h} is the half-thickness of the specimen metal base.

Table 3.8 shows the results of the static bending tests, the relative elongations of the metal substrate at which the fibers of the glass cloth broke and wrinkles and folds appeared on its external surface. The flexural tests indicated that in the case of the pulled fibers, the threads of the glass cloth broke; in the case of the pushed fibers, wrinkles and folds appeared on the external layer of the glass cloth while the inner layers were firm on the metal surface. The static tests of the specimens with the metal base extension showed that the glass-reinforced plastic peeled off the metal for stress in the metal not less than 355 MPa.

For comparison, tests on biplates made of steel and glass-reinforced plastic with VAK as binder were carried out along with tests of similar specimens with UP 5-177-1 epoxy compound as binder. Impact resistance of the VAK specimens exceeded that of the UP 5-177-1 specimens by 14–19%; for the bending resistance the difference was 33–34%. The plastic with UP 5-177-1 binder peeled off the metal at a stress in the metal of 302 MPa, which is 15% less than that for the plastic with VAK binder. The glass-reinforced plastic was formed in air at 20°C.

TABLE 3.8 Results of the Biplate Flexural Tests

Metal specimen thickness (mm)	Failure deformation	
	$\varepsilon_{\text{pull}}$	$\varepsilon_{\text{push}}$
3	0.37–0.41	0.39–0.43
10	0.36–0.41	0.39–0.42
14	0.35–0.40	0.38–0.42

The adhesion strength of the glass-reinforced plastic–steel system was determined using steel dollies and layers of the glass cloth impregnated with the adhesive were placed between the dollies or plates. The glass-reinforced plastic in the adhesive-bonded joint was formed in air, in water, and in oil with subsequent placing of the cemented specimens into water.

The results of tests at 10–12 days after cementing and holding in the corresponding medium are presented in Table 3.9. The adhesion strength decreases with increase of the number of glass cloth layers, although in all cases the rupture is of cohesive character (for the adhesive). The drop in strength seems to be due to the glass-reinforced plastic formed under contact pressure having many faults in its structure, such as uncemented sections, non-uniformities of the adhesive layer, and so on.

3.3.2 Controlling the properties of adhesives based on epoxy rubber polymeric mixtures

The properties of existing epoxy adhesives fail to meet a number of requirements of current engineering, bringing the problem of their modification to the forefront because, despite the good adhesion and dielectric properties of the epoxy polymers, their elasticity and resistance to impact loads are not high.

Among the most effective modifiers allowing an increase in the impact viscosity and cracking resistance of epoxy polymers are the oligobutadiene acrylonitrile rubbers with reactive end groups. The structures of the reactive rubbers that are commonly used to increase the impact viscosity of the epoxy resins are as follows. CTBN is a

TABLE 3.9 Adhesion Strength of the Metal–Glass-Reinforced Plastic System

	Failure stress σ_n (MPa) for rupture of cemented specimens				Shear failure stress τ_{ad} (MPa) for cemented specimens in air
Specimen type	In air	In water	In air then in water	In oil	
Without glass cloth	26.0	16.8	24.5	12.0	21.3
With glass cloth					
1 layer	22.8	13.4	21.5	7.2	16.2
2 layers	19.0	11.4	17.4	4.0	—
4 layers	13.6	7.8	13.0	—	—

copolymer of butadiene with acrylonitrile with a carboxyl end group (in this case $x = 5$, $y = 1$, $z = 9$):

$$\mathrm{HOOC}\,[-(\mathrm{CH_2}-\mathrm{CH}=\mathrm{CH}-\mathrm{CH_2}-)_x-(-\mathrm{CH_2}-\underset{\displaystyle \mathrm{C}\equiv\mathrm{N}}{\mathrm{CH}}-)_y-]_z\,\mathrm{COOH}$$

CTBNX is a copolymer of butadiene with acrylonitrile with end and side carboxyl groups ($x = 5$, $y = 7$, $t = 3$, $z = 9$):

$$\mathrm{HOOC}\,[-(\mathrm{CH_2}-\mathrm{CH}=\mathrm{CH}-\mathrm{CH_2}-)_x(-\mathrm{CH_2}-\underset{\displaystyle \mathrm{COOH}}{\mathrm{CH}}-)_t-(-\mathrm{CH_2}-\underset{\displaystyle \mathrm{C}\equiv\mathrm{N}}{\mathrm{CH}})_y-]_z\,\mathrm{COOH}$$

HYCAR 1312 is a copolymer of butadiene with acrylonitrile ($x = 3$, $y = 1$, $z = 24$):

$$-[(\mathrm{CH_2}-\mathrm{CH}=\mathrm{CH}-\mathrm{CH_2})_x-(\mathrm{CH_2}-\underset{\displaystyle \mathrm{C}\equiv\mathrm{N}}{\mathrm{CH}}-)_y]_z-$$

A copolymer of butadiene with the acrylonitrile with end mercaptan groups ($x = 5$, $y = 1$, $z = 9$) was also considered but not used:

$$\mathrm{SH}-[(\mathrm{CH_2}-\mathrm{CH}=\mathrm{CH}-\mathrm{CH_2})_x-(\mathrm{CH_2}-\underset{\displaystyle \mathrm{C}\equiv\mathrm{N}}{\mathrm{CH}}-)_y]_z-\mathrm{SH}$$

These chemical formulas show that the modifiers are low-molecular weight copolymers of acrylonitrile and butadiene that variously contain or do not contain groups capable of reacting with epoxy groups (the side-groups in the chain can also be reactive).

Modification of epoxy resins by such modifiers is usually achieved in two ways: by integrating the rubber into the epoxy polymer or by copolymerizing the epoxy and rubber oligomers simultaneously with the process of epoxy resin curing at 393 or 433 K.

The best modifiers are the rubbers containing reactive groups —COOH, —OH, or —SH. For rubbers with carboxyl groups, the following chemical reactions can occur:

$$-\underset{\underset{O}{\|}}{C}-OH + \underset{\diagdown O \diagup}{H_2C-CH} \rightarrow \sim\underset{\underset{O}{\|}}{C}-O-CH_2-\underset{\underset{OH}{|}}{CH}\sim \quad (3.13)$$

$$-\overset{\overset{O}{\|}}{C}-OH + \sim\underset{\underset{OH}{|}}{CH}\sim \rightarrow \sim\underset{\underset{O-\underset{\underset{O}{\|}}{C}\sim}{|}}{CH}\sim + H_2O \quad (3.14)$$

$$-\underset{\underset{OH}{|}}{CH}\sim + \underset{\diagdown O \diagup}{H_2C-CH}\sim \rightarrow \sim\underset{\underset{O-CH_2-CH\sim}{|}}{CH}\sim \quad \overset{OH}{|} \quad (3.15)$$

Reaction (3.13) proceeds up to the moment of complete consumption of all the carboxyl groups, and the remaining epoxy groups react according to mechanism (3.15).

With the appropriate curing agent present, both epoxy rings open, forming the centers of branching in the molecular chains; that is, the molecule of the bisphenol A diglycidyl ester (DGEBA) is tetrafunctional. The reaction of the epoxy group with primary amines was shown to proceed according to mechanism (3.16), while with secondary amines and acids it runs according to the mechanisms (3.17) and (3.18) [126].

$$RNH_2 + \underset{\diagdown O \diagup}{H_2C-CH}- \rightarrow RNH-CH_2-\underset{\underset{OH}{|}}{CH}- \quad (3.16)$$

$$R_2NH + \underset{\diagdown O \diagup}{H_2C-CH}- \rightarrow R_2N-CH_2-\underset{\underset{OH}{|}}{CH}- \quad (3.17)$$

$$\mathrm{R{-}\underset{\underset{O}{\|}}{C}OH + H_2C\underset{O}{\overset{}{-}}CH{-} \longrightarrow R{-}\underset{\underset{O}{\|}}{C}{-}CH_2{-}\underset{\underset{OH}{|}}{CH}{-}} \tag{3.18}$$

When epoxy resins modified by rubbers are cured with the tertiary amines, the following processes occur:

$$\mathrm{R'{-}\overset{\overset{O}{\|}}{C}{-}OH + R_3N \longrightarrow R'{-}\overset{\overset{O}{\|}}{C}O^{-}{-}\overset{+}{N}HR_3} \tag{3.19}$$

The salt formed then reacts quickly with the epoxy group:

$$\mathrm{R'{-}\overset{\overset{O}{\|}}{C}{-}O^{-}{-}\overset{+}{N}HR_3 + H_2C\underset{O}{-}CH{-} \longrightarrow R'{-}\overset{\overset{O}{\|}}{C}{-}O{-}} \tag{3.20}$$

$$\mathrm{{-}CH_2{-}\overset{|}{C}H{-}\bar{O}{-}NHR_3}$$

As the cited paper [126] notes, the product seems to be similar to an ABA block polymer in which A consists of the DGEBA fragments and a curing agent, and B is represented by the butadiene–acrylonitrile chain of one OBDANC molecule (an oligomeric butadiene–acrylonitrile copolymer with the end carboxyl groups) with the following structure:

$$\mathrm{HOOC\left[{-}(CH_2{-}CH{=}CH{-}CH_2{-})_x(CH_2{-}\underset{\underset{C\equiv N}{|}}{CH}{-})_y{-}\right]_z COOH}$$

The usual process of epoxy resin curing runs alongside the above reactions.

The principal requirement of the modifier is usually considered to be solubility in the epoxy oligomer, but when the epoxy is cured the modifier must precipitate into a separate phase. This requirement is met by OBDANC because the difference in the solubility parameters of OBDANC and DGEBA is quite small, which makes its solution in the epoxy oligomer possible and its subsequent phase separation in the course of curing.

Integration of such modifiers as SKD–KTRA oligobutadiene rubber with the carboxyl end groups into an epoxy resin increases the adhesion strength, with a noticeable tendency to phase separation on mixing the epoxy and rubber oligomers that makes for difficulty in the uniform distribution of the modifier molecules in the epoxy oligomer. A noticeable improvement of compatibility of epoxy and rubber oligomers is observed in the oligobutadiene–acrylonitrilic copolymers (CTBM or OBDANC). Exact coincidence of the solubility parameters of DGEBA and OBDANC is achieved at 28.5% of the acrylonitrile in the rubber. In this connection, the majority of studies aiming to obtain impact-resistant epoxy compounds have been devoted to elastomers based on OBDANC and to epoxy resins based on DGEBA.

Now let us analyze what the properties of epoxy rubber compounds (ERC) depend on. As the authors of [128] note, the properties are determined by the epoxy oligomer used, by the type and concentration of the curing agent, by the content of the acrylonitrile in the rubber, and by its molecular weight. According to various authors, the optimal content of the rubber is from 0.25% to 20%. The region in which the maximum modifying effect is displayed is determined mainly by the reactivity of the rubber. Thus, when using chemically inert rubbers they are precipitated as a separate phase; when reactive rubbers are used, the epoxy rubber copolymer is precipitated into a separate phase. The molecular weight of rubbers with similar reactivities influences the extent of the modifying effect as well as the limit of the components' compatibility.

Studies of the dependence of ERC properties obtained as above on the formation method showed that the precipitated finely dispersed rubber that is not bonded to the epoxy matrix does not endow it with high mechanical characteristics. When reactive oligomers are applied with high-temperature cure, there is a sharp enhancement of the modifying efficiency, accounting for the more intensive separation of the phases in the system and for the increase of the yield of the epoxy rubber copolymer. Thus, the structure and properties of epoxy resins modified by rubbers are mainly determined by the mode of system curing, but the quantitative relationships between curing mode and final properties have not been sufficiently studied.

Of special interest are investigations reported in [129] in which the epoxy resin was modified by rubbers using a third method—preesterification reaction (PER) between the epoxy and carboxyl groups of the rubber and the resin and the subsequent curing of the copolymer obtained at room temperature. In particular, it was shown that with thorough blending of the epoxy and rubber oligomers and subsequent heating of the mixture at 423 K for 2 h there is a noncatalytic interaction of the carboxyl and epoxy groups that results in formation of the hydroxyester; the resin obtained has the following structure:

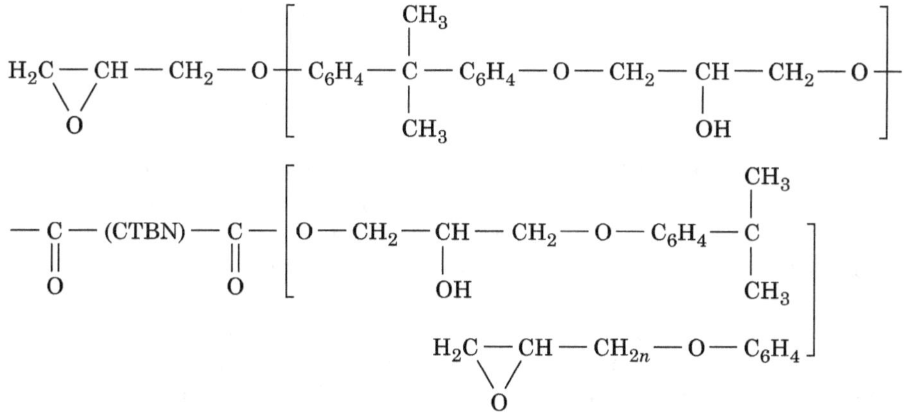

$$
\begin{array}{l}
\underset{\diagdown\,\diagup\atop O}{H_2C-CH}-CH_2-O\left[-C_6H_4-\underset{CH_3}{\overset{CH_3}{\underset{|}{\overset{|}{C}}}}-C_6H_4-O-CH_2-\underset{OH}{\underset{|}{CH}}-CH_2-O-\right] \\
-\underset{O}{\underset{\|}{C}}-(CTBN)-\underset{O}{\underset{\|}{C}}-\left[O-CH_2-\underset{OH}{\underset{|}{CH}}-CH_2-O-C_6H_4-\underset{CH_3}{\overset{CH_3}{\underset{|}{\overset{|}{C}}}}\right. \\
\left.\qquad\qquad \underset{\diagdown\,\diagup\atop O}{H_2C-CH}-CH_{2n}-O-C_6H_4\right]
\end{array}
\tag{3.21}
$$

The etherification between the epoxy and rubber oligomers can occur also at 353 K but requires such catalysts as chromium salts, diisopropylsalicylic acid, triphenylphosphine, betaine, triethanolamine, triethanolamine borate, benzyldiethylamine, or acetyltriethylene ammonium bromide.

For modification of epoxy resins with rubbers containing double bonds, the principal difficulty lies in the formation of hydroperoxide groups. These in turn can fragment into radicals, which cause undesirable and uncontrolled side-reactions.

3.3.2.1 Phase separation in epoxy rubber compounds and their morphological structure. The most important factor that ensures phase separation of ERC is the difference between the solubility parameters of the resin and of the rubber. If this difference is small, the phase separation appears to be incomplete. On the other hand, with a large difference the elastomer does not form a homogeneous solution in the initial stage of the modification process. The molecular weight of the rubber also affects the efficiency of the phase separation because at high molecular weight the rubber is poorly dissolved in the epoxy oligomer; at low molecular weight the phase separation does not occur and as a result an elasticized epoxy resin is formed. The use of rubber with average molecular weight of 3800 results in satisfactory phase separation and in good resistance to failure.

The efficiency of the phase separation cannot be always predicted from thermodynamic considerations; in some systems it does not occur at all. Thus, in the case of fast-curing epoxy resins the separation can practically cease due to the increase of viscosity (this is observed in cases of a curing agent with high reactivity such as triethylenetetramine, especially at higher content and with no mixing). The use of a less active curing agent such as piperidine ensures better phase separation.

The approach taken by the authors of [131] is of interest; they studied the problem of the dependence of the phase separation in ERC on thermodynamic and kinetic factors. They show that in epoxy material modified by rubber there is an upper temperature limit—the solubility temperature T_s above which the rubber and the resin are sufficiently compatible with one another that the phases do not separate before gel formation.

The efficiency of phase separation in ERC has been controlled by the integration of bisphenol A. In this case the resin molecular weight increases as a result of interaction of the phenol and epoxy groups. Optimal phase separation can be expected only when the difference $\delta_1 - \delta_2$ is very small and negative at the beginning of the reaction. Only under these conditions does the rubber dissolve in the epoxy oligomer; it precipitates into an independent phase as the molecular weight of the epoxy resin increases. From this it was concluded that the poor phase separation in the mixture can be improved by allowing for the increase of the molecular weight of the resin or rubber (i.e., for the increase of the difference of the solubility of the epoxy and rubber oligomers).

Increase of the phase separation efficiency in ERC can be achieved by various additives [133]. The effect of the rubber particle dispersion in the epoxy resin is enhanced by application of surfactants with active groups able to participate in the formation of hydrogen bonds between the structural elements. Similar regularity is noted with some aromatic and aliphatic resins (for example, resins with the brand name Structol 60NS) and ABC-copolymer used for this purpose. With the ABC-copolymer, which contains a large quantity of polybutadiene, the elastomer represents a dispersed phase of particles of a defined size. The advantage of such a modification is the higher glass-transition temperature T_g compared with the OBDANC rubber.

Study of the morphological features of ERC indicate precipitation in the course of the epoxy resin cure of particles of the discrete phase of the rubber, the dispersion of which in the epoxy polymer matrix has a considerable effect on the mechanical properties of the latter. If the rubber particles are large enough, they can be detected by optical microscopy, but the most important results are obtained by electron microscopy.

The morphological structure of ERC has been studied [134] by applying osmium tetroxide, which is frequently used for the detection of unsaturated bonds. In this case it reacts with the double bonds of the rubber, with the formation of the osmium cyclic ester (see (3.22)).

This results in additional curing of the rubber that in its turn makes it easy to prepare with enhanced surface contrast.

$$O_2OsO_2 + \begin{matrix} | \\ CH \\ \| \\ CH \\ | \end{matrix} \longrightarrow O_2Os\begin{matrix} O-CH \\ | \\ O-CH \end{matrix} \qquad (3.22)$$

In studies of the ERC heterogeneous structure obtained in copolymerization between the epoxy and rubber oligomers that occurs along with the epoxy resin cure (i.e., hot cure compounds), the average size of the rubber particles was found to be dependent on its polarity and on the level of compatibility with the epoxy resin as well as on the chemical composition of the modifier [132]. The author also showed that the size of the rubber inclusions decreases and the structure of the composition becomes more homogeneous with increase of the polarity of the elastomers and the curing temperature of the ERC.

Studies of the thermodynamic compatibility of the epoxy and rubber oligomers by the interferometric micromethod also showed [135] that the region of homogeneous mixing expands noticeably with increase of the rubber polarity. In this case the nucleation mechanism of heterophase formation is realized in which growth of the nuclei is the limiting stage. With active curing agents, occlusions of the epoxy resin in the rubber are formed during phase separation, which results in increase in the volume of the rubber particles. This was said [132] to be due to reaction of the epoxy resin with the rubber.

Many of the particles, especially the smallest ones, are homogeneous. Although this does not mean that there is no epoxy resin in the elastomer phase, it indicates that the epoxy resin is possibly dispersed at the molecular level as a result of the etherification reaction between the epoxy and carboxyl groups of the resin and the rubber. The usual size of the rubber particles is 0.1–0.8 μm. The phase separation in the mixture of the epoxy and rubber oligomers is assumed to proceed in two stages. In the initial stage the usual phase separation with formation of the rubber particles (0.8 μm) occurs. The level of phase separation is determined by the interphase tension, by the viscosity of the epoxy and rubber oligomers, by the difference of the parameters of their solubility, and other factors. The finer particles (0.1 μm) are formed after the epoxy resin is cured due to the formation of the block polymer of the resin and the rubber. Varying the modification conditions alters the degree of cross-linking within rather narrow limits, while the size of the rubber particles can change from 0.5 to 5 μm.

Correlation of the size distribution curves of the particles of the rubber dispersed phase with the compatibility of the initial ERC components showed [135] that the widest particle distribution is observed for the rubber that is of poor compatibility with the epoxy oligomer. Using a rubber with better compatibility, the particle size distribution becomes noticeably narrower. From this observation it was concluded that there is a correlation between the thermodynamic compatibility of the ERC components at the mixing stage and the phase state of the cured compound.

3.3.2.2 Relaxation properties of epoxy rubber compounds and mechanisms of the rubber strengthening effect. Usually ERC displays two regions of relaxation. The first is at a temperature much higher than room temperature and depends on the glass-transition of the matrix (epoxy) component of the compound. The glass-transition of the rubber phase is related to the second relaxation region, which is detected at temperatures below room temperature.

The basic methods of detecting the secondary relaxation transitions are dynamic mechanical spectroscopy (DMS) and differential thermal analysis; in some cases information can be obtained by evaluation of the dielectric characteristics, by NMR, and by DSC.

Studies of ERC by the DMS [136] found that the rubber dissolved in the matrix of the epoxy polymer plasticized it, with a decrease of the glass transition. The content of dissolved rubber was evaluated by means of the Gordon–Taylor equation:

$$\frac{1}{T_g} = \frac{1}{W_1 + kW_2}\left(\frac{W_1}{T_{g1}} + \frac{kW_2}{T_{g2}}\right) \tag{3.23}$$

where W_1 and W_2 are the contents (in %) of the epoxy resin and the rubber; T_{g1}, T_{g2}, and T_g are the glass-transition temperatures of the initial epoxy resin, the rubber, and the homogeneous matrix phase enriched with the rubber; and k is the normalizing constant.

The authors [136] showed that, within the range of small rubber additions, the glass-transition temperature drop is determined by the position of the main maximum of the mechanical losses. With further increase of the modifier content in the system, the glass-transition remains practically constant.

Four regions of ERC mechanical relaxation related to types of molecular mobility were found. One study [137] determined a direct dependence between the ERC impact resistance and the height of peaks of the mechanical losses. From the experimental data obtained in that study, the authors affirm that the physical basis of this correlation is that the relaxation mechanism makes a contribution to the energy dissipation in the material. It was also found that in combining the epoxy resin with the rubbers the conditions of molecular motion change in both the primary and the secondary regions of the mechanical relaxation (in this case both transitions shift to lower temperatures), which is related to the formation of a number of neighboring regions of mechanical relaxation due to the structural heterogeneity of the ERC.

Investigation of ERC by dielectric and ultrasound methods [138] detected three relaxation transitions. The low-temperature transition (220–246 K at 10^2–10^5 Hz) is related to the molecular mobility of the polyacrylonitrile butadiene. Relaxation transitions related to molecular mobility, either in the boundary layer between the rubber and the resin or in the regions of structure formed in the course of the crosslinking, correspond to higher temperatures. But, as this paper points out, the higher-temperature relaxation transitions do not appear in the frequency dependences. They can be detected only in the dependence of the modulus of elasticity on the temperature.

In the heterophasic system comprising poorly compatible components, the degree of the heterophasicity has a noticeable effect on the dynamics of molecular mobility. In this case the homogeneous system of properly compatible components displays their strong mutual effect on the molecular mobility, while the poorly compatible rubber that forms an independent phase has no effect on it. Properly compatible oligomeric rubber has a plasticizing effect, and a poorly compatible one serves as a structural modifier or an antiplasticizer.

Let us consider the possible mechanisms of strengthening of epoxy polymers by rubbers. The failure of a polymeric material, as of any solid, proceeds through the development of cracks. In the failure of linear polymers the material structure can change under the influence of mechanical stresses in the tip of the growing crack, which in turn results in orientation and consequently in the strengthening of the polymer in the area of the crack tip. In network polymers (epoxy in particular) the possibility of plastic deformation in the crack tip is lower due to the additional limits on the chain mobility applied by the network chemical crosslinks. This is why one observes a low value of the effective failure surface energy compared with linear polymers.

Hypotheses have been advanced to explain the increase of the impact strength of the epoxy resins by the integration of finely dispersed rubbers. There is an assumption [139] that the rubber particles in the epoxy polymer matrix absorb the impact energy in the manner of mechanical damping. In this case the dissipation of the excess energy occurs mainly inside the formed interphase layer with decreased density of the cohesion energy. In addition, to provide maximum impact strength of ERC a strong adhesion interaction is needed at the interface between the epoxy polymer and the rubber (in this case the cracks formed penetrate into the rubber particle and do not pass over it). The decrease of the brittleness of the epoxy resin observed with similar modification occurs as a result of the increase of elasticity, insofar as the looseness of the epoxy polymer increases with integration of rubber into the matrix of the epoxy polymer. The role of the rubber particles consists in providing for a free volume, which is necessary for the formation of cold flow, as well as in disturbing the chain regularity, which facilitates the course of relaxation and decreases the internal (residual) stresses.

Of definite interest is the finding of two flow mechanisms for epoxy resins containing elastomeric particles: these are the formation of crazes and shear yield [130] similar to that found for linear polymers containing rubber particles [126]. The role of the latter consists in the fact that they can both initiate and limit the growth of crazes. With the rubber present, a large number of small crazes rather than a small number of large crazes are formed and in this case the energy dissipates in a much greater volume. The rubber particles also facilitate shear deformation of the epoxy polymer matrix.

There are many other hypotheses regarding the mechanism of the rubber's influence on the strengthening of the epoxy polymers, but the matter is not yet resolved. Generally one can say that the ERC mechanical properties depend on the properties of the initial components of the compound and that their change under the mutual effect, both in bulk and at the interface, depends on the morphological structure of the mixture and on the properties of the interface between the components. When ERC are used as adhesives or coatings, the properties of the polymeric mixture depend on the substrate.

Determination of the properties of the system components and their interactions that affect the mechanical behavior of ERC is at present quite a complicated problem. In this connection it is assumed that the most convenient method of analyzing processes that occur during ERC formation is the thermodynamic method [141–149]. At the same time, the thermodynamic approach is the most rigorous because it allows one to evaluate completely all kinds of interactions in the system. However, this approach has not so far acquired wide application because of difficulties related to the nonequilibrium nature of the process of ERC formation, which is determined not only by thermodynamics but also by kinetic considerations. As the conversion level of the system increases, its viscosity increases, which prevents the reaching of thermodynamic equilibrium.

To characterize the effect of the degree of the intermolecular interaction or, equivalently, the level of thermodynamic compatibility (stability) of the initial components (the epoxy and rubber oligomers) on the final properties of ERC, at the equilibrium conditions, highly viscid systems with subsequent high cure rate were studied. In this case the rapid formation of the fragments of the three-dimensional network of the epoxy polymer "freezes" the structure of the reactive mixture of the eopxy and rubber oligomers.

3.3.2.3 Thermodynamic compatibility of epoxy and rubber mixtures with crosslinking agents. The thermodynamic compatibility of the epoxy oligomers of various structures with the rubbers containing various quantities of acrylonitrile groups was evaluated by the RGC method by determining the thermodynamic interaction parameter $\chi_{2,3}^*$. In the same way, the compatibility of mixtures containing the product of interaction of the epoxy resins with the rubbers was determined. The viscosity values of the initial components and the gel formation times of the systems are presented in Tables 3.10 and 3.11.

The mixture of UP-637 and SKD-KTRA oligomers is thermodynamically compatible up to 5% rubber at 298 K. The increase in the rubber polarity results in this case in widening of the region of mutual solubility of the epoxy and rubber oligomers, with two regions of composition with less stable thermodynamic equilibrium for the UP-637 and SKN-3KTR mixture below 10% and 50% of the rubber in the mixture.

Figure 3.10 displays dependences of the $\chi_{2,3}^*$ parameter on the composition of mixtures of ED-20 epoxy resin with rubbers of various polarity. For a mixture of ED-20 with SKD-KTRA nonpolar rubber, positive values of $\chi_{2,3}^*$ are observed practically throughout the con-

TABLE 3.10 Initial Viscosity of the Components of the ERG

Epoxy oligomer or (and) modifier		Curing agent	
Brand name	Viscosity at 298 K (Pa·s)	Brand name	Viscosity at 293 K (Pa·s)
ED-20	17.0	PEPA	7.0
Modified by 20% SKD-KTRA	125.5	UP-583	15.0
UP-652	6.0		
Modified by 20% SKD-KTRA	100.0	UP-0639	20.0
UP-637	5.0		
Modified by 20% SKD-KTRA	90.5	PO-300	35.0
SKD-KTRA	21.0		
SKN-1KTR	50.5		
SKN-3KTR	73.5		

TABLE 3.11 Gel Formation Time of Epoxy and Epoxy–Rubber Compounds at 296 K

	Gel formation time (min)			
Compound	PEPA	UP-583	UP-0639	PO-300
Ed-20	20	15	35	150
Modified by SKD-KTRA (20%)	5	30	50	180
UP-652	15	12	28	201
Modified by SKD-KTRA (20%)	30	25	40	130
UPK-637	16	12	30	110
Modified by SKD-KTRA (20%)	30	25	40	150

centration range that shows thermodynamic incompatibility of the epoxy and rubber oligomers. Analysis of the temperature dependence reveals that $\chi_{2,3}^*$ decreases with temperature elevation, which is the criterion for a top critical temperature of blending (TCTB) for the given mixture of the oligomers. Change of the rubber polarity results in widening of the region of homogeneous blending; the mixture of ED-20–SKN-3KTR polar rubber oligomers is compatible up to 40% of the rubber at 298 K.

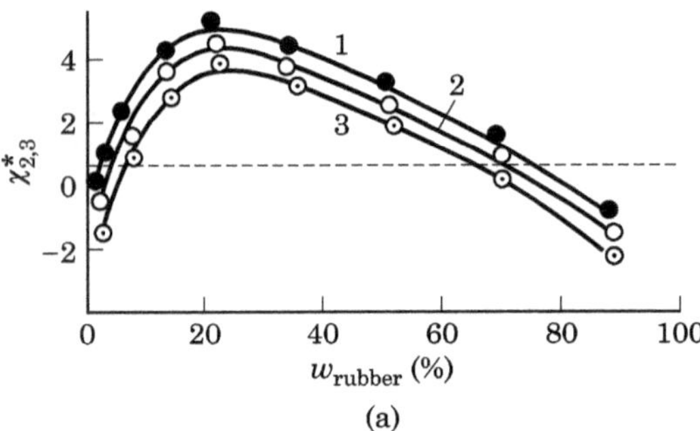

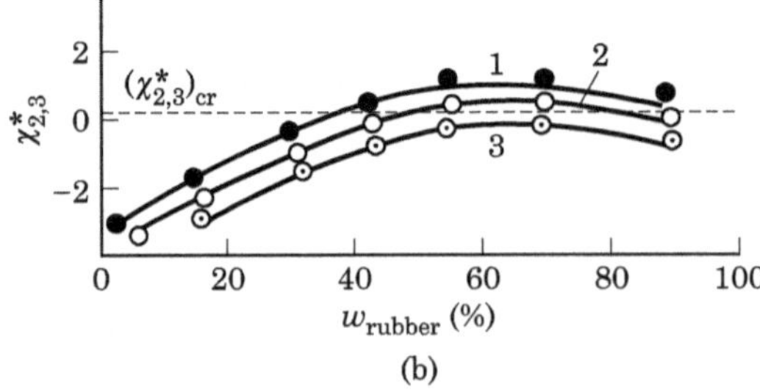

Figure 3.10 Concentration dependence of the parameter $\chi_{2,3}^*$ of the systems (a) ED-20–SKD-KTRA and (b) ED-20–SKN-30 KTR at (1) 298 K, (2) 323 K, and (3) 333 K.

Let us consider the reasons for the improvement of the thermodynamic compatibility of the epoxy and rubber oligomers that is observed with increase of the polarity of the rubber. One of the factors that facilitate the compatibility of the compound components is the ability of their molecules to form various bonds between each other; in particular, the molecules of the rubber and the epoxy resins under study have functional groups that can form hydrogen bonds (Table 3.12).

In the spectrum of SKN-3KTR there are two bands of equal intensity: in the region of the carbonyl absorption at 1710 cm^{-1} (hydrogen-bonded C=O) and at 1740 cm^{-1} (C=O groups free of the hydrogen bonding). When SKN-3KTR is mixed with ED-20 there is no noticeable absorption in the region of 1710 cm^{-1}. Only the band of C=O groups free of the hydrogen bonding appears at 1740 cm^{-1}. In the region of 3300–3600 cm^{-1} there are bands that indicate that all the OH-groups of the compound are linked by hydrogen bonds. These observations indicate that all the OH-groups of the SKN-3KTR molecules free of the hydrogen bonding with the C=O groups can form hydrogen bonds with the ED-20 acceptor groups; that is, there is redistribution of the intermolecular hydrogen bonds in the mixture, which enhances the compatibility of the components.

Proceeding from this concept of the solubility parameter one can expect the oligomer mixture to be compatible if the components have close values of this parameter. Table 3.13 shows that the parameter of the rubber solubility δ changes with acrylonitrile content from 1.84×10^{-3} J/kg$^{1/2}$ for SKD-KTRA to 2.07×10^{-3} J/kg$^{1/2}$ for SKN-3KTR. The solubility parameter of the epoxy oligomer is 2.41×10^{-3} J/kg$^{1/2}$.

The solubility parameter was calculated from the equation

$$\delta^2 = \frac{\sum \Delta E_i}{N_A} \sum \Delta V_i \qquad (3.24)$$

where ΔE_i is the contribution of each atom and of each type of intermolecular interaction to the value of the cohesion effective molar energy; N_A is the Avogadro number; $\sum \Delta V_i$ is the van der Waals volume of the molecule, equal to the sum of the van der Waals volumes of the constituent atoms. Thus, in this case there is a correlation between the difference of the solubility parameters and the region of mutual blending determined by the RGC method. IR spectroscopy data confirm the relationship.

Let us consider the thermodynamic compatibility of the crosslinking agents with the modifiers in use, which occurs at the stage of the ERC formation. The mixture of the PEPA–SKD-KTRA oligomers is thermodynamically compatible up to 2% of SKD-KTRA at 298 K, while for 5%

TABLE 3.12 Characteristics of Components of Epoxy and Epoxy–Rubber Compositions

Type	Brand name	Chemical formula
Epoxy oligomers		
Epoxide–diane resin (MW = 410)	ED-20	$H_2C(O)CH—CH_2—[—OC_6H_4—C(CH_3)_2—C_6H_4—O—CH_2—CH—]_n—O—C_6H_4—C(CH_3)_2—C_6H_4—O—CH_2—HC(O)CH_2$
1,2-Epoxypropoxydiphenyl ester of ethylene glycol (MW = 304)	UP-652	$H_2C(O)CH—CH_2—O—C_6H_4—O—CH_2—CH_2—OC_6H_4—OCH_2—CH(O)CH_2$
Diglycidyl ester of resorcin (MW = 210)	UP-637	$H_2C—CH—CH_2—[—O—C_6H_4—O—]_n—CH_2—CH(O)CH_2$

Modifying agents		
Oligobutadiene rubber (MW = 2880)	SKD-KTRA	$HOOC[-CH_2-CH=CH-CH_2-]_nCOOH$
Oligobutadiene–acrylonitrile rubber (MW = 2900)	SKN-10 KTR ($x = 5; y = 1; z = 10$)	$HOOC-[-(CH_2-CH=CH-CH_2)_x-(CH_2-\underset{\displaystyle C\equiv N}{CH})_y-]_z-COOH$
(MW = 3200)	SKN-30 KTR ($x = 5; y = 1; z = 30$)	
Laproxide (MW = 490)	503 M	$\begin{array}{l} CH_2-O-(-CH_2-\underset{CH_3}{CH}-O-)_n-CH_2-\overset{O}{HC-CH_2} \\ CH-O-(-CH_2-\underset{CH_3}{CH}-O-)_n-CH_2-\overset{O}{HC-CH_2} \\ CH_2-(-O-(CH_2-\underset{CH_3}{CH}-O-)_n-CH_2-\overset{O}{HC-CH_2} \end{array}$
(MW= 700)	703 M	

(Continued)

TABLE 3.12 Characteristics of Components of Epoxy and Epoxy-Rubber Compositions *(Continued)*

Type	Brand name	Chemical formula
	Modifying agents *(Continued)*	
Oxyethylated ester of alkylphenol (MW = 790)	OP-10	$R—C_6H_4—O—(CH_2—CH_2O)_nH$ $R═C_9—C_{10}; n = 10–12$
Monoalkyl ester of polyethylene glycol on base of primary fatty alcohols (SAS) (MW = 709)	DS-10	$C_nH_{2n+1}O(CH_2CH_2O)_mH$ $n = 18; m = 10$
	Linking agents	
Polyethylenepolyamine (MW = 120)	PEPA	$H_2N—[—(CH_2—CH_2)NH]CH_2—CH_2—NH_2$ $H_2N—[—(CH_2—CH_2)NH]_2—CH—CH_2—NH_2$
Diethylenetriaminomethylphenol (MW = 210)	UP-583	$HO—C_6H_4—CH_2—N(H)—(CH_2—CH_2)—N(H)—CH_2—CH_2—NH_2$
Imidazoline on base of sebacic acid (MW = 336)	UP-0639	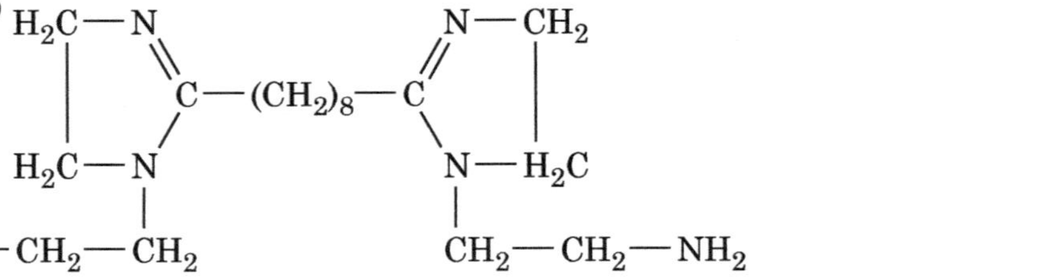 $H_2C—N \qquad\qquad N—CH_2$ $C—(CH_2)_8—C$ $H_2C—N \qquad\qquad N—H_2C$ $H_2N—CH_2—CH_2 \qquad CH_2—CH_2—NH_2$

TABLE 3.13 Values of the Solubility Parameter δ of Epoxy and Rubber Oligomers of Various Brands

Epoxy oligomer		Rubber oligomer	
Brand name	$10^3\,\delta$ (J/kg$^{1/2}$)	Brand name	$10^3\,\delta$ (J/kg$^{1/2}$)
ED-20	2.41	SKD-KTRA	1.84
UP-652	2.36	SKN-1KTR	1.93
UP-638	2.5	SKN-3KTR	2.07

of the rubber in the mixture at the same temperature maximum thermodynamic instability is when parameter $\chi_{2,3}{}^*$ is maximal. The higher rubber polarity in the case of SKN-40KTR results in some widening of the region of the homogeneous blending. For the UP-583–SKD-KTR mixture the values of $\chi_{2,3}{}^*$ are lower than the critical value that exhibits the thermodynamic compatibility.

Note a somewhat unusual dependence of the parameter $\chi_{2,3}{}^*$ on temperature. With up to 40% of rubber in the mixture the value of $\chi_{2,3}{}^*$ increases with falling temperature, while at high rubber concentrations $\chi_{2,3}{}^*$ falls with falling temperature. The thermodynamic compatibility becomes poorer with increase of the rubber polarity in these mixtures. For the UP-583–SKN-30 oligomers the magnitude of $\chi_{2,3}{}^*$ is much higher than the critical value.

The mixture of the UP-0639–SKD-KTRA oligomers has limited thermodynamic compatibility up to 10% of SKD-KTRA at 298 K. With increasing temperature $\chi_{2,3}{}^*$ decreases, which indicates the existence of a blending top critical temperature in this mixture of oligomers. The increase of the rubber polarity in the case of SKN-30KTR results in decline of the thermodynamic compatibility, and the values of $\chi_{2,3}{}^*$ increase with increasing temperature, indicating the existence of a blending bottom critical temperature for this mixture of oligomers.

The concentration dependence of the parameter $\chi_{2,3}{}^*$ for the mixture of PO-300–SKD-KTRA oligomers indicates the complex thermodynamic behavior of the mixture. The function $\chi_{2,3}{}^* = f(\phi_3)$, where ϕ_3 is the volume fraction, is of bimodal character with the maximum of the thermodynamic instability observed at 5% and 45% of SKD-KTRA; that is, the above mixture has two regions of composition with the least stable thermodynamic equilibrium. In this case the stability is itself a function of temperature and increases with temperature, indicating the existence of a blending bottom critical temperature for this mixture of oligomers.

Thus, mixtures of epoxy resins or curing agents with rubbers are systems with strong specific interactions, which ultimately determine all the features of the thermodynamic behavior of these systems. The mutual solubility of oligomer mixtures is affected by the epoxy oligomer composition, by the modifier's polarity, by the compound composition, and by temperature. The structure of the composite can be controlled by varying the temperature during curing of the epoxy resins by various crosslinking agents as well as with different quantities of the oligobutadiene rubber or OBDANC present.

3.3.2.4 Mechanical properties of epoxy rubber compounds. The shape of the phase diagram can be judged to some extent by the type of concentration dependence of the thermodynamic interaction parameter $\chi_{2,3}{}^{*}$ of the cured polymeric compound. The parameter $\chi_{2,3}{}^{*}$ for the UP-637 system (cured by the UP-0639 crosslinking agent) with SKD-KTRA rubber is positive and essentially does not change when 5–40% of the rubber is added. But the evaluation of the fraction of the rubber (which is in the form of a dispersion in the epoxy matrix) by determination of the enthalpy component of $\chi_{2,3}{}^{*}$ ($\chi_H = -T\,\partial\chi_{2,3}{}^{*}/\partial/T$) showed that this fraction does not have a constant increase with rubber content growth but has a maximum at 20% of the rubber.

Let us consider the effect of the ERC thermodynamic state on the mechanical properties. Figure 3.11 presents the temperature relationships of the dynamic modulus of elasticity E' and the tangent of the angle of the mechanical losses $\tan\delta$ for the epoxy polymers based on

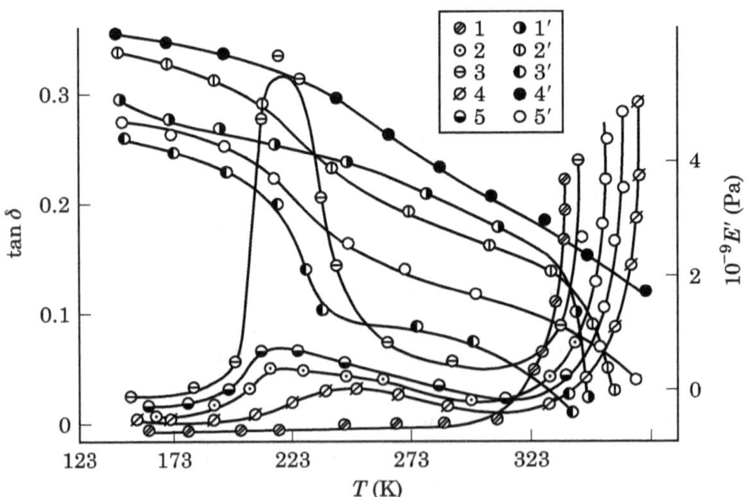

Figure 3.11 Temperature dependences of $\tan\delta$ (1–5) and E' (1′–5′) for epoxy compounds based on UP-637 (1–3, 1′–3′) and ED-20 (4, 4′, 5, 5′) oligomers modified by SKD-KTRA rubber; curing agent is UP-0639; the ratio of the oligomer to the rubber is 100:0 (1, 1′, 4, 4′); 90:10 (2, 2′); 70:30 (3, 3′); and 80:20 (5, 5′).

UP-637 and ED-20 cured by UP-0639 with and without addition of the low-molecular-weight rubber. As in systems based on diene oligomers, the integration of the rubber into oligomers based on UP-637 results in a peak in the temperature relationship of $\tan\delta$ at 213 K. Increase of the rubber content produces increase of the peak intensity only, without any practical change in the temperature at which $\tan\delta$ is maximal.

Note that whereas a broad region of β-relaxation within the temperature range 203–303 K is observed for the epoxy polymer based on ED-20 unmodified oligomer, no β-relaxation is observed for the polymer based on UP-637. In this case the β-relaxation is related [140] to the motion of the group of atoms

$$—CH_2—\underset{\displaystyle OH}{\underset{|}{CH}}—CH_2—O—$$

In connection with the facts that the peak at 213 K corresponds exactly to the glass transtion of the SKD-KTRA rubber and that system softening at 328 K coincides with the glass transition of the epoxy matrix, one can affirm the incompatibility of the rubber and epoxy components of the above ERC. Electron-microscopic data also indicate the existence of the ERC two-phase structure only in the case of SKD-KTRA as a modifier, which is thermodynamically incompatible at the stage of blending with the epoxy oligomer.

Thus, the phase separation that runs in parallel with the reaction of formation of the three-dimensional epoxy polymer network in ERC, obtained by the PER method with application of SKN-3KTR rubber that is thermodynamically compatible with the epoxy oligomer, does not occur, indicating the conservation of the phase structure of the compound formed at the blending stage and in the cured state. This supports the possibility of using the thermodynamic approach for describing the physical-mechanical properties of ERC obtained by the PER method.

Let us consider the effect of the ERC phase state upon its crack resistance. The maximum modifying effect from addition of the rubber into the epoxy resin is achieved for rubber particles in the form of the individual phase being available in the epoxy polymer matrix. As was shown for ERC obtained by the PER method, this condition is observed when using SKD-KTRA as a modifier.

Consider the critical value of the stress intensity coefficient, which like the classical mechanical characteristics (strength and yield limit, ultimate elongation per unit length, etc.) reflects an objective property of the material—its capacity to resist the development of cracks.

For mechanically mixed resin and rubber, the SKD-KTRA rubber concentration has essentially no effect upon the value of K_{IC}. For thermally treated mechanical mixtures, the value of K_{IC} increases about twofold. If the thermal treatment of the specimens was performed after the compound was completely cured (degree of conversion not less than 93%), the heating does not effect the level of phase separation but it alters the interphase interaction between the components not only by increasing the strength of the resin itself but also through the abrupt increase of the effect of the rubber upon K_{IC}.

Comparison of the concentration dependence of the K_{IC} values of the compounds obtained by the above two methods with the concentration dependence of the $\chi_{2,3}^*$ thermodynamic interaction parameter of oligomer mixtures (epoxy resin–modifier and curing agent–modifier) displays the same dependences. The maximum value of K_{IC} of $1.1\,\text{MPa}\cdot\text{m}^{1/2}$ is observed at 5% content of the SKD-KTRA rubber. This is the region of maximal thermodynamic instability of the oligomer mixture.

For compounds obtained with PER, the maximal modifying effect is observed at 20% of SKD-KTRA ($K_{IC} = 1.45\,\text{MPa}\cdot\text{m}^{1/2}$). The concentration dependence of $\chi_{2,3}^*$ (see Fig. 3.10) shows that this region is characterized by the maximal thermodynamic instability of the mixture of the epoxy resin with the modifier.

Thus, from the above data one could conclude that to achieve the maximal values of K_{IC} the method of modification with PER is most effective. Additionally, this method is more technologically applicable because it can be used in formation of cold-cure compounds. Special attention must be paid to the correlation of the K_{IC} characteristic with $\chi_{2,3}^*$ for mixtures such as epoxy resin–modifier and curing agent–modifier. The presented characteristics of the ERC cracking resistance indicate the strong dependence of K_{IC} on the structure of these materials, which enables the use of this characteristic in optimizing the compositions of new and already known epoxy rubber materials.

Now let us consider how the addition of the SKD-KTRA rubber influences the ultimate strength properties of ERC based on ED-20 and UPR-637 oligomers obtained by the method with PER and cured by UP-0639.

With increase of the rubber content in a compound, the specimens become more elastic and can develop larger deformations, but the strength of the compounds decreases (this decrease becomes insignificant at high concentrations of the rubber). In this context there is much more work to be done on their failure due to the increase of deformability. Note that in the course of deformation of the modified specimens turbidity is observed that seems to be a direct consequence of the appearance of microcracks in the matrix adjacent to the rubber particles at the start of the development of high elastic deformation.

The rupture strength of the initial epoxy polymers and of those modified by the rubber decreases monotonically with increase of the test temperature for all the specimens under study (Fig. 3.12). For polymers based on the UP-637 oligomer, the drop in strength is more precipitous. As a result, the strength properties of the UP-637 oligomer-based polymers exceed those of the ED-20-based polymers only within a narrow range of low temperatures. At higher temperatures the diene polymers are somewhat stronger. In addition, the strength of the UP-637-based modified polymer is lower than that of the unmodified polymer within the whole range of temperatures studied, while for the polymer based on the diene oligomer the modified specimen (beginning at about 340 K) becomes stronger due to the lower rate of decline of σ_p with the temperature drop.

The temperature dependence of the rupture deformation ε_p shows that the modified polymers have higher deformability compared with the initial polymers throughout the whole temperature range mentioned above. This difference is especially noticeable at low temperatures (up to about 323 K), which correspond to the glass-transition state of the polymers under investigation. The rupture elongation ε_p

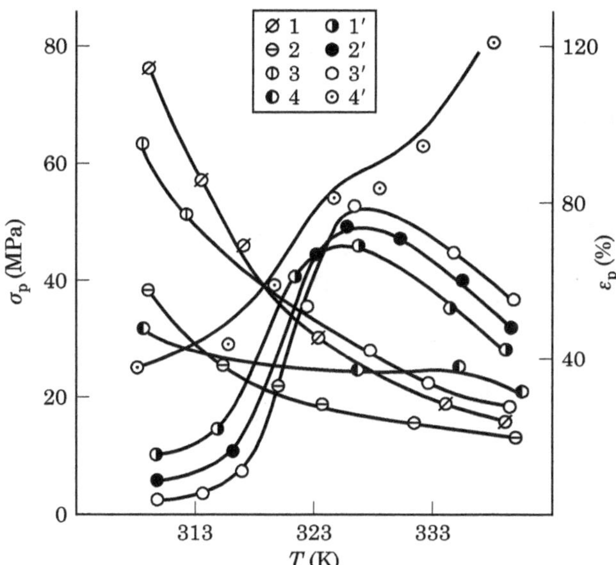

Figure 3.12 Temperature dependences of the strength σ_p (1–4) and the deformations ε_p (1′–4′) for polymers based on the initial oligomers (1, 1′, 3, 3′) and oligomers modified by SKD-KTRA rubber (2, 2′, 4, 4′). The ratio of the oligomer to the rubber in the modified oligomer is 80:20.

increases sharply at temperatures above T_g due to the development of high-elasticity deformation, and then runs through a maximum at about 333 K. The exception is the specimen based on the modified diane oligomer, for which the enhanced deformability is observed at high temperatures as well. Consequently, in epoxy oligomers based on the diane oligomer the modification by the rubber facilitates improvement of their deformation features at both low and high temperatures.

It is interesting to compare the strength σ_p and the elongation ε_p during extension of the film and block specimens. Figure 3.13 shows the dependences of σ_p and ε_p on the rubber content in compounds based on ED-20 and UP-637 oligomers cured by UP-0639. It is evident that in UP-637-based compounds the values of σ_p for the film specimens (Fig. 3.13, curve 1) and for the block specimens (curve 2) almost coincide for any quantity of the rubber. For compounds based on the ED-20 modified oligomer and containing up to 4% of the rubber, the block specimens have smaller σ_p (curve 4) than the film specimens (curve 3). With increase of the rubber content, σ_p for the film specimens decreases monotonically, while for the block specimens an extremal dependence is observed, and in the case of the rubber content above 4% the breaking strength of the block specimen becomes higher

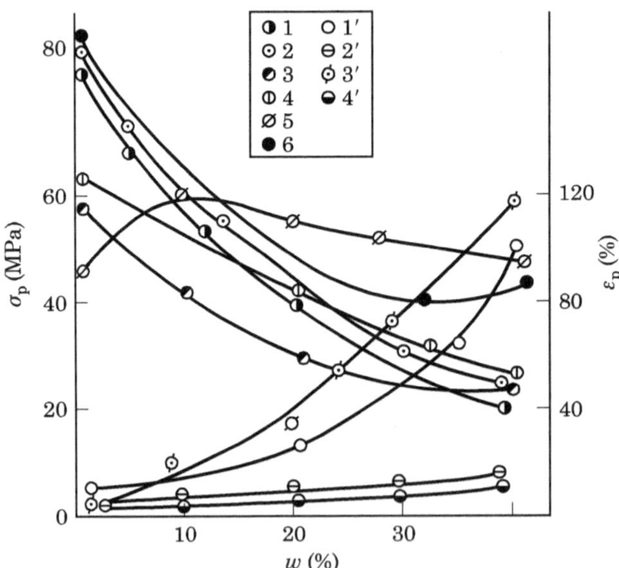

Figure 3.13 Dependence of σ_p (1–6) and ε_p (1′–4′) on the SKD-KTRA rubber content w(%) for epoxy polymers based on ED-20 (2–5) and UP-637 (1, 2, 6) oligomers; (1, 3, 5, 6) are film specimens; (2, 4) are specimens in block. The value of σ_p is calculated for the initial (1–4) and true (5, 6) cross-sections of the polymer specimen.

than that of the film ones. The rupture deformation for both types of specimens increases monotonically with increase of the rubber content. Hence, for the film specimens ε_p is essentially higher than for the block specimens, especially when the quantity of rubber is above 10%. It should also be noted that for the block specimens σ_p is higher for the compound based on the UP-637 oligomer (curve 2), while for the films this parameter is higher when the specimens are based on the diane oligomer. The comparatively small deformations at rupture of the block specimens seem to be governed by the higher probability of defect formation compared with the fine films. Note in particular that, although σ_p for the film specimens decreases with the increase of the quantity of added rubber (this decrease becomes insignificant for large quantities of rubber), the abrupt increase of the system deformability is more important.

It should be noted that the values of σ_p (Fig. 3.13, curves 1–4) are calculated for the value of the initial cross-section of the specimen, but in the course of deformation the specimen cross-section decreases. As Fig. 3.13 shows, the value of the true stress at rupture of the polymeric films decreases with increasing rubber content to a lesser extent than σ_0. At large rubber content, $\sigma_{p\ true}$ increases somewhat (possibly due to orientation strengthening).

The mechanical characteristics and the moduli of elasticity of UP-652-based polymers modified by the SKD-KTRA and SKN-3KTR rubbers and cured by the UP-0639 curing agent are presented in Table 3.14. These data indicate that the decrease of the tensile strength σ_p

TABLE 3.14 The Effects of Addition of SKD-KTRA and SKN-3KTR Rubbers upon the Mechanical Properties of Epoxy Oligomers Based on UP-652 Epoxy Resin Cured by UP-0639

Oligomer to rubber ratio	σ_p (MPa)[a]	ε_p (%)[a]	E (MPa)[a]
	SKD-KTRA		
100:00	30.6/60.8	10.2/12.9	580/890
95:5	30.8/52.3	10.8/13.5	520/840
90:10	30.1/42.2	11.8/15.0	460/765
80:20	28.2/31.5	26.5/20.3	211/396
70:30	26.5/27.8	72.3/56.4	123/165
60:40	18.5/20.3	100.8/89.5	82/104
	SKN-3KTR		
95:5	29.5/52.3	10.5/12.9	608/858
90:10	27.5/45.1	10.8/13.0	610/812
80:20	21.5/39.7	14.7/16.5	515/703
70:30	14.8/31.8	53.2/28.5	352/574

and of the modulus of elasticity E (about 2.5-fold with addition of 40% of the modifier) is accompanied by much greater (about 8-fold) increase of the films' breaking deformation.

So far as $\sigma_p \varepsilon_p/2$ gives an approximate characteristic of the material failure work, the considerable increase of the deformability of the modified polymers with the comparatively small drop in strength must improve the serviceability of the polymeric material. This is also true for the heat-treated specimens, which have increased σ_p and E as a result of heating. The value of ε_p increases somewhat after heat treatment for the unmodified polymer containing comparatively small amounts (up to 20%) of the rubber, and decreases insignificantly for large quantities of the modifier. The increase of the deformability as a result of the heat treatment seems to be due to the formation of the looser structure of the glassy polymer, that is, to the free volume increase. Thus, mesurement of the density of the modified polymer revealed the decrease of this parameter following heat treatment (to 1.192 from 1.208 kg/m^3 for the unheated specimen).

The integration of liquid rubbers into the epoxy polymers is used mainly to increase the impact strength. For systems based on the ED-20, UP-652, and UP-637 epoxy oligomers, the addition of SKD-KTRA rubber with UP-0639 as curing agent results in ~3-fold increase of the polymer's specific impact viscosity. This is observed most clearly with addition of small quantities of the rubber (less than 10%). The polymer based on the unmodified diane oligomer is characterized by a comparatively small value of the impact viscosity of about 6 kJ/m^2, but increasing the quantity of the integrated rubber results in rapid increase in the impact strength, as a result of which the specific impact viscosity of the polymer is 26 kJ/m^2 at 25% content of the modifier. With addition of more than 5% of the rubber into an epoxy polymer of any type there is a practically linear growth of the impact viscosity with the increase of the elastomer content (up to 30%).

Let us consider the dependence of the bending strength σ_{bend} of the systems based on ED-20, UP-652, and UP-637 epoxy oligomers cured by UP-0639 on the concentration of the SKD-KTRA modifier. In the case of the UP-637-based polymer σ_{bend} decreases monotonically with the increase of the quantity of added rubber, this decrease being especially considerable at small (less than 5%) and large (over 20%) contents of the rubber. In the intermediate region, the decrease of σ_{bend} is comparatively small. In the case of the ED-20- and UP-652-based polymers, σ_{bend} increases along with the growth of the rubber content. For the ED-20-based polymers, a small maximum is observed at 5% content of the rubber; as the rubber content increases, σ_{bend} begins to decrease monotonically. For the UP-652-based polymers, the σ_{bend} maximal values are observed at 10% of rubber with rather substantial increase of σ_{bend}. This amount of rubber seems to be optimal for obtaining impact-resistant engineering materials based on UP-652.

Modification of epoxy polymers by rubbers results in decrease of the residual stresses σ_{dresid} cured in. Thus, for the polymers based on UP-637 and cured by UP-0639, σ_{dresid} falls monotonically from 5.5 to 2.0 MPa as the SKD-KTRA rubber content increases from 0% to 30%. For the compounds based on the diane oligomer, the stresses decrease under the same conditions from 10.2 to 5.5 MPa because of the increase of the deformation characteristics and of the acceleration of relaxation.

The above considerations suggest that modification of epoxy polymers of various chemical structures (based on diene oligomer and resorcinol diglycidyle ethers) by the carboxyl-containing oligobutadiene and oligobutadiene-acrylonitrile rubbers by the method with PER produces ERC with valuable properties. In this case the relative breaking elongation ε_r, specific impact viscosity a, and failure work A_p increase, and the residual stress σ_{dresid} decreases. Note also that despite the lower efficiency of the modifying effect of the rubbers on the impact viscosity and the bending strength, the UP-637-based polymers have better mechanical characteristics, accounting for better properties of the initial (unmodified) polymer in comparison of the UP-637 oligomer with the ED-20 diane oligomer within a very broad and practically important range of rubber content (up to 25%).

Let us consider the static relaxation properties of ERC. There are numerous publications dedicated to the investigation of the morphology and the extreme mechanical characteristics of ERC obtained by applying high-temperature cure. But there is no information on the static relaxation properties of epoxy polymers modified by the low-molecular weight rubbers obtained by the method with PER. Such characteristics of polymers are of interest in so far as the polymeric materials may be unusable not only because of failure but also due to intolerable changes of shape and size. As Fig. 3.14 shows, the integration of different quantities of rubber (up to 30%) does not produce a great decrease of the glass-transition temperature. For small additions (of about 10%) some increase of T_g is observed compared with the unmodified oligomers. A distinguishing feature of the UP-637-based polymers is the large temperature range of the high-elasticity state (about 323–523 K) compared with that of the ED-20-based compound.

A further peculiarity of the thermomechanical behavior of the studied compounds is important. The temperature at which the deformation increases abruptly is not very sensitive to the rubber content (Fig. 3.14).

The effect of addition of SKD-KTRA and SKN-3KTR rubbers on the heat resistance of the UP-652 based epoxy polymer cured by UP-0639 is similar. The addition of SKN-3KTR does not practically change the

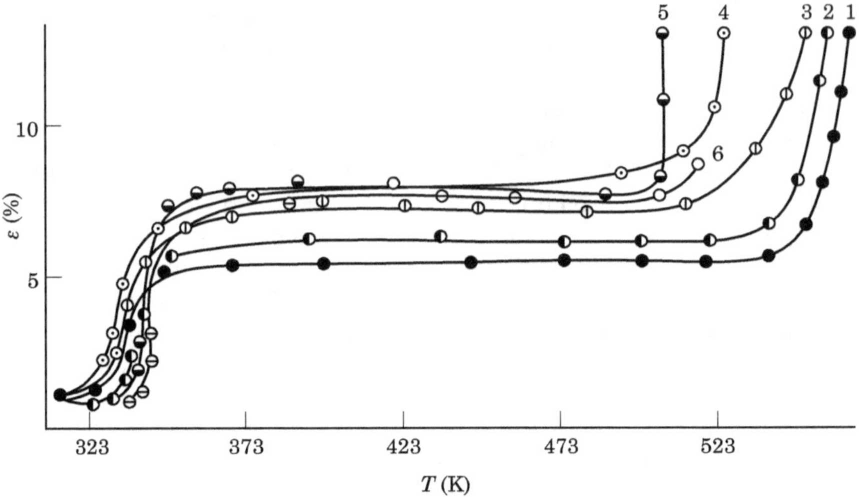

Figure 3.14 Thermal-mechanical curves for epoxy polymers based on the UP-637 (1–4) and ED-20 (5, 6) oligomers. The ratio of the oligomer to the rubber is 100:0 (1, 5), 90:10 (2, 6), 80:20 (3), 70:30 (4).

glass transition temperature of the specimen within a quite broad range (up to 30%). At the same time, the addition of only 10% of SKD-KTRA produces a sharp decrease of the polymer's heat resistance. Heat treatment at 393 K for 3 h enhances the heat resistance due to the increase of the chemical crosslinking density as a result of the system postcure. The conversion level of the epoxy groups at ERC heat-up was 99%, with the molecular weight of the chain section between the adjacent crosslinks of the space network (M_C) decreasing from 780 to 630.

The effect of addition of SKD-KTRA rubber on the static relaxation properties of the epoxy polymers was determined by studying the creep under conditions of single-axial extension at all possible levels of the mechanical stress (up to film failure) within the temperature range 298–353 K. For the initial polymer and that modified by a small (10%) amount of rubber, the creep rates coincide at similar loads when the polymers are in the glassy state. But at stresses close to the strength limit the creep of the modified polymer is substantially higher than that of the initial specimen. In the case of the polymer containing 30% rubber, a fairly large creep deformation substantially exceeding those of the initial specimens develops at comparatively small loads (10 and 20 MPa).

If the temperature is raised to 323 K rates of creep for the modified and initial polymers increase but to different extents for specimens with different rubber contents (Fig. 3.15). Under these conditions the

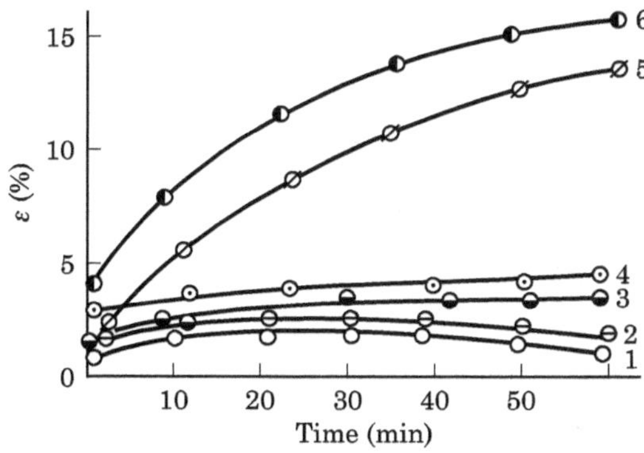

Figure 3.15 Alteration of the creep of epoxy polymers based on UP-637 initial oligomer (2, 5) and that modified by 10% (1, 4) and 30% (3, 6) of SKD-KTRA plotted at stresses of 5 MPa (1–3) and 10 MPa (4–6) and at a temperature of 323 K.

development of the creep process is resisted most by the specimen modified with 10% rubber. For the initial epoxy polymers and for those modified by 30% rubber, the creep deformation and the rate of its development are substantially higher, especially for much higher levels of the mechanical stress. Consequently, the introduction of small quantities of the modifying additive (about 10%) not only decreases the creep deformation compared with the unmodified polymer but widens somewhat the serviceable temperature range of the material.

This conclusion is confirmed by the generalized pliability curves (see Fig. 3.16) in the range of 298–353 K using a reduced temperature (T/T_0) relative to 298 K. The generalized pliability curves show that the value and the rate of creep of the epoxy polymer modified with 10% rubber are less than those of the initial polymer. For the specimen with 30% modifier, these parameters are much higher, especially at high levels of mechanical stress.

The generalized pliability curves determined for two levels of the mechanical stress (5 and 10 MPa) for the initial polymer and that modified by 10% rubber within the region of small duration of the creep process practically overlap one another, indicating the linearity of their mechanical behavior. In the case of the longer process, the linearity is disturbed for both types of polymers, with larger deviations observed for the initial (unmodified) polymer.

The increase of the glass-transition temperature and the decrease of creep of epoxy polymers with 10% SKD-KTRA demonstrate the effect of the rubber upon the epoxy matrix itself. Such abnormal behavior of ERC can be explained thermodynamically. An SKD-KTRA content of 10% in ERC ensures thermodynamic conditions for separation of the

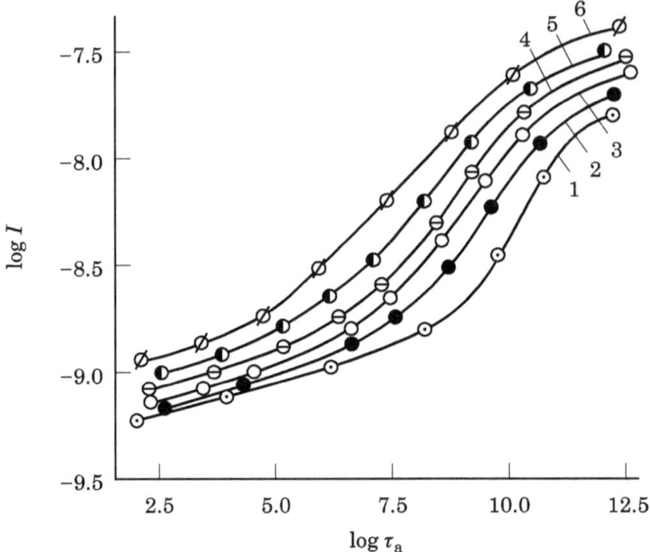

Figure 3.16 Generalized curves of the pliability I (Pa^{-1}) for film specimens based on UP-637 initial oligomer (3, 4) and that modified by 10% (1, 2) and 30% (3, 6) of SKD-KTRA obtained for stresses of 5 MPa (1, 3, 5) and 10 MPa (2, 4, 6). The reduced temperature is calculated relative to 298 K.

system into two phases, which results in a fairly developed interface of epoxy polymer–rubber.

As the authors of [16] note, polymer boundary layers of up to 0.05 μm on a low-energy surface are characterized by higher packing density, by a glass transition temperature 5–7°C higher than that of the polymers in block, and by other characteristics. The most probable reason for these findings seems to be the restriction of the macromolecules' conformation set and their adsorption interaction with a solid at the interface. Alteration of the properties of the epoxy matrix results also in the improvement of a number of other physical and mechanical characteristics of ERC containing 10% SKD-KTRA, particularly in enhancement of the flexural and tensile strength.

Let us consider the adhesion properties of ERC. The mechanism of the effect of the low-molecular-weight additives on the properties of the polymers in block has been studied in detail, though their effect on the strength of adhesive-bonded joints has not yet been studied in sufficient detail.

Figure 3.17 presents the concentration dependences of τ_{ad} for ED-20-based ERC modified with SKD-KTRA rubber and cured by UP-0639 and produced in different ways. It is evident that the maximum

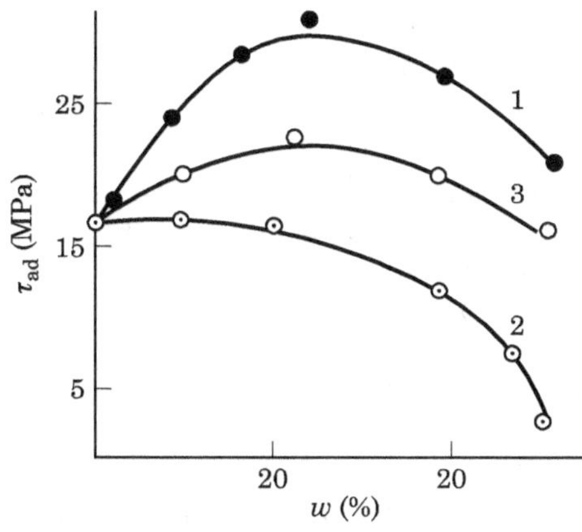

Figure 3.17 Effect of the quantity of SKD-KTRA rubber on τ_{ad} for ERC based on ED-20 cured by UP-0639 obtained by the method with PER (1) and by mechanical blending with heating for 2 h at 372 K (3).

increase of the adhesion strength is observed when obtaining ERC by the PER method with less than 20% rubber. For ERC obtained by mechanical blending of the epoxy and rubber oligomers, the adhesion strength changes little for small quantities of the modifier, while at greater additive concentration its adhesion strength decreases. Some improvement of the ERC properties after the heat treatment can be gained through the formation of chemical links between the molecules of the modifier and the epoxide, accounting for etherification of the groups that have not reacted in the course of curing at room temperature. In addition, the annealing enhances the homogeneity of the epoxy matrix structure due to the decrease of the number of defects and internal stresses.

For compounds of epoxy resins based on epoxy-diane oligomers and resorcinol diglycidyle esters set with amine curing agents, the dependence of τ_{ad} on the content and type of the modifying additive displays distinct maxima. The adhesion strength is enhanced most when the SKD-KTRA rubber is applied at 20% (see Figs. 3.17 and 3.18). With SKN-KTR as modifier the effect is less.

The adhesion properties of ERC are directly related to their phase structure. When comparing the values of $\chi_{2,3}{}^{*}$ that characterize the thermodynamic stability of the oligomer mixtures (epoxy resins–rubbers and curing agents–rubbers) with the values of τ_{ad}, a similar dependence is observed for ERC obtained by the method with PER. The maximum value of the ERC adhesion strength corresponds to that quantity of the rubber at which $\chi_{2,3}{}^{*}$ is maximal.

For compounds containing SKD-KTRA rubber incompatible with the epoxy oligomer, the higher adhesion strength can be explained both by the increase of the ERC cohesion strength and by the possible concentration of impurities at the interface between the epoxy matrix and the rubber, as a result of which the probability of their sorption on the boundary between the adhesive and the substrate decreases, a weak boundary zone appears, and the adhesion strength is reduced [151]. Note that for the compounds containing SKN-3KTR rubber the adhesion strength also increases in the region of concentrations in which the thermodynamic compatibility of the epoxy and rubber oligomers is observed to decrease. For compounds cured by PO-300 (Fig. 3.18) the adhesion strength has maxima at 5% and 20% of the modifier. The first corresponds to the maximal thermodynamic incompatibility of the curing agent–modifier mixture; the second maximum corresponds to that of the epoxy oligomer–modifier mixture.

Thus, the formation of the two-phase system by the method with PER ensures optimal values of some of the physical-mechanical characteristics. These conclusions are confirmed by the results of electron-microscopic studies. With chemical bonds between the epoxy polymers and the rubber particles, the developing crack does not pass round the rubber particle but penetrates it, which causes delocalization of stresses at the tip of the crack and results in improvement of the physical and mechanical properties of the ERC.

These findings allow control of the modification of epoxy resins by rubbers. For example, to obtain ERC with improved adhesion strength it is necessary to use a quantity of the rubber at which thermodynamic incompatibility with the epoxy oligomer is observed.

ERC with improved cracking resistance can be obtained using 20% of SKD-KTRA rubber by the method with PER; for hot-cure compounds the maximal values of the cracking resistance can be achieved by adding 5% of the rubber.

When using UP-0639 as the curing agent, the addition of 10% of SKD-KTRA to the epoxy oligomers results in compounds with lower

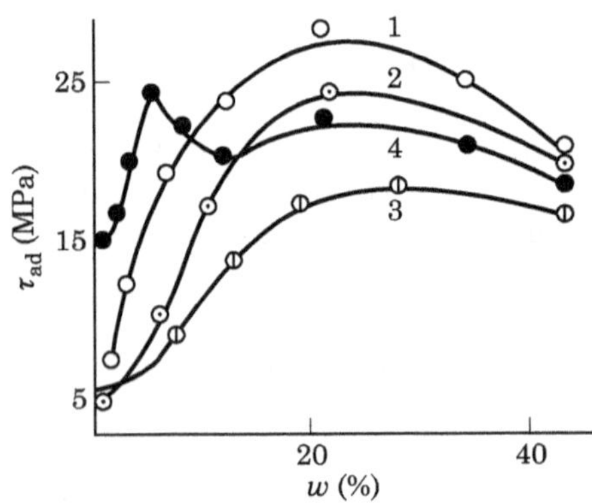

Figure 3.18 Dependence of τ_{ad} for ED-20 cured by PEPA (1–3) and PO-300 (4) on the quantity of rubber: (1) SKD-KTRA; (2) SKN-10 KTR; (3) SKN-30 KTR; (4) SKD-KTRA.

creep rate than the unmodified polymer and with improved heat resistance and bending strength, which are related to the thermodynamic state of the ERC components.

To obtain compounds that have improved impact resistance and are characterized by high failure work as well as high values of relative elongation before break-off, it is necessary to add not more than 30–40% of the SKD-KTRA rubber.

3.3.2.5 Effect of surface-active substances on the properties of epoxy rubber compounds. ERC have a long interface boundary between the system components, and surfactant concentrating here can alter both the components' thermodynamic stability and the interface energy characteristics. In addition, the addition of surfactant changes the properties of the epoxy matrix itself [152–155]. Consequently, the addition of surfactant to ERC must change its thermodynamic and physical-mechanical properties.

As Fig. 3.19 shows, the maximal values of K_{IC} are obtained when SKN-3KTR rubber is used as a modifier, which has a similar solubility parameter (see Table 3.13) to that of the epoxy oligomer. Addition of surfactant (OP-10) to ERC obtained by mechanical blending of the ED-20 epoxy oligomer with the SKD-KTRA and SKN-3KTR rubber oligomers gives considerable improvement of their cracking resistance. The critical value of the stress intensity coefficient increases twofold when SKD-KTRA rubber, which differs from the epoxy oligomer in the solubility parameter, is used as a modifier.

Let us consider the results in detail. OP-10 is thermodynamically incompatible with SKD-KTRA rubber but is compatible with SKN-3KTR. It exerts a structurizing effect on the epoxy polymer when

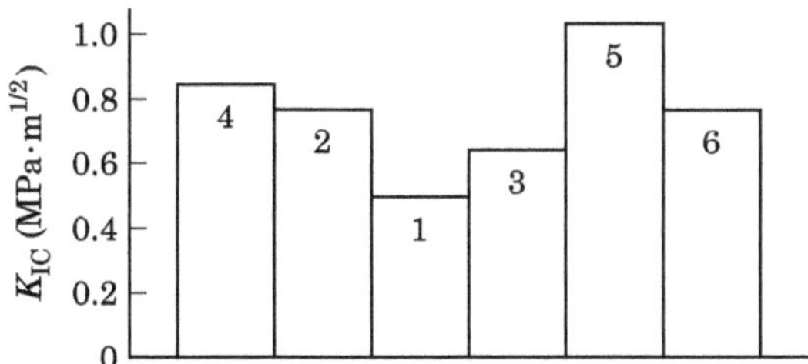

Figure 3.19 Effect of surface-active additives (%) on the cracking resistance K_{IC} of ERC based on ED-20 cured by PEPA: (1) ED-20; its mechanical mixtures with (2) 20% of SKN-30 KTR, (3) 20% of SKD-KTRA, (4) 20% of SKN-30 KTR and OP-10 additive, (5) 20% of SKD-KTRA and OP-10 additive, (6) ED-20 with OP-10.

added at 6%. In this connection, it can be assumed that for ERC based on ED-20 epoxy resin and SKD-KTRA rubber the maximal values of K_{IC} are due to the fact that surfactant spreads uniformly inside the epoxy oligomer, affecting its formed structure. In addition, OP-10 is surface active at the interface between the epoxy resin and the low-energy surface, which alters the intensity of interphase interactions between the epoxy polymer and the rubber and changes the dispersion level. This assumption is confirmed by electron-microscopic analysis.

From the above, one can conclude that for ERC with SKN-3KTR the conditions are met for maximal adhesion interaction between the epoxy polymer and the rubber due to the components' compatibility. The maximal values of K_{IC} are determined for ERC obtained by the method with PER (Fig. 3.20). It should also be noted that these conditions ensure the higher adhesion strength of ERC (Fig. 3.21). Small additions of surfactant to ERC obtained by the method with PER allow additional improvement of their adhesion characteristics.

3.3.3 Modification of EP-20 epoxy-diane resin by epoxided polypropylene glycol (Laproxides 503M and 703)

Rubberlike materials are formed in the course of curing of Laproxides by amine-type crosslinking agents. In connection with the fact that the relaxation capacity of the Laproxides is much lower than that of ED-20

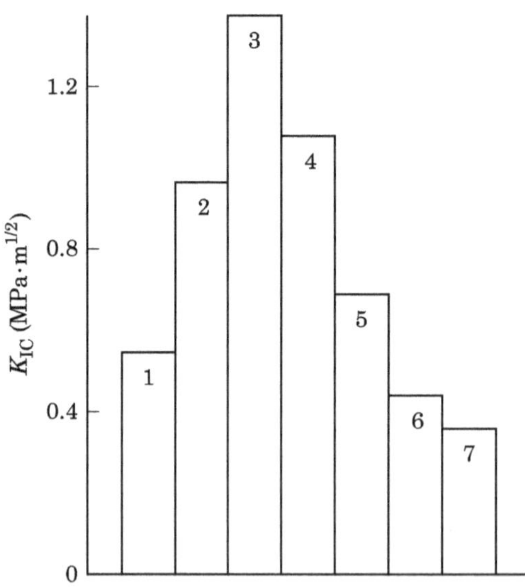

Figure 3.20 Effect of the method of ED-20 epoxy resin modification on the cracking resistance K_{IC}: (1) mechanical mixture with 20% SKN-30 KTR; (2) modification with 20% SKN-30 KTR; (3) modification with 20% SKD-KTRA with preformed PER; (4) heat-treated mixture with 20% SKD-KTRA; (5) heat-treated resin; (6) mechanical mixture with 20% SKD-KTRA; (7) cured resin without heat treatment.

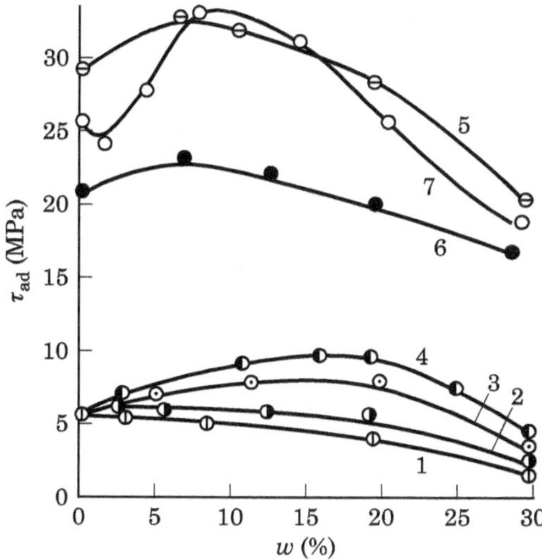

Figure 3.21 Dependence of τ_{ad} on the rubber content (1–4) and surfactant content (3–7) for epoxy resins based on ED-20 cured by PEPA (1–6) and UP-583 (7). Mechanical mixtures with (1) SKD-KTRA, (2) SKN-30 KTR, (3) SKN-30 KTR and 6% OP-10, (4) SKD-KTRA and 6% OP-10; (5) modified with 20% SKD-KTRA with PER; (6) ED-20 modified with 20% SKN-30 KTR; (7) UP-652 modified with 20% SKD-KTRA.

resin, their mutual cure results in polymeric mixtures similar to the epoxy rubber ones.

Before considering the particularities of ED-20 epoxy resin modification by Laproxides of different molecular weight, it is necessary to note that low-viscosity systems, with more time for formation of the three-dimensional network fragments, are formed with application of low-activity curing agents (like UP-0633). This is why in such systems the formation of the polymeric mixture structure must keep step with the thermodynamic changes that occur during curing of these two crosslinked polymers. Consequently, in this case one should not expect the correlation between the properties of the polymeric mixture and the thermodynamic stability of the initial components of the compound that is observed at the stage of its formation. The molecular weight of the modifier (epoxide oligoester) affects the thermodynamic compatibility of ED-20 mixtures with Laproxides. The mixture of ED-20–Laproxide 503M oligomers is thermodynamically compatible within the whole range of concentrations studied, while the mixture of ED-20 and Laproxide 703M is compatible only up to 40% of the modifier content at 298 K. In this case $\chi_{2,3}{}^{*}$ increases as temperature decreases for these mixtures of oligomers, indicating the existence of a top blending critical temperature.

The phase separation in systems based on the high-molecular weight compounds is known to occur not only due to alteration of composition, temperature or pressure but also as a result of chemical reaction [19]. In the course of curing of these epoxy oligomers (containing two or more reactive groups) two crosslinked polymers are formed. The formation of their fragments is accompanied by the appearance of thermodynamic incompatibility, which determines the initiation of the microheterogeneous structure. Because of the difficulties in studies of systems of this type, there are few papers [156, 157] at present that deal with the study of processes of phase separation by small-angle X-ray scattering. Let us consider how the thermodynamic analysis (RGC proximate method) can be used to study the changes of the thermodynamic stability of the ERC components that occur as a result of chemical reaction (the setting of two epoxy oligomers by one and the same curing agent up to a degree conversion not less than 85%).

For systems based on the above two epoxy polymers, the concentration dependence of the parameter $\log V_q$ represents a nonmonotonically increasing deviation from the additive value. Hence, for small additions of one polymer to another a decrease of V_q is observed; in terms of the absorption process this can be explained by the decrease of the system sorption activity, probably due to closer packing of the polymer mixture.

As seen from Fig. 3.22, the concentration dependence of the thermodynamic interaction parameter $\chi_{2,3}^*$ of the above mixtures is nonmonotonic and is bimodal in character (the maxima are observed at 20% and 70% of Laproxide 703M). The existence of two maxima indicates that these systems have two regions of composition with least-stable thermodynamic equilibrium, which is a function of temperature. It follows from the temperature dependence of $\chi_{2,3}^*$ that the thermodynamic compatibility of these mixtures of polymers declines as the temperature drops—there is the top mixture critical temperature. Whereas at the blending stage there is thermodynamic compatibility of the ED-20 oligomers and Laproxide 703M for modifier contents up to 40% at 298 K, the parameter $\chi_{2,3}^*$ of the mixture of the two epoxy

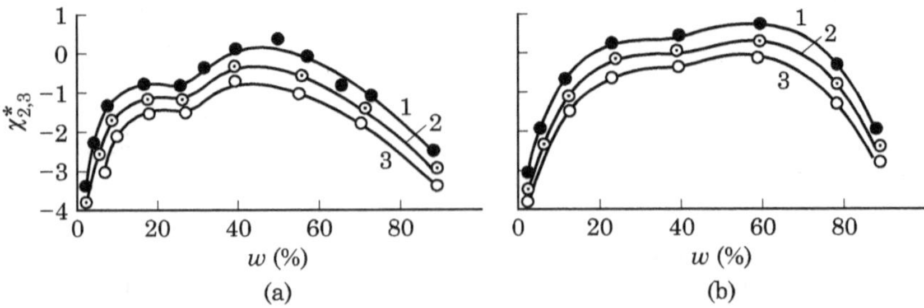

Figure 3.22 Concentration dependence of the $\chi_{2,3}^*$ parameter of ED-20 systems (with Laproxides 503M (a) and 703M (b)) cured by UP-0633 at (1) 363 K, (2) 373 K, and (3) 383 K.

polymers is negative for low contents of the polymeric modifier, up to approximately 20% of Laproxide 703M at 393 K.

Figure 3.23 presents physical-mechanical properties of ERC based on two epoxy polymers. The concentration dependence of the adhesion strength and of the breaking strength has its maximum at 5% of Laproxide 703M. Evidently, the enhancement of the breaking and adhesion strength after adding the second component can be explained by a high level of energetic interactions between unlike molecules, which is confirmed by enhancement of the packing of these polymer mixtures and by increase of the parameter $\chi_{2,3}{}^{*}$ at low contents of the polymeric modifier, approximately up to 5%.

Compounds of this type are characterized by the fact that enhancement of the adhesion strength (Fig. 3.23a) and increase of the breaking strength (Fig. 3.23b) are accompanied by decrease of the impact viscosity (Fig. 3.23c) and vice versa. Apparently, this effect can be related to the beginning of microphase separation of two crosslinked polymers, which occurs at more than 5% of the modifier (Laproxide 703M) content. This assumption is also confirmed by electron-microscopic analysis and by the positive values of the thermodynamic interaction parameter $\chi_{2,3}{}^{*}$ (see Fig. 3.22).

Summing up the results regarding the modification of ED-20 epoxy resin by oligoester epoxides of different molecular weights, thermodynamic incompatibility of the formed epoxy polymers appears during the curing of the homogeneous mixtures of the epoxy oligomers by one and the same curing agent. For ERC of the given type (low-viscosity, slowly formed systems) there is no correlation between the degree of thermodynamic compatibility of the oligomers' initial mixtures and

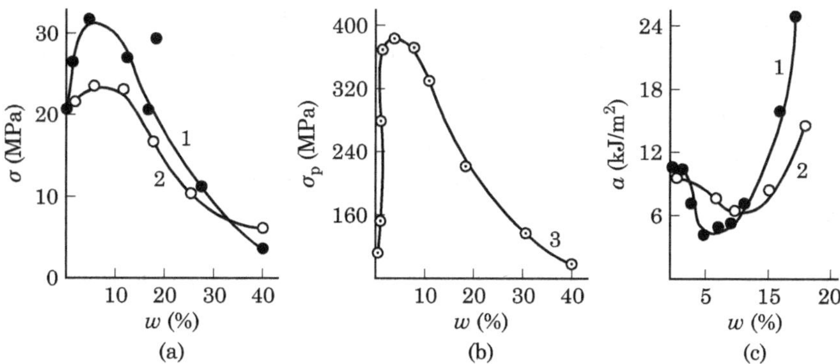

Figure 3.23 Concentration dependence of the adhesion strength (a), tensile strength (b), and impact viscosity (c) for ED-20 systems with (1) Laproxide 703M and (2) Laproxide 503M cured by UP-0633; (3) reactive oligomer.

the physical-mechanical properties of the completely cured polymeric compound. The optimal physical-mechanical characteristics were obtained for ERC using an oligoester epoxide of higher molecular weight (Laproxide 703M) as modified, which ensures the maximal degree of heterogeneity during curing. Use of Laproxide 503 M also results in improvement of the physical-mechanical properties, but to a lesser extent.

3.3.4 Influence of surfactants on structure and properties of polyurethanes based on oligomer mixtures

3.3.4.1 Investigation of thermodynamic compatibility of initial oligoglycols. In the selection of oligomer pairs featuring the compatibility required for investigation, reversed-phase gas chromatography was used [132], with an accuracy of 10% in parameter determination. The compatibility criterion was taken as a negative value of free energy of mixing (ΔG_{m}) for thermodynamically compatible systems. The oligomer pair polyoxypropylene glycol (PPG)–polyoxytetramethylene glycol (PTMG) is characterized by $\Delta G_{\mathrm{m}} \sim 0$, with a feeble temperature dependence. Excess enthalpy of mixing is low and has positive sign, and thus it may be assumed that mixture formation requires additional energy and is accompanied by heat absorption. ΔS_{m} at different temperatures is characterized by values close to zero and increasing weakly as temperature decreases. These findings indicate that this oligomer pair is compatible and that this mixture is close to ideal.

Mixtures of PPG with oligodiethyleneglycol adipate (ODA) are characterized by entirely different thermodynamic conditions. ΔG_{m} is positive and increases with temperature, characterizing the mixture as incompatible; the incompatibility increases with increase in the temperature. Negative values of excess mixing entropy demonstrate the high tendency of the components to self-association. The mixing process is accompanied by evolution of heat.

Another typically incompatible system is the mixture of PTMG and ODA [133]. Hence, one of the three systems considered is ideally compatible and two pairs of oligoglycols are incompatible.

3.3.4.2 Thermodynamic functions of surfactant surface layers in the initial oligomer. Investigation of the thermodynamic functions of the surfactant surface layers in the initial oligomers is based on experimental values of the surface tension measured for solutions of KEP-2 surfactant of various concentrations at various temperatures by the method of Wilgelmi [126]. Temperature dependences of surface pressure (P) for various molar concentrations of surfactant are determined from experimental isotherms. The thermodynamic functions of the surfactant surface layers are defined through an equation relating surface pressure to the entropy change and the enthalpy of the surface layers [134]:

$$\Delta H_s = P + \frac{dP}{dT} \tag{3.25}$$

where ΔH_s = change of the surface layer enthalpy [cal/(mol·K)]
P = surface pressure, presenting the difference between surface tension of the solvent and solution (mN/m)
dP/dT = rate of change of the surface layer entropy [cal/(mol·deg^2)]
T = temperature (K)

See Table 3.15 for isotherms of surface tension of KEP-2 (polyorganosiloxane and polyoxyalkylene block copolymer) solutions in PTMG measured in the temperature interval of 323–353 K. The data presented in the table demonstrate that the critical concentration of micelle formation in such solutions is observed for a surfactant fraction around 2.5×10^{-4}.

If the KEP-2 solubility from the isotherms of surface tension of surfactant solutions in PTMG is compared with the data in [11], which reports that the KEP-2 fraction in PPG corresponding to the critical concentration of micelle formation (CCMF) is 2.5×10^{-3}, it is evident that KEP-2 solubility in these simple oligoglycols differs tenfold. Temperature dependences of the surface pressure of KEP-2 solutions in PTMG obtained from the experimental isotherms of surface tension at all the investigated mass fractions of KEP-2 are linear.

TABLE 3.15 Surface Tension–Temperature Dependences of Surfactant Solutions in PTMG

Surfactant mass fraction in the solution (±3%)	Surface tension, γ (mN/m) (±0.2) at the temperature			
	323 ± 0.1 K	333 ± 0.1 K	343 ± 0.1 K	353 ± 0.1 K
0	38.90	38.00	37.10	30.50
2×10^{-5}	34.30	33.30	32.50	31.50
3×10^{-5}	30.80	30.30	29.10	28.50
5×10^{-5}	29.30	28.20	27.60	27.06
1×10^{-4}	27.00	26.40	25.90	28.50
1.8×10^{-4}	23.60	23.20	22.90	22.60
3×10^{-4}	23.40	22.90	22.60	22.30
0.99×10^{-3}	23.40	22.90	22.60	22.30

Thus the change of the surface pressure with the temperature is a constant for each solution, whose values are presented in Table 3.16, under ΔS_s. The table also gives ΔH_s values calculated from Equation (3.25). It is clear from the surface entropy values that the concentration change of the surface layer entropy is temperature-invariant. The enthalpy of the surface layer of the surfactant depends on the temperature of investigation, though ΔH_s concentration dependences appear to be of the same type as the $\Delta S_s = f(c_{\text{surf}})$ (Table 3.16) dependence with the minimum exhibited at the same surfactant content. From the surface entropy and enthalpy values the most well-ordered adsorption layers are formed at temperatures of 333–353 K and a KEP-2 mass fraction in PTMG of 1.8×10^{-4}. We refer to high contents of surfactant molecules in the surface layer because CCMF of these KEP-2 solutions is mass fraction 2.5×10^{-4} (see Table 3.15). Thus, with KEP-2 solutions in PTMG more and more ordered adsorption layers are formed with increase of the surfactant mass fraction ($\Delta S_s < 0$ for almost all the surfactant contents studied) and are also characterized by negative values of the surface enthalpy. Furthermore, the concentration dependences of ΔH_s and ΔS_s at different temperatures are of the same character, unlike the thermodynamic functions of similar solutions of PPG [11].

Let us consider the results of similar investigation of KEP-2 solutions in ODA. Table 3.17 shows that the introduction of this surfactant into a complex oligoester can reduce the surface tension by approximately 40% and that a mass fraction of KEP-2 of the order $(5\text{–}7) \times 10^{-5}$ is sufficient. The system reaches a constant value of surface tension starting at 7.5×10^{-5} surfactant mass fraction and this amount of KEP-2 is the critical concentration for micelle formation. Consequently, the solubility of this oligomer surfactant in ODA is the lowest of all three of the oligoglycols investigated. The surface pressure changes linearly with temperature although for different surfactant concentrations the sign of the slope varies and accordingly the surface entropy (Table 3.18) changes depending on temperature and concentration of the surfactant. It follows from these data that surface entropy changes with concentration are described by an extremal dependence with a maximum around 3×10^{-5} mass fraction of KEP-2. As in PTMG solutions, the concentration change of ΔS_s is expressed as a single curve for all investigated temperatures, i.e., ΔS_s does not depend on temperature. The significant difference is that the ΔS_s of the surface layer is negative only for a surfactant mass fraction $\sim 1.7 \times 10^{-5}$.

Thus, it is at the extremum that the most disordered state of adsorption (surface) layer is observed. The positive extremum of ΔH_s values (Table 3.18) is observed in the same region of surfactant mass frac-

TABLE 3.16 Values of Surface Enthalpy and Entropy of KEP-2 Solutions in PTMG at Different Temperatures

KEP-2	$323 \pm 0.1\,K$		$333 \pm 0.1\,K$		$343 \pm 0.1\,K$		$353 \pm 0.1\,K$	
mass fraction ($\pm 3\%$)	$10^5 \Delta S_s$ ($J/m^2 \cdot K$)	$10^3 \Delta H_s$ ($J/m^2 \cdot K$)	$10^5 \Delta S_s$ ($J/m^2 \cdot K$)	$10^3 \Delta H_s$ ($J/m^2 \cdot K$)	$10^5 \Delta S_s$ ($J/m^2 \cdot K$)	$10^3 \Delta H_s$ ($J/m^2 \cdot K$)	$10^5 \Delta S_s$ ($J/m^2 \cdot K$)	$10^3 \Delta H_s$ ($J/m^2 \cdot K$)
2×10^{-5}	1.20	8.40	1.20	8.60	1.20	8.60	1.20	9.10
3×10^{-5}	0.0	8.10	0.0	7.70	0.0	8.00	0.0	8.00
5×10^{-5}	−0.90	6.90	−0.90	6.90	−0.90	6.50	−0.90	6.50
1×10^{-4}	−3.00	2.20	−3.00	1.6	−3.00	1.00	−3.00	0.40
1.8×10^{-4}	−5.3	−0.80	−5.3	−1.8	−5.3	−2.20	−5.3	−3.70
3×10^{-4}	−4.3	0.4	−4.3	−0.25	−4.3	−1.70	−4.3	−2.30
9.9×10^{-4}	−4.0	1.0	−4.0	0.20	−4.0	−0.90	−4.0	−1.60

TABLE 3.17 Surface Tension Temperature Dependences of KEP-2 Solutions in ODA

Surfactant mass fraction (±0.3%)	Surface tension, γ (mN/m) (±0.2) at the temperature			
	323 ± 0.1 K	333 ± 0.1 K	343 ± 0.1 K	353 ± 0.1 K
0	38.3	37.7	37.1	36.6
1×10^{-5}	33.5	33.2	32.9	32.6
1.5×10^{-5}	31.1	30.8	30.2	29.7
2×10^{-5}	30.1	29.42	28.8	28.2
2.8×10^{-5}	26.6	26.0	24.7	23.9
3.5×10^{-5}	25.0	24.4	23.2	22.6
5×10^{-5}	24.3	23.7	22.9	22.3
7.5×10^{-5}	23.8	23.3	22.6	22.0
1.0×10^{-4}	23.8	23.3	22.6	22.0

tions. Accordingly, KEP-2 surface layers in ODA solutions are mainly characterized by high degree of disorder and maximum values of the power component of total surface energy.

From comparison of the concentration dependences of surface entropy and enthalpy of KEP-2 solutions in PTMG, ODA, and PPG [10, 11] it may be concluded that in all cases considered they have extremal character, although these extrema are observed for KEP-2 mass fractions differing tenfold and more for different oligoglycols. Additionally, there are two specific surfactant concentrations in PPG: apart from the extremal region there is another at which ΔS_s is observed to be independent of temperature. Thermodynamic analysis of the surface layers of KEP-2 surfactant in ODA and PTMG showed that entropy is temperature-invariant for practically all concentrations considered. Since hydrogen bonding is a very temperature-sensitive factor, it can be supposed that its role for KEP-2 solutions in PPG is demonstrably greater than in the other cases.

3.3.4.3 Investigation of interactions in oligomer–surfactant systems by spectral methods. Infrared (IR) spectroscopy was used for this part of the investigation. Surfactant concentrations for spectral comparison were chosen according to their solubility in different oligomers. It was found that KEP-2, regardless of its concentration, does not practically influence the state of the hydroxyl groups of PTMG and ODA oligomers. A different pattern was observed for KEP-2 in PPG. For mole fractions of 3×10^{-4} and 1.5×10^{-4} the surfactant does not influence the OH-group band. With other surfactant concentrations, regular reduction of the band maximum below 3460–3480 cm^{-1} is observed (Table 3.19). Furthermore, at those surfactant contents the band at 3590 cm^{-1} disappears, i.e., the profile of this part of the spectrum changes. The band

TABLE 3.18 Values of Surface Enthalpy and Entropy for KEP-2 and ODA Solutions at Different Temperatures

KEP-2 mass fraction (±3%)	323 ± 0.1 K		333 ± 0.1 K		343 ± 0.1 K		353 ± 0.1 K	
	$10^5 \Delta S_s$ (J/m^2·K)	$10^3 \Delta H_s$ (J/m^2·K)	$10^5 \Delta S_s$ (J/m^2·K)	$10^3 \Delta H_s$ (J/m^2·K)	$10^5 \Delta S_s$ (J/m^2·K)	$10^3 \Delta H_s$ (J/m^2·K)	$10^5 \Delta S_s$ (J/m^2·K)	$10^3 \Delta H_s$ (J/m^2·K)
1×10^{-5}	−3.2	−5.4	−3.2	−6.0	−3.2	−6.6	−3.2	−7.2
1.5×10^{-5}	−1.5	2.40	−1.5	2.1	−1.5	1.8	−1.5	1.5
2×10^{-5}	0.0	8.00	0.0	8.3	0.0	8.3	0.0	8.3
2.8×10^{-5}	5.5	26.50	5.5	28.4	5.5	31.3	5.5	32.1
3.5×10^{-5}	5.5	29.4	5.5	30.0	5.5	32.8	5.5	34.0
5×10^{-5}	3.7	26.0	3.7	26.3	3.7	27.0	3.7	27.6
7.5×10^{-5}	0.0	14.6	0.0	14.6	0.0	14.6	0.0	14.6
1.0×10^{-4}	0.0	14.6	0.0	14.6	0.0	14.6	0.0	14.6

TABLE 3.19 Band Parameters of the Individual Oligoglycol Hydroxyl Groups and Their Mixtures Without Surfactant and in the Presence of KEP-2

Oligoglycol	KEP-2 concentration mole fraction (±3%)	Band integral intensity g (±10%)	Rated integral intensity of the band g (±10%)	Band half-width (cm^{-1})
PPG	0	0.1635	—	1.62
PPG	2×10^{-3}	0.1203	—	1.65
PTMG	0	0.2124	—	1.90
PTMG	2×10^{-4}	0.2116	—	1.90
ODA	0	0.4077	—	2.50
DA	2×10^{-5}	0.4077	—	2.50
PPG, PTMG	1×10^{-3}	0.0812	0.1660	1.55
PPG, PTMG	0	0.1368	0.1729	2.00
PG, ODA	0	0.2032	0.2978	2.00
PPG, ODA	1.1×10^{-3}	0.2784	0.2783	2.00

at 3590 cm^{-1} characterizes intramolecular hydrogen bonding in five cycles, which is nonreactive in uncatalysed reactions of urethane formation, as shown in [135].

Comparing the IR spectral results with data from thermodynamic analysis of adsorption layers of the surfactant, it is concluded that when hydrophobic parts of the surfactant molecules, regardless of their conformation, are characterized by extremely dense packing in the surface layer, the surfactant does not influence the state of the hydroxyl groups of the solvent oligomers.

3.3.4.4 Physical and mechanical properties of polyurethane networks and surfactant additives. The cohesion and adhesion properties of the cured polymers were investigated to elucidate the influence of KEP-2 on the physical and mechanical properties of polyurethane networks based on PPG in the presence of individual oligomers.

Results for polyurethanes based on PPG with various surfactant concentrations are shown in Table 3.20. When these data are analysed it is seen that introduction of a certain surfactant concentration increases cohesion and adhesion strength but that changes in σ and σ_p for the same surfactant concentrations are inverse: if σ increases, σ_p falls, and vice versa. It was found that the extremal character of the breaking strength with concentration (Table 3.20) corresponds essentially to the dependences of the thermodynamic functions [10]. Therefore, for very dense surface layer packing, cohesion strength increases and adhesion strength decreases. Growth of adhesion strength is observed for those surfactant concentrations at which the

TABLE 3.20 Physical and Mechanical Properties of Polyurethane Networks Based on PPG, Trimethylolpropane, and TDI at 343 ± 1 K

KEP-2 mass fraction (±3%)	Tensile strength σ (MPa) (±10%)	Ultimate strength under normal fracture σ_p (MPa) (±10%)	Percent elongation (%) (±10%)
0	3.0	4.0	400
7.5×10^{-5}	3.5	4.0	420
1.5×10^{-4}	3.6	3.9	450
3×10^{-4}	2.6	6.0	380
5×10^{-4}	2.5	6.2	400
7.5×10^{-4}	4.6	3.0	400
1.0×10^{-3}	4.6	3.0	420
1.5×10^{-3}	4.5	2.9	420

structure of the surface layer is characterized at the given temperature by constancy of ΔH_s and ΔS_s.

The physical and mechanical properties of polyurethane networks synthesized using ODA oligoesters at 343 ± 1 K were also studied; the results are presented in Table 3.21. Polyurethane specimens were synthesized with various KEP-2 concentrations, including that of mass fraction 3.5×10^{-5} at which the extremum in the dependences of thermodynamic functions was observed for oligoglycol solutions. It is especially notable that KEP-2 results in considerable (twofold) worsening of cohesion strength, significant reduction of elasticity, and subsequent insensitivity of the ultimate strength under normal fracture to the presence of surfactant in the system. Thus, the cohesion and adhesion characteristics of polyurethane networks synthesized using ODA do not depend on the structure of KEP-2 adsorption layers in the initial oligomer.

Analogous results for polyurethane networks based on PTMG are also presented in Table 3.21. Polyurethane specimens were synthesized with various concentrations of KEP-2 in the oligomer, including the mass fraction 1.8×10^{-4} at which extrema are observed in the thermodynamic functions in PTMG solutions. Clearly, KEP-2 produces considerable worsening of cohesion strength, reduction of the percent elongation of polyurethane specimens, and constancy of the ultimate strength value under normal fracture. Thus, as with ODA, the cohesion and adhesion characteristics of polyurethane networks do not depend on the structure of KEP-2 adsorption layers in the initial oligomer.

Hence, if there is no specific interaction between the initial oligomer and the surfactant, there is also no connection between thermodynamic functions characterizing the surface layers at the start of the

TABLE 3.21 Physical and Mechanical Properties of Polyurethane Networks Based on Different Oligoglycols (ODA, PTMG), Trimethylolpropane, and TDI at (343 ± 1) K

KEP-2 mass fraction (±3%)	ODA			PTMG		
	Tensile strength (MPa) (±10%)	Ultimate strength under normal fracture (MPa) (±10%)	Percent elongation (%) (±10%)	Tensile strength (MPa) (±10%)	Ultimate strength under normal fracture (MPa) (±10%)	Percent elongation (%) (±10%)
0	4.0	3.6	650	5.0	5.0	500
3.5×10^{-5}	3.9	3.6	600	5.0	5.0	400
7×10^{-5}	2.6	3.5	530	4.0	5.0	350
1.8×10^{-4}	2.6	3.5	400	3.5	4.8	300
5.0×10^{-4}	2.5	3.45	500	3.5	4.8	350
1.0×10^{-3}	2.4	3.5	500	3.4	4.9	380
1.5×10^{-3}	2.4	3.5	500	3.4	5.0	350

reaction and the properties, and therefore the structure, of the polyurethanes in the cured condition.

3.3.4.5 Kinetics of formation of polyurethanes based upon oligoethers and oligoesters. A promising direction in the production of polyurethane (PU) adhesives is the creation of one-component resilient adhesives based on prepolymers with free isocyanate (NCO) groups and without the use of solvent; these represent highly efficient bonds that are low in toxicity and form strong adhesive joints with different materials [85, 136, 137]. The adhesion strength of these adhesives can be increased by the use of appropriate polymer–polymer compositions [138]. Polymer mixtures are obtained without solvents by mixing their melts [139] or by matching of polymerizing monomers to form network polymers. Properties typical of crosslinked three-dimensional networks can be achieved as a result of only physical interactions in systems based on linear or branched polymers [140].

This part of the work is devoted to the study of adhesive joint formation based on urethane polymer mixtures with free NCO groups. The materials under investigation were adhesives obtained from toluylene diisocyanate (TDI) and hydroxyl-containing polyesters. The hydroxyl-containing components were oligooxypropylenetriols (L-3003 and P-2200 oligoester, the product of reaction of diethylene glycol, trimethylolpropane and adipic acid interaction).

From kinetic investigation of the curing of macro-triisocyanates (MTI) based on L-3003, P-2200, and their mixtures, kinetic curves of the polymerization process were obtained at 298 K. Kinetic curves for the curing of polyurethane prepolymer mixtures obtained by reaction of TDI with L-3003, P-2200, and various mixtures with a double excess of NCO groups with respect to OH groups are linear. These dependences show that NCO group reactions continue until high fractional conversions comply with the second-order equation

$$\frac{1}{1-P} = KC_0t \tag{3.26}$$

where P = extent of reaction; K = rate constant; C_0 = initial concentration of NCO groups in the reaction mixture; t = time.

The rate constants calculated for total conversion of NCO groups in PU adhesive compositions are shown in Table 3.22. An obvious and significant difference (fivefold) is observed in the reactivity of prepolymer NCO groups based on each oligoester and oligoether separately. This allows the use of an approach originally devised to describe parallel reaction behavior with similar functional groups [142] for the kinetic evaluation of the curing of these incompatible

TABLE 3.22 Experimental and Rated Values of Curing Reaction Rate Constants for Reactive Oligourethanes of Various Compositions at 298 K

Component mole ratio I:II	Volumetric fraction of the component based on oligomer I	$10^7 K_{eff}$ [l/(mol·s)]	$10^7 K_r$ [l/(mol·s)]
0:1	0.0	7	—
1:1	0.61	43	23
2:1	0.76	40	27
5:1	0.89	31	31
40:1	0.98	35	34
1:0	1.0	35	—

reagents. As was shown in [133], when thermodynamically compatible reagents have the same functional groups, their interaction with chemically active substances is described by reactions running in parallel provided that their reactivities differ. In accordance with [141] and [142], the effective rate constant in the mixture and the observed rate constant for the components are related by

$$K_{eff} = K_1 + (K_2 - K_1)\frac{[C]_2}{[C]_0} \tag{3.27}$$

where K_1 and K_2 are the rate constants of the less and more reactive groups respectively, and $[C]_2$ and $[C]_0$ are concentrations of the more active NCO groups and of all isocyanate groups in the system, respectively. Rate constants calculated from Equation (3.27) are also presented in Table 3.22. Comparison of the rated and experimental values shows these values are properly correlated when the volumetric content of the component based on L-3003 oligomer exceeds 0.88. With the further increase of dispersed phase in the mixture (i.e., with P-2200 concentration) a discrepancy is encountered: the experimental values increase while calculated values decrease noticeably. This behavior is connected with existence of not only two internal phases but also a developed interphase layer with altered properties. The larger the contribution this layer makes to the kinetics, the greater is the volumetric fraction of the component base on P-2200 and, consequently, the longer is the interphase boundary. Hence, description of the process using only concepts of parallel reactions in each phase does not provide a complete picture of the curing kinetics for the oligourethanes investigated, since the phenomena at the boundary are not taken into account, although comparison of the experimental and rated values provides a qualitative characterization of the role of reactions occurring close to the interphase boundary.

3.3.4.6 Cohesion and adhesion characteristics of polymer networks based on polyurethane mixtures. The curing of macro-polyisocyanates between the bonded surfaces provides adequately strong adhesive joints between different materials. Investigation of the adhesion strength to steel of compositions based on oligourethanes [143] showed that increase of strength at normal fracture occurs over time and is especially rapid during the first 10 days. Prepolymer compositions based on oligoethers possess good strength characteristics (4.9–5.3 MPa). As the breaking of adhesive joints has cohesive character and is determined by the strength of the cured polymer, reduction of adhesion strength is to be expected with increase of molecular mass of the polyol component due to the decrease of concentration of urethane groups in the oligourethane [144]. With the tri-functional polyol L-3003, despite the increase of the initial polyester molecular mass, a strength increase is observed, connected with the formation of branched macromolecules of oligourethane capable of stronger interphase interaction with the metal. The strength of MTI based on P-2200 polyester is somewhat higher, due to stronger intermolecular interaction, than is found with oligoethers [145]. A considerable increase in strength (8.0–8.5 MPa) is noted with adhesives containing an MTI mixture based on P-2200 and L-3003 combined during synthesis. This is apparently connected with formation of a more compact spatial grid.

The strength increase for specimens cured at higher temperature (333 K, in comparison with 293 K) probably occurs because temperature increase provides polymerization by NCO groups as well as the chemical and permolecular crosslinking at the polymer–substrate boundary.

The use of UP-606/2 [2,4,6-tri(dimethylaminomethyl)phenol] is of interest as a catalyst for acceleration of the system curing because of its influence on the physical and mechanical indices of the corresponding adhesives. As has been shown [143], introduction of the catalyst reduces the strength of the adhesive joint, probably due to foaming in the adhesive seam because of intense liberation of carbon dioxide upon interaction of NCO groups with moisture. The drop in adhesion strength in the system investigated depends on the quantity of the catalyst used: only small quantities (0.05–0.1 mass%) of catalyst can be introduced without considerable reduction of adhesion strength.

Study of the influence of water on the adhesion strength of PU adhesives based on individual isocyanate-containing oligourethanes has revealed low water resistance [45]. Increase of strength and water resistance in adhesives based on mixtures of branched oligourethanes with free isocyanate groups, as well as of these mixtures modified by perchlorvinyl resin (PPCV), may be expected. The materials investigated were PU compositions with free NCO groups based on toluylene diisocyanate (TDI) and hydroxyl-containing oligoesters and oligoethers: oligo-oxypropylenetriol (L-3003), P-2200, ODA-800, and their mixtures in 1:1 mole ratio at a total NCO/OH ratio of 2. One of the compositions investigated contained predried perchlorvinyl resin (PPCV-LS) as the additive introduced into the system during synthesis.

The adhesion properties of the compositions and their water resistance were studied by observing the changes over time of the adhesion strength of joints cured by the ambient moisture during 30 days of exposure under water. Macro-triisocyanates based on L-3003 and P-2200 are characterized by similar strength values. In mixtures of oligourethanes with these polyols a sharp increase in adhesion strength is observed, which may be explained by more compact spatial packing of polar groups on the substrate surface. Introduction of PPCV-LS into the system leads to sharper increase of adhesion strength as a result of "bond in the net" structure formation [146] and due to the increase in the number of polar groups at the interphase boundary, providing structual ordering and strengthening of interphase polymer–metal interaction. Maximum adhesion strength reaches 12 MPa and is obtainable in 30 days.

The water resistance of these systems of oligourethanes under conditions of increased humidity has been studied in connection with their possible use as adhesives. The low water resistance of prepolymers obtained through TDI reaction and of individual polyols, especially of oligourethanes based on ODA-800 and P-2200, may be explained by the high interphase tension, which facilitates the conditions of selective water sorption at the interphase boundary of adhesive and substrate [145].

For mixtures of macro-polyisocyanates, there is typically an increase of adhesion strength before and after the soaking of adhesive joints in water that is more significant with introduction of PPCV-LS into them, and a subsequent reduction of adhesion strength in water for all the mixtures investigated within 30 days such that $\sigma/\sigma_{30} =$ 2.0–2.5. That is, there is no noticeable increase of water resistance of the adhesive joint by means of PPCV-LS additives. Lower water resistance of adhesive joints based on isocyanate-containing oligourethane than with films of the same material was found in [146] and was explained by faster water penetration along the adhesive–substrate boundary due to defectiveness of the permolecular structure of the polymer and selective water sorption at the metal surface, as well as by formation of carbon dioxide bubbles on reaction of NCO groups with this moisture. This explanation is confirmed by studying the influence of PU-oligomer curing catalysts on the adhesion and water resistance. Introduction of tin into the system of octoanate and dibutyl dilaurate accelerated interaction of the isocyanate groups with moisture and, correspondingly, caused more intensive liberation of carbon dioxide, producing a very sharp fall of adhesion strength and water resistance of the adhesive contact with the metal. The reduction of adhesion strength and water resistance of compositions using catalysts exceeds the positive effect from introduction of PPCV-LS when no catalyst is used.

Enhancement of the interphase interaction of adhesive with metal and ordering of the structure of the mixtures results in considerable improvement of their physical and mechanical properties and, correspondingly, considerable increase of adhesion strength, which remains higher after long-term soaking in water, in comparison with individual oligourethanes.

3.3.4.7 Investigation of the structure of prepolymers with isocyanate end groups based on mixtures of incompatible oligoesters and of their polymers by contrast optical microscopy. As has been shown, oligourethanes based on oligoester mixtures and cured by moisture in air and with isocyanate end groups exhibit an increase of adhesion strength at normal fracture in comparison with corresponding adhesives based on the individual polyesters. The reasons for this change in the properties of the cured composition remains uncertain. The regularities characterizing the formation of polymer compositions with improved properties based on self-cured oligourethanes obtained from mixtures of oligoethers and oligoesters were therefore of interest.

Oligourethanes with the NCO end groups based on oligoester mixtures were investigated. L-3003 and P-2200 oligoester were used as the hydroxyl-containing components and the ratio of hydroxyl groups in oligoether to those in oligoester was varied from 100:1 to 1:1.

Using contrast microscopy [147] it was established that the oligourethanes generated are mutually insoluble under the conditions studied. Mutual insolubility is the result of the different polarity of prepolymers based on oligoethers and oligoesters, represented by values of surface tension of, respectively, 28 ± 0.4 and 40 ± 0.4 mN/m. This results in formation of either kinetically stable emulsions (from 100:1 up to 5:1) or emulsions that partially split when stored (above 5:1). When the drop sizes of emulsions formed in oligourethanes of different composition are compared it appears that the structures that are stable over time form drops of 1×10^{-5} m diameter or less. Thus, the synthesis of oligourethanes from polyether and polyester mixtures is a process of formation of stable emulsions without introduction of special stabilizing additives. It should be noted that by mixing of macro-triisocyanates based upon each polyester separately, it is impossible to obtain a nondelaminating mixture. Hence, the substances exerting stabilizing action are formed during the oligourethane synthesis. The appearance of substances able to sorb at the interphase boundary and reducing the value of interphase tension is demonstrated by experimental data.

3.3.4.8 Determination of the interphase tension of reaction mixtures during synthesis of oligourethanes. The kinetics of interphase tension change at the boundary with air and mercury during the formation of prepolymers with isocyanate end groups was investigated as follows. The charge of hydroxyl-containing component was placed into a reactor, dried at 373 ± 5 K, and mixed intensively using a magneic mixer for 5–6 h. Then, at 323 ± 0.1 K, a defined quantity of a distilled mixture of 65% 2,4- and 35% 2,6-isomers of TDI was added and was intensively mixed for 5–10 min and a measuring cell located in a thermostated chamber of the unit for measuring interphase tension was filled with the reaction mixture at 323 ± 0.1 K. If the interphase tension was measured at the boundary with mercury, then a mercury drop was placed onto the backing of the measuring cell. If the measurement was performed at the boundary with air, a drop of the mixture under investigation was placed onto the backing of the measuring cell. The drop was photographed during 240–300 min at defined time intervals. The photographed drop dimensions were measured with a UIM-21 measuring microscope. To study the influence of different order of mixing of the initial components on the properties of the systems obtained, the reagents were also mixed in reverse order: the distilled mixture of 2,4- and 2,6-isomers of TDI was placed into the reactor and the charge of dried hydroxyl-containing component was added, with subsequent mixing for 5–10 min. The remainder of the experiment was performed as described.

It is apparent from Table 3.23 that for both orders of mixing the process of system formation is accompanied by reduction of the interphase tension, though when macro-triisocyanate (MTI) is formed this reduction is greater with diisocyanate in excess.

For the interphase tension change at the boundary with air during prepolymers formation based on individual polyesters and mixtures, the surface tension displayed no change and was 28 mN/m for the prepolymer based on L-3003 polyether, 40 mN/m for that based on P-2200 polyester, and 32 mN/m for that based on a 40:1 mixture of polyether and polyester (Table 3.23). For the interphase tension at the boundary with mercury, it was shown that formation of prepolymers based on individual oligoethers and oligoesters and on mixtures is accompanied by significant reduction of interphase tension, this

TABLE 3.23 Kinetics of Interphase and Surface Tension Change in the Process of Formation of Prepolymers Based on Individual Complex Oligomers and Their Mixtures at 323 ± 0.1 K

	Interphase tension (mN/m) (±2%) of MTI formation process based on				Surface tension (mN/m) (±1%) of MTI formation process based on		
Time (min)	TDI and L-3003	L-3003 and TDI	TDI and P-2200	TDI and L-3003 + P-2200 40:1	TDI and L-3003	TDI and P-2200	TDI and L-3003 + P-2200 40:1
25	—	—	—	347	—	—	—
40	302	302.0	299.0	310	27.8	42.0	32.0
60	254	253.0	297.0	240	28.2	40.6	31.6
90	222	246.0	295.0	222	29.0	40.4	31.6
120	226	241.0	282.0	214	28.6	40.0	31.6
150	206	237.0	247.5	215	27.0	40.0	31.8
180	224	235.0	231.0	217	27.2	40.0	31.8
210	214	233.5	226.5	217	27.6	40.0	31.8
230	211	233.0	225.0	215	28.0	40.0	32.0
250	210	233.0	225.0	215	28.0	40.0	32.0

reduction being of the same order (100–110 mN/m) and character for all the systems investigated (Table 3.23).

As well as interphase phenomena in polyurethane compositions, their adhesion properties were also studied; bonding of steel dollies was done immediately after the synthesis and two days later. Tests for normal fracture were performed 30 days after bonding. Table 3.24 shows that specimens bonded immediately after synthesis are not distinguished by high adhesion strength. The adhesion strength of prepolymers based on oligoesters and oligoethers is somewhat higher than that of prepolymers based on mixtures. Another picture is observed in bonding of specimens two days after synthesis. In this case considerable increase of the strength characteristics (to 9.0 MPa) of prepolymers containing MTI mixtures and those based on P-2200 and L-3003 is observed. Fivefold reduction of the amount of polyester in MTI based on oligoether and oligoester mixture does not alter the adhesion properties of the compositions investigated. The adhesion strength data of as-synthesized polyurethane compositions agreed with the results for interphase tension at the formation stage. The values of interphase tension are the same and the adhesion strength is of the same type for all the systems considered.

IR spectroscopic investigation of MTI specimens as-prepared or held for several days showed a gradual shift of carbonyl group absorption from 1740 cm^{-1} (free C=O) to 1720 cm^{-1}, demonstrating [148] the accumulation of hydrogen-bonded carbonyls, i.e., associated and self-associated urethane groups. Thus, the increase of adhesion strength observed may be subsequent to ordering due to formation of hydrogen bonds in the system structure during storage with exclusion of moisture. This creates conditions preventing delamination (components are

TABLE 3.24 Adhesion Properties of Specimens Bonded by Polyurethane Compositions Based on Individual Oligoethers and Mixtures

Initial components			Strength under normal fracture of specimens bonded directly after synthesis (MPa) (±10%)	Strength under normal fracture of specimens bonded 2 days after synthesis (MPa) (±10%)
NCO	P-2200	POPT L-3003		
2:1	0	2	5.1	4.7
2:1	2	0	6.0	5.6
2:1	1	1	5.0	9.0
2:1	1	2	4.6	8.5
2:1	1	3	4.6	8.7
2:1	1	5	5.5	9.1
2:1	1	10	3.1	—

mutually insoluble) and also induces the composite in the adhesive joint to act as a single unit.

3.3.4.9 Rheological behaviour of polyurethane mixtures. The colloid chemistry of oligourethanes as stable heterogeneous systems is of interest and estimation of the rheological behavior of the prepolymers may be useful. Oligourethanes based on oligoethers and on mixtures with mass fraction of L-3003 of more than 85% (by hydroxyl groups) are Newtonian liquids. At the same time, oligourethanes with a higher content of oligoester are characterized by non-Newtonian behavior as regards low shift speeds and established flow pattern at other shift voltages. Thus, typical rheological behavior is observed during the transition from diluted to concentrated emulsions [149]. In the latter case there is the possibility of dispersed phase particles with solvation spheres in the continuous phase interacting with each other (structure-forming interaction), as a result of which the system acquires viscoelastic properties. The concentration dependence of the dynamic viscosity (η) of oligourethanes under established flow has extremal character (Fig. 3.24). Figure 3.24 shows that the viscosity of oligourethanes based on polyester mixtures is always less than that of oligourethanes based on P-2200. Moreover, there is a region of polyether and polyester ratios (from 5:1 up to 100:1) where the viscosity of oligourethanes of mixed structure is less than for minimal initial oligourethane (obtained from L-3003 only). Experimental data also

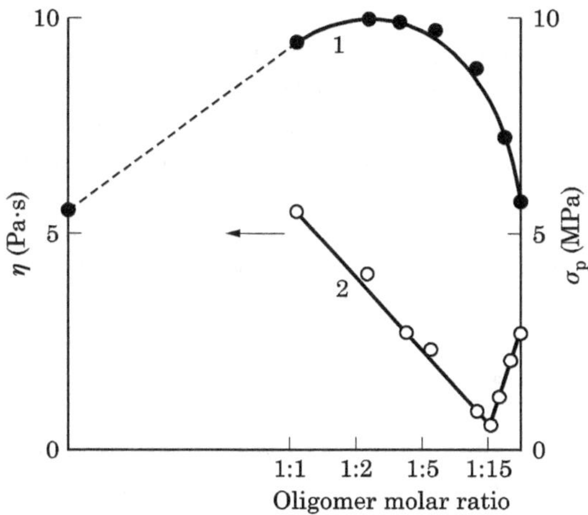

Figure 3.24 Dependence of the cured adhesive joint strength (1) and viscosity of initial isocyanate-containing oligourethanes (2) on the polyether and polyester ratio.

show reduction of density in oligourethanes based on mixtures ($1.04 \pm 0.04\,\text{kg/m}^3$) compared with those based on each polyester separately (1.35 ± 0.04 and $1.11 \pm 0.04\,\text{kg/m}^3$, respectively, for simple and complex ones). The extremal character of the dependence of viscosity on oligourethane composition is evidence of the inversely acting factors displayed in rheological behavior. A factor promoting viscosity increase is the introduction of a high-viscosity component based on P-2200 ($\sim 60 \pm 0.1\,\text{Pa·s}$). Its contribution is reduced with the decrease of the amount of this substance in the mixture. The second factor is formation of a layer promoting stability of the emulsion, a so-called structural–mechanical barrier [150], at the boundary of two liquid phases, since it is known that long-term existence of the emulsions is provided by stabilization connected with formation of an adsorption layer at the interphase boundary [150] and stability is provided when this layer is located outside the drop only, that is, in the dispersion medium. Apparently, this intermediate layer possesses a looser structure due to conformational difficulties appearing near the boundary [138]. The more developed is the boundary between existing phases, the greater is the contribution to the rheology of interphase layer properties. As shown in Fig. 3.24, this influence is maximally displayed at a polyether to polyester ratio of 15:1. As specific interactions [151] play a very significant role in polyurethanes, a change in the character of hydrogen bonding [152] may be expected in the components' incompatibility conditions for polyurethane structure formation. The rheological behavior of polyurethanes is sensitive to such changes. Figure 3.24 shows that for curing of oligourethanes with polyester ratio between 1:1 and 5:1 there is maximum increase of adhesion strength compared with the initial compositions. Thus, the optimal adhesion characteristics are achieved for compositions at which the oligourethanes are characterized by non-Newtonian behavior. This may be connected with microstructural ordering typical of dispersed systems and, as a result, with the appearance of oriented macrostructures. The distances between the drops of emulsified phase are comparable with the size of dispersed liquid particles at high mass fractions of oligourethanes based on P-2200. The effect of dispersion medium structuring is lost with decrease in the number of dispersed particles in unit volume.

3.3.4.10 Microstructure of polyurethane networks based on oligomer mixtures. The morphological structure of cured MTI based on L-3003 and mixtures of oligomers of varied structure has been studied by electron microscopy. The morphological pattern is practically the same for a polymer surface formed in air and for bulk polymer, but it is expressed more clearly in the latter. Cured compositions based on oligomer mixtures with larger mass fractions of P-2200 are characterized by groups of globules having diameters of 30–100 nm in chains of varied length. With increase in the composite of the initial MTI product based on L-3003 and random location of permolecular formations, globule sizes increase up to 200 nm, approaching those of polyurethanes based on a single L-3003, and regrouping of permolecular formations from chain structures to a cellular arrangement of globules also occurs. Thus, microstructure of the cured MTI is characterized by the occurrence of order in the form of chains or cells according to the oligomer mixture used; for stronger structure formations and better adhesion characteristics, the former is preferable.

It should thus be possible to synthesize macrodiisocyanates in the form of stable emulsions from mixtures of mutually insoluble polyesters without special stabilizing additives. In this case oligourethanes with various (including low) viscosity may be obtained according to the oligoester ratio. It was shown that the observed adhesion strength increase depends on the amount of the substance present in dispersed state and that the maximum adhesive joint strength is achieved with curing of oligourethanes characterized by non-Newtonian behavior, i.e., structured systems featuring a microstructure with enhanced ordering.

3.3.4.11 Phase division in polymers obtained by oligourethane-triisocyanate curing. It is known that the mechanical properties of dispersed systems are determined by four main factors: the volumetric fraction of the dispersed phase; the size and shape of the particles; the mechanical properties of the phases; and the strength of the bond at the interphase boundary between dispersed phase and dispersed medium. The role of the interphase layer in polymer structure formation was studied in the present investigation. The glass-transition temperature was used for characterization of polymer structures; for mixing of incompatible polymers the individual phases maintain their T_g, as a rule, and change in this parameter is evidence of change of chemical composition or phase structure [153]. Differential scanning calorimetry (DSC) was used in this investigation.

To outline the features of the glass-transition temperature of the polymers investigated, the change in their thermal capacity in the temperature interval 140–300 K was studied. Polymer specimens cured with MTI based on individual oligomers were compared. As seen in Fig. 3.25 (curves 2, 4), specimens based on L-3003 and P-2200 are both characterized by curves with one thermal capacity jump and have glass-transition temperatures of 226 and 243 K, respectively. A polymer based on MTI of an oligomer mixture (curve 1) is

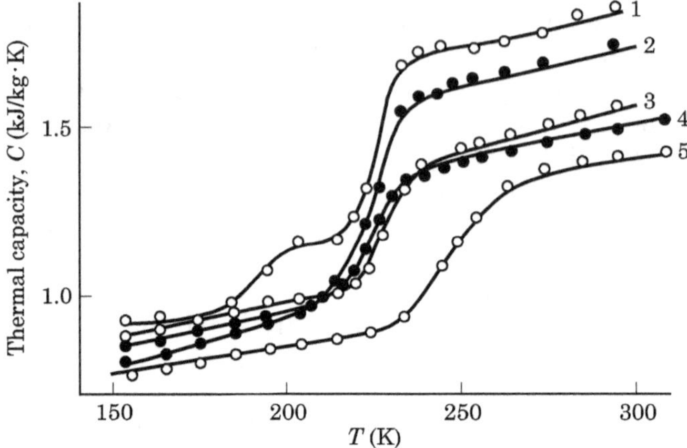

Figure 3.25 Thermal capacity/temperature change curves for curing of compositions based on: (1) MTI of TDI and 5:1 L-3003 and P-2200 oligomer mixture; (2) MTI of TDI and L-3003; (3) MTI of L-3003 and TDI; (4) mechanical mixture of polymers based on L-3003 and P-2200; (5) MTI of TDI and P-2200.

characterized by a temperature dependence with two jumps, at temperatures of 190 and 228 K. However, a mechanical mixture of these polymers representing a 5:1 quantitative ratio exhibits a thermal capacity curve with a single temperature transition at 227 K (curve 3): in other words, the second oligoester-based component provides no contribution. Consequently, DSC data characterize polymers based on an oligomer mixture having the same main glass-transition temperature as a specimen based on oligoether plus an additional transition at lower temperature. As has been noted [154] for mixtures of nonmixing polymers, various mixing levels are possible—from a distinct two-phase system with interphase layers of various lengths up to such a degree of dispersion of one component in the other that the dispersed phase does not display the properties of a separate phase and either T_g shows values typical of or close to those of the individual phases or new structures exhibit new T_g values. In the case considered, judging from DSC data the cured composition is a polymer serving as dispersed medium in which a loose interphase layer with a less densely crosslinked network appears characterized by T_g. The second component displays no properties of a separate phase, though its existence is manifest as formations less than 1×10^{-5} m in size under contrast microscopy. Besides this, it follows from the DSC data that curing of oligourethane isocyanates based on L-3003 (synthesized with various orders of mixing of the components) produces polymers with the same glass-transition temperature but different values of the thermal capacity jump (Fig. 3.25, curves 2, 4). The full jump in thermal capacity at the glass-transition is determined by the change of two molecular states, i.e., their oscillation spectrum and configuration behavior [155]. Thus, the method of initial MTI synthesis defines this characteristic in the finished polymer.

Dynamic spectroscopy (this method is more sensitive to T_g modifications) characterized cured MTI as polymer networks. The existence of the high-elasticity plateau typical for cross-linked organic polymers makes it possible to determine values of high elastic modulus in the pseudoequilibrium zone. Systems based on oligoethers and mixtures with oligoesters are characterized by E values of 5.0 and 8.5 MPa, respectively, which comply with adhesion strength data.

Data from dynamic mechanical spectroscopy also define more precisely the behavior of polymers at the transition from glasslike to a highly elastic condition. Polymers based on L-3003 and oligoester mixtures in different ratios only are characterized by various $\tan\delta$ temperature dependences. Two maxima at temperatures of 201 and 245 K are observed in the cured MTI based on 5:1 oligoether–oligoester, while the polymer network based on L-3003 is described by a curve with one maximum at 239 K. An even larger $\tan\delta$ maximum shift to 249 K is experienced for the cured MTI with 1:1 oligomer ratio. On the assumption that this method is also insensitive for resolution of T_g values of small separation ($T_{gc} - T_{gn} = 15°C$), curing of MTI with 1:1 ratio was used at increased temperature (333 K) as in the study of curing kinetics [145]; it was found that MTI based on oligoethers and oligoesters are cured with different temperature coefficients. It was found that for a 5:1 ratio the initial MTI is a stable emulsion and at 1:1 ratio of components the latter delaminates on storage.

Thus, polymers cured at less than 298 K based on MTI of oligoethers and two areas of mechanical losses characterize oligoester mixtures. The position of the main maximum and, consequently, T_g depends on the ratio of polymers based on simple and complex oligomers and a new area of structural transition may be referred to the loosened interphase layer. Hence, polymers of heterogeneous composition may behave as single phases in curing, which explains the improvement of properties in comparison with individual polymers.

3.3.4.12 Thermodynamic functions of surfactant surface layers in oligomer mixtures. For task-oriented investigation of the influence of surfactant on the formation of polyurethane networks, information concerning the structure of the surfactant surface layers is required for various surfactant contents as well as identification of the concentration interval in which the surfactant is fully soluble in the oligomer mixtures. For this reason, surface tension isotherms of surfactant solutions were studied and thermodynamic analysis was undertaken.

Mixtures of a single compatible oligomer pair (PPG–PTMG) and of two incompatible ones (PPG–ODA and PTMG–ODA) were chosen for investigation. KEP-2 was used as surfactant. The principal condition for thermodynamic analysis and qualitative determination of surface functions is that the systems should be binary [133]. Accordingly, surfactant solutions cannot be analyzed in thermodynamically incompatible oligomer pairs (ODA–PPG and ODA–PTMG). We thus limit ourselves to analysis of thermodynamic surface functions of KEP-2 solution mixtures of compatible oligoglycols. This research is based on experimental values of the surface tension measured for KEP-2 solutions of various concentrations at different temperatures by the Wilgelmi method [126].

Thermodynamic functions of the surfactant surface layers for simple oligoglycol mixtures were derived from surface tension data presented in Table 3.25. The results show that KEP-2 solubility in PPG–PTMG is halved in comparison with that in a single PPG oligomer. Temperature dependences of surface pressure P for various KEP-2 mass fractions in oligomer mixtures are determined from experimental isotherms of surface tension (Fig. 3.26). As is obvious from Fig. 3.26, surface pressure depends nonmonotonically on temperature for practically all mass contents of KEP-2 despite the P–T dependences observed earlier in PTMG or ODA solutions. This dependence also determines the complex character of the concentration and temperature dependences of ΔS_s shown in Table 3.26. This table also gives enthalpy values ΔH_s. From Table 3.26 it follows that in solutions with KEP-2 mass fraction of 2.25×10^{-4}–5×10^{-4} the values of ΔH_s and ΔS_s are temperature-invariant and the entropy change does not depend on surfactant content in the solution, as it is equal to zero. Thus, for these KEP-2 contents at all the studied temperatures the surface layer is characterized by high structural stability.

Furthermore, concentration changes of enthalpy and entropy (Table 3.26) with increase of temperature are characterized by regular transformation of the dependences observed, i.e., gradual transition of the curves to negative values of entropy and enthalpy and the appearance of two extrema for solutions with KEP-2 mass fractions of 1.5×10^{-4} and 7.5×10^{-4}. In this case, the intensity of the second maximum increases with increase in temperature. The dependences obtained for thermodynamic functions can be interpreted from the point of view of the phenomena considered in Section 2.8.1.

The maximum values of the adsorption constant for KEP-2 solutions in PPG–PTMG mixtures were determined by the transformed Gibbs equation [160]:

TABLE 3.25 Surface Tension Temperature Dependences in Systems Based on PPG–PTMG–KEP-2

KEP-2 mass fraction (±3%)	Surface tension γ (mN/m) (±0.2) at the temperature					
	328 ± 0.1 K	333 ± 0.1 K	338 ± 0.1 K	343 ± 0.1 K	348 ± 0.1 K	353 ± 0.1 K
0	33.0	32.7	32.4	32.1	31.5	30.9
7.5×10^{-5}	28.8	28.2	27.6	27.3	26.7	26.1
1.5×10^{-4}	26.4	25.9	25.6	25.1	24.7	24.3
2.25×10^{-4}	25.9	25.6	25.3	25.0	24.4	24.1
3×10^{-4}	25.6	25.3	25.0	24.7	24.1	23.8
5×10^{-4}	24.7	24.4	24.1	23.8	23.2	22.9
7.5×10^{-4}	24.1	23.8	23.5	23.2	22.9	22.3
1.0×10^{-3}	23.7	23.2	23.2	22.8	22.5	22.1
1.25×10^{-3}	23.3	23.0	22.9	22.6	22.3	22.0
1.5×10^{-3}	23.3	23.0	22.9	22.6	22.3	22.0

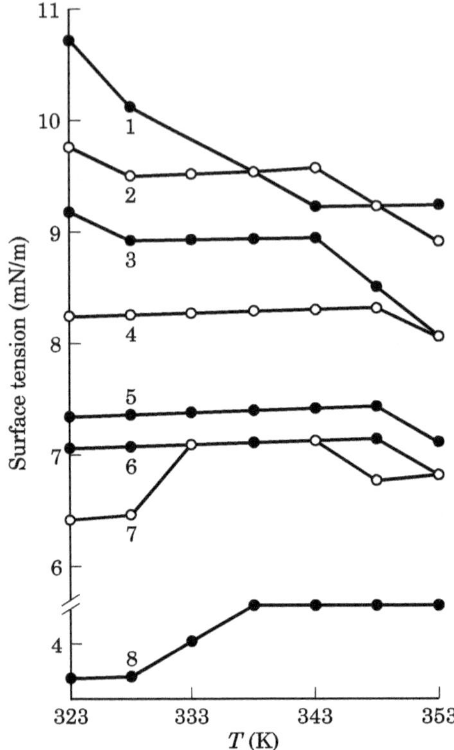

Figure 3.26 Dependence on temperature of surface tension of KEP-2 solutions in PPG and PTMG equimolar mixture. Surfactant mass fraction in the solution: (1) 0.0015; (2) 0.00125; (3) 0.00075; (4) 0.0005; (5) 0.0003; (6) 0.000225; (7) 0.00015; (8) 0.00075.

$$\Gamma_\infty = \frac{1}{RT} \cdot \frac{d\gamma}{d \ln C} \tag{3.28}$$

where Γ_∞ = maximum surfactant adsorption (mol/m^2), γ = surface tension (J/m^2), C = surfactant concentration (mol/m^3), T = temperature (K), and R = gas constant.

Γ values and calculated areas for a molecule of surfactant in a saturated sorption layer are given in Table 3.27.

From the results presented in Table 3.27 and in accordance with the results of [157, 158], for our case at 333 K, the surface layer at maximum adsorption (surfactant mass fractions from 2.25×10^{-4} to 7.5×10^{-4}) is an extremely compressed monolayer, whereas the majority of dimethylsiloxane units lie on the surface of the liquid. At higher temperatures (343–353 K) but in a narrower concentration range (from 5×10^{-4} to 7.5×10^{-4}) the polydimethylsiloxane chain tears off the surface, forming spiral turns of 6–8 elements, as witnessed by a single element area of 0.09–0.11 nm^2, which may be referred to one spiral turn [157]. From the concentration and tempera-

TABLE 3.26 Surface Enthalpy and Entropy Values for Solutions with Various KEP-2 Mass Fractions in PPG–PTMG Mixtures at Various Temperatures

KEP-2 mass fraction (±3%)	328 ± 0.1 K		333 ± 0.1 K		338 ± 0.1 K		343 ± 0.1 K		348 ± 0.1 K		351 ± 0.1 K	
	$10^3 \Delta H_s$ (J/m^2) (±5%)	$10^5 \Delta S_s$ (J/m^2 · K) (±5%)	$10^3 \Delta H_s$ (J/m^2) (±5%)	$10^5 \Delta S_s$ (J/m^2 · K) (±5%)	$10^3 \Delta H_s$ (J/m^2) (±5%)	$10^5 \Delta S_s$ (J/m^2 · K) (±5%)	$10^3 \Delta H_s$ (J/m^2) (±5%)	$10^5 \Delta S_s$ (J/m^2 · K) (±5%)	$10^3 \Delta H_s$ (J/m^2) (±5%)	$10^5 \Delta S_s$ (J/m^2 · K) (±5%)	$10^3 \Delta H_s$ (J/m^2) (±5%)	$10^5 \Delta S_s$ (J/m^2 · K) (±5%)
7.5×10^{-5}	21.1	5.2	17.0	3.8	12.4	2.3	4.7	0	4.8	0	4.8	0
1.5×10^{-4}	25.7	5.8	25.2	5.4	7.1	0	−1.1	−2.4	−2.6	−2.7	−2.7	−2.7
2.25×10^{-4}	7.1	0	7.1	0	7.1	0	7.1	0	7.1	0	6.8	0
3×10^{-4}	7.4	0	7.4	0	7.4	0	7.4	0	7.4	0	7.1	0
5×10^{-4}	8.3	0	8.3	0	8.3	0	8.3	0	8.3	0	8.0	0
7.5×10^{-4}	8.9	0	8.9	0	2.4	−1.9	−9.0	−5.2	−9.0	−7.4	−30.0	−10.9
1×10^{-3}	9.5	0	9.5	0	1.5	−2.4	−2.9	−3.6	−2.9	−5.5	−18.7	−7.9
1.25×10^{-3}	9.5	0	9.5	0	1.5	−2.4	−2.9	−3.6	−2.9	−5.5	−18.7	−7.9
1.5×10^{-3}	−22.7	−10.0	−10.5	−6.1	−6.2	−4.6	−0.2	−2.7	−0.2	0	9.2	0

TABLE 3.27 Rated Values of Surfactant Maximum Adsorption and Area Occupied by a Surfactant Molecule and One Siloxane Element for KEP-2 Solutions in PPG–PTMG Mixtures

T (K) (±0.1)	Γ (mol/m^2)	A_m (nm^2)	A_{el} (nm^2)
333	5.3×10^{-7}	3.13	0.156
343	7.4×10^{-7}	2.24	0.112
353	9.2×10^{-7}	1.82	0.091

ture dependences of surface tension and other derived characteristics of the surfactant adsorption layer the structure could be characterized for various concentrations and temperatures to elucidate the effects of these factors.

3.3.4.13 Study of interactions in polyurethane mixtures in the presence of surfactants by spectral methods. NMR spectroscopic investigation was undertaken to discover whether there was interaction between that part of the adsorption layer formed by oxyalkylene lateral extensions and molecules of the solvent (oligoglycol). The integrated intensities (II) of proton signals in equimolar mixtures of compatible and incompatible oligomer pairs without surfactant and with KEP-2 at mass fraction 1×10^{-3} were compared. It was found that in both oligomer pairs the introduction of KEP-2 results in changes of the II of proton signals, but in compatible systems the surfactant leads to II reduction, and in mixtures of incompatible oligomers the reverse is observed. It is of paramount importance that in the range of temperatures investigated the influence of the surfactant for each oligomer pair does not depend on temperature. Preliminary results obtained by NMR are qualitative, but they form the basis for more directed IR spectral investigation of oligomer mixtures with surfactant additives.

It was shown that in KEP-2 solutions in equimolar mixtures of PPG and PTMG containing 7.5×10^{-4} and 1.5×10^{-4} surfactant mole fractions the presence of surfactant does not influence the OH group band. At other surfactant concentrations the normal reduction of the band maximum to less than 3460–3480 cm^{-1} is observed. In addition, at these surfactant contents the band at 3590 cm^{-1} disappears, i.e., the spectral profile contour changes. The 3590 cm^{-1} band characterizes intramolecular hydrogen bonding in five-element cycles, which are nonreactive in uncatalyzed reactions of urethane formation. A similar influence of KEP-2 is observed in pure PPG.

The data show that KEP-2 solutions in equimolar mixtures and in pure PPG show the same behavior for surfactant concentrations differing twofold, which may be a result of differing KEP-2 solubility in

PPG and PTMG. From the chemical composition of the oligomers and of KEP-2 it may be supposed that PPG is the surfactant solvent and the second oligomer acts as diluent. Therefore, the integral intensity of OH groups in PPG containing 2×10^{-3} mole fraction of KEP-2 (see Table 3.19) was taken into account in calculation of the integral intensity of the OH group band in the PPG–PTMG mixture. Data given in Table 3.19 show that in the PPG–PTMG system strong specific interactions arise between molecules of these oligomers, as the intensity of the OH group band in the mixture is 1.5 times less than the additive value in the presence of KEP-2, which may be evidence of participation of both glycol molecules in specific interactions with KEP-2. This may be due to formation of hydrogen bonds with a surfactant not by separate molecules but by associations with hydrogen bonds of oligomer molecules. The fact that at certain surfactant concentrations the intramolecular five-cycles are broken in PPG itself and in PPG–PTMG mixtures may be related to the change of conformation of the polyoxypropylene glycol molecules under the influence of hydrophilic fragments of KEP-2 molecules located in the boundary layer.

Let us compare the results from IR spectral studies with data from thermodynamic analysis of the surfactant adsorption layers. It follows that when hydrophobic parts of surfactant molecules, regardless of their conformation, are characterized by extremely compact packing in the surface layer, the surfactant does not influence the state of the hydroxyl groups of solvent oligomers. In all the other cases hydrophilic lateral extensions of the surfactant change the character of specific interactions in the oligomer mixture: the OH group band contour is changed and its intensity is reduced.

IR spectra of mixtures of incompatible PPG and ODA oligoethers with various KEP-2 contents show that at certain surfactant concentrations (1.7×10^{-4} and 8.5×10^{-4} mole fraction) KEP-2 has no effect on the form or magnitude of the OH group band in the spectrum of the oligomer mixture, while other surfactant concentrations cause the growth of the hydroxyl group band. As seen from Table 3.28, in PPG and ODA the band integral intensity is 1.5 times less than the additive value without surfactant, and with KEP-2 at 1×10^{-3} mole fraction it shows essentially additive behavior of the hydroxyl group mixture. This may be considered evidence of participation of some of the PPG and ODA molecules in the formation of intermolecular hydrogen bonds, despite their thermodynamic incompatibility. Surfactant at 1×10^{-3} mole fraction results in independent contributions of the OH groups of each oligomer to the common band, i.e., ODA and PPG essentially do not form mutual hydrogen bonds at these KEP-2 concentrations. The following observation deserves attention in relation

TABLE 3.28 Physical and Mechanical Properties of Polyurethanes Based on Compatible Oligoethers, Trimethylolpropane, and Toluylene Diisocyanate

	Curing temperature (K) (±1)					
KEP-2 mass fraction (±3%)	343	333	343	333	343	333
	Tensile strength (MPa) (±10%)		Ultimate strength under normal fracture (MPa) (±10%)		Elongation at breaking (%) (±10%)	
0	3.0	2.0	9.0	9.0	800	470
7.5×10^{-5}	4.0	1.5	8.0	9.0	820	600
1.5×10^{-4}	5.0	0.7	7.0	12.3	850	960
3×10^{-4}	3.0	4.2	16.0	14.0	440	600
5×10^{-4}	3.8	3.5	17.0	13.0	600	450
7.5×10^{-4}	6.0	3.6	9.0	9.0	750	410
1.0×10^{-3}	4.5	3.5	10.0	11.0	400	510
1.5×10^{-3}	4.0	5.0	13.0	13.0	500	640

to this oligomer system, as in the compatible one. In the solutions containing 1.7×10^{-4} and 8.5×10^{-4} mole fraction of KEP-2 it appears that the surfactant additive does not participate in specific interactions with oligomers (or more precisely, as investigation showed, with one of the oligomers, namely, PPG) since ODA hydroxyl groups do not show specific interaction with KEP-2 at any concentration. As the oligomers investigated are incompatible and only PPG is related to KEP-2 by chemical composition (KEP-2 solubility in ODA is 70 times lower than in PPG), the surfactant is predominantly dissolved in PPG. Thus, surfactant additive behavior is determined by its concentration of PPG. This concentration is 1.8 times higher than specified. Taking the above into account, the results in [10, 11] may be used for discussion of IR spectral data, from which it follows that KEP-2 "indifference" is observed at solution concentrations for which the structure of the surfactant surface layer in PPG is in precritical condition. At low surfactant concentration, reorientation with respect to the surface occurs; and under high concentration, there is disordering of the structures formed by spiral fragments of molecules.

Thus, in both types of oligomer systems the surfactant influences specific interactions at all concentrations in addition to those corresponding to maximally ordered surface layer structures.

3.3.4.14 Microstructure of polyurethane networks based on oligomer mixtures. As KEP-2 is a chemically unreactive surfactant, it may be supposed that one of the essential factors controlling modification of polyurethane properties under its influence is their morphological structure. From comparison of micrographs of the morphological structure of polyurethane networks based on compatible oligomer mixtures cured at $343 \pm 1\,K$ and $333 \pm 1\,K$, it is obvious that the microstructures of polyurethanes based on individual oligomers of PPG and of PTMG have different characters. In the first case these are disordered permolecular formations of irregular shape with two types of microstructural heterogeneity; in the second case the structure is fairly homogeneous with elements of order. These polyurethanes are characterized by corresponding tensile strengths of 3.0 and 7.0 MPa. Morphological elements typical of the polymer based on a single PTMG prevail in the microstructures of polyurethanes based on PPG–PTMG mixtures. Polymer morphological structure becomes homogeneous and is characterized by less globule-like form and dimensions with the introduction of KEP-2 at 1.5×10^{-4} mass fraction at the stage of polyurethane synthesis. Further increase of surfactant content in the system leads again to the formation of irregular morphological structure with numerous large, irregularly shaped elements. A polyurethane network obtained in the presence of KEP-2 at 7.5×10^{-4} mass fraction is characterized by fairly homogeneous microstructure similar to that of polymers based on a single PTMG. Increase of the KEP-2 concentration causes generation of irregular structure with areas of larger structural elements and smaller globules. Thus, microstructural elements in the polyurethane system are smaller than in the same systems with surfactant at mass fractions of 1.5×10^{-4} and 7.5×10^{-4}.

Investigation of microstructure in polyurethanes cured at 333 ± 1 K has shown that two levels of heterogeneity characterize polyurethanes based on mixtures of compatible oligomers when smaller globule-like structural elements are grouped into macroglobules. Introduction of 1.5×10^{-4} mass fraction of KEP-2 results in these macroglobules breaking and the appearance of larger formations that are 2–3 times larger and have irregular shape at the macro level. Further increase of the surfactant content in polyurethane leads to complete disappearance of structural elements, typical of the initial polyurethane without surfactant, from the morphological structure of specimens.

Morphological analysis of polyurethane networks based on mixtures of PPG and ODA incompatible oligomers shows that polymers based on ODA are distinguished by a regular microglobular structure. Typically, an emulsion structure with a drop size of order 100 μm and having regular shape is seen in the microstructure of polyurethanes based on mixtures of oligoethers and oligoesters. The micro level structures of this polyurethane are characterized by elements typical of polyurethanes based on PPG but are smaller in size and are more uniform. Introduction of 1.7×10^{-4} and 1.1×10^{-3} mass fraction of surfactant results in formation of fairly homogeneous microstructure consisting of smaller globules even showing elements of order. For a KEP-2 content of 5.5×10^{-4} mass fraction in polyurethane based on an incompatible oligomer mixture, the microstructure does not essentially differ from the morphology of the polymer without surfactant. Heterogenization of the microstructure with the formation of irregularly shaped macroglobules from quite large microformations is typical for polymers with 8.5×10^{-4} mass fraction of KEP-2. All the polyurethanes obtained in the presence of surfactant are characterized by the disappearance of the emulsion structure observed in the initial polyurethane.

3.3.4.15 Swelling of polyurethane networks. Proceeding from the general conception of the role of surfactants in the formation of the secondary structure of polyurethanes, studies of the influence of KEP-2 on the distribution of hydrogen bonds in polyurethane networks were required. The literature on the thermodynamic properties of polymer solutions and gels was searched and, in particular, the work of the authors of [117] is of direct interest in this context. This work considers values of the heat and free energy of polymer mixing with solvent for rubbers and the energy of polymer cohesion and the liquid phase (solvent) with derivation of equations. The equations derived show good correlation with experimental data. This work has served as the conceptual basis for development of ideas of the swelling of completely insoluble vulcanized rubbers. In conformity with these conceptions, the equality of specific cohesion values of the polymer and solvent represents the thermodynamic condition for maximum swelling of crosslinked polymers. As the value of the specific cohesion is equal to the square of the solubility parameter (σ), it is more convenient to use the latter. Thus, the ideas of [117], whose application to polar polymers was demonstrated in [118], may be used to reveal specific interactions in polyurethanes of various compositions. The essence of the method is that solubility parameters (σ_n) may be determined by the position of the maximum in the dependence of equilibrium swelling (Q_p) on the solvent solubility parameter (σ_p), even for crosslinked polymers. In [119] this approach was used for characterization of redistribution of hydrogen bonds in polyurethane networks under the influence of fillers. This qualitative evaluation proved to be possible because it was shown that the polyurethanes investigated had two extremal swelling areas differing in energy. It is known that there are two types of proton acceptors in polyurethanes, namely, the oxygen atom of the glycol fragment and the oxygen atom of the urethane group, allowing for the formation of two types of hydrogen bond with differing energies [163]. Qualitative correlation of the two types of hydrogen bonds is determined by the particular polyurethane's ability to self-associate [164, 165]. In accordance with data on relaxation behavior, the localization of various structure types formed by homogeneous hydrogen bonds should be observed in polyurethane nets [165], regardless of the nature of the initial simple oligoglycol. Thus, all the findings reported support the contention that application of the ideas of [117] to polyurethanes allows us to distinguish types of hydrogen bonds with differing energies. The first maximum in the $Q_p = f(\sigma_p)$ dependence may be attributed to weaker hydrogen bonds of the NH group with the oxygen atom in the glycol chain; the second maximum characterizes the breaking of both types of bonds [165]. For systematic studies of specific interactions in polyurethanes by the swelling method it is essential to use a number of solvents with various solubility parameters but having a common type of interaction with polymer [162], which is accomplished by the use of ethers and esters in the present investigation. The required values of the solvent solubility parameters are taken from [118, 119, 120].

The results of equilibrium swelling studies on polyurethanes based on oligomer mixtures are presented in Figs. 3.27 and 3.28. Q_p values in various solvents are correlated with the Q_p of each polymer in toluene to clarify the contribution of specific interactions to the equilibrium swelling curves. This reduction is effected by proceeding from the fact that swelling touches on all kinds of polymer systems' physical interactions, and use of toluene permits nonspecific (van der Waals, dispersion, etc.) inter- and intramolecular interactions to be taken into account. This solvent is characterized by a three-dimensional solubility parameter close to the maximum for polyurethane thermodynamic affinity [18–19 $(MJ/m^3)^{1/2}$] [118]. The dispersion component [122] is the main contribution to its value. Thus, curves in Figs. 3.27 and 3.28 demonstrate the role only of specific interactions in the process of polyurethane swelling in a number of solvents differing in cohesion energy. The difference between the equilibrium swelling values for the given solvent and for toluene (ΔQ_p) is plotted as the ordinate for this purpose.

Let us consider the correspondingly processed experimental results for the system based on oligoglycol mixtures. Figure 3.27 depicts the dependences obtained for polyurethane networks based on mixtures of compatible oligomers in the presence of KEP-2. Swelling of polymers based on the individual oligomers composing the mixture of PPG and PTMG in toluene gives respective values of 2.0 and 1.0 for equilibrium

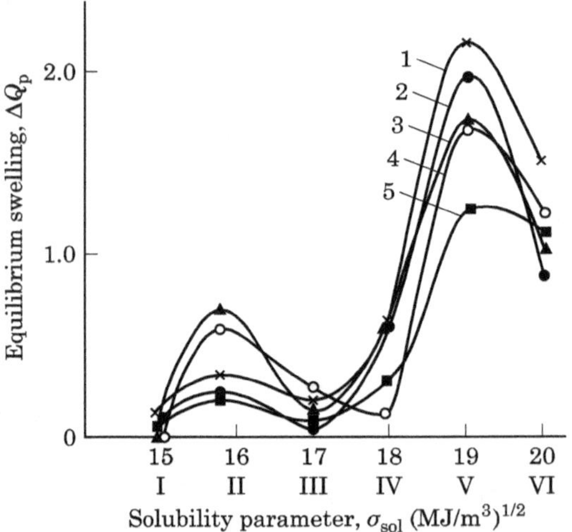

Figure 3.27 Dependence of equilibrium swelling on the solubility parameter in a number of solvents: (I) diethyl ether; (II) diethylene glycol diethyl ether; (III) butyl acetate; (IV) ethyl acetate; (V) tetrahydrofuran; (VI) dioxane. Polyurethane networks based on PTMG and PPG mixture without additives (4) and containing various KEP-2 mass fractions: (1) 1.5×10^{-4}; (3) 5×10^{-4}; (2) 7.5×10^{-4}; (5) 1.5×10^{-3}.

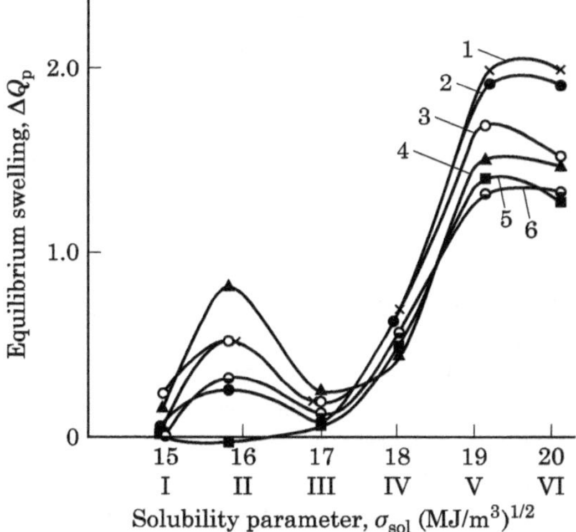

Figure 3.28 Dependence of equilibrium swelling on the solubility parameter in a number of solvents: (I) diethyl ether; (II) diethylene glycol diethyl ether; (III) butyl acetate; (IV) ethyl acetate; (V) tetrahydrofuran; (VI) dioxane. Polyurethane networks based on ODA and PPG mixture without additives (6) and containing various KEP-2 mass fractions: (1) 1.7×10^{-4}; (3) 5.5×10^{-4}; (2) 1.1×10^{-3}; (4) 8.5×10^{-4}. (5) Polyurethane based on ODA.

swelling; for polymers based on mixtures with and without surfactant added, values close to the average rated values of order 1.4–1.5 are observed. Figure 3.27, curve 4, shows that the $Q_p = f(\sigma_p)$ dependence for polyurethane based on an oligomer mixture is characterized by a curve with two well-expressed maxima at 15.8 and 19.0 (MJ/m^3)$^{1/2}$. Other curves indicate that introduction of 5×10^{-4} mass fractions of surfactant (curve 3) into the polyurethane system does not significantly influence the number and types of hydrogen bonds and their balance in the system. The contribution of different types of hydrogen bonds to the polyurethane secondary structure remains close to that observed in the initial polymer when the surfactant content is 1.5×10^{-3} mass fraction (curve 5), but the number of hydrogen bonds due to the ether group oxygen is reduced. Nevertheless, curves 1 and 2 indicate that the polyurethane systems are characterized by important redistribution of the physical structures in favor of self-associated urethane domains at two surfactant concentrations of 1.5×10^{-4} and 7.5×10^{-4} mass fractions.

Let us consider the data from swelling studies in polyurethanes based on two incompatible oligoglycols. Figure 3.28 shows dependences found for polymer networks synthesized from PPG and ODA mixtures. The equilibrium swelling of polyurethanes based on the corresponding individual oligomers in toluene varies tenfold and is 0.2 for the complex oligomers. All the mixtures in this solvent swell equally and Q_p is characterized by a value of 1.0, close to the rated value. The role of specific interactions in the formation of this polyurethane mix structure is demonstrated in Fig. 3.28, which shows that, in contrast to polyurethanes based on the single ODA (curve 5), all the other polymers investigated exhibit two maxima (curves 1–4) whose intensity ratio depends on the quantities of surfactant in the system. The occurrence of the first maximum in $Q_p = f(\sigma_p)$ (Fig. 3.28) demonstrates the involvement of hydrogen bonds, due to the oxygen atom of the PPG ether group, in the polyurethane network structure based on mixtures of simple and complex oligoglycols. Comparing the curves corresponding to various surfactant concentrations, there is no direct relationship between increase in KEP-2 content and its influence on the contribution of hydrogen bonds to the polymer structure. Combining the intensity of maxima with the number of broken hydrogen bonds of a given strength suggests that an approximately 1.5-fold increase of hydrogen bonds due to self-associated urethane segments occurs for KEP-2 concentrations of 1.7×10^{-4} and 1.1×10^{-3} mass fraction (curves 1, 2). The lower of these concentrations promotes formation of the greater number of hydrogen bonds with lower energy (curve 1). At the same time, the introduction of 8.5×10^{-4} mass fractions of KEP-2 into the polyurethane system results in proportional (1.4-fold) reduction in the number of hydrogen bonds typical for higher polyurethane cohesion energy and an increase in the number of hydrogen bonds whose breaking is described by the low-energy maximum (curve 4). The data on specific interactions in polyurethane systems based on PPG and ODA mixture show that for the same surfactant contents the growth of self-association and formation of more ordered microstructure is observed as in polyurethane systems based on compatible oligomer pairs.

Specific interaction studies on the polyurethane nets based on ODA and PTMG incompatible oligomers shows that, similarly to the polyurethane system based on a single ODA (Fig. 3.28, curve 5), those based on ODA–PTMG mixtures are characterized by swelling curves with one extremum in the area of 19–20 $(MJ/m^3)^{1/2}$. Therefore, these polyurethanes retain the tendency to segregation of the chain urethane fragments in mixtures with PTMG, with and without the surfactant present. In all cases the introduction of surfactant results in a considerable decrease of hydrogen bonds formed in self-associations. KEP-2 mass fractions of 1.9×10^{-4} and 1.5×10^{-3} in the cured polyurethane system cause a two-fold reduction of the swelling intensity maximum. It can be supposed that macromolecules based on both glycols participate in self-association of the urethane segments in these polyurethanes; and that introduction of surfactant prevents the possibility of such interaction in principle. The effect is strongest in the range of surfactant concentrations for which adsorption layers with maximally ordered structure are formed of large KEP-2 associations.

3.3.4.16 Influence of surfactant on physical and mechanical properties of polyurethane networks based on oligomer mixtures. Table 3.28 shows results of investigations on the cohesion and adhesion properties of polyurethanes based on compatible oligomer mixture in the presence of different amounts of surfactant. As was shown in Table 3.26, the characters of the concentration dependences of entropy and enthalpy in PPG and PTMG equimolar mixture at $333 \pm 1\,\mathrm{K}$ and $343 \pm 1\,\mathrm{K}$ are essentially different. It was thus of interest to investigate their influence on the physical and mechanical properties of cured polyurethanes. Analysis of the data in Table 3.28 indicates that introduction of surfactant at certain concentrations increases the cohesion and adhesion strength of polyurethane networks and alters their elasticity at both curing temperatures. The trends of change in cohesion and adhesion strength for a given surfactant concentration are inversely related—if cohesion strength increases, adhesion strength falls. The extremal character of the concentration dependence of the breaking strength characteristics (Table 3.28) actually corresponds to the same dependences of the thermodynamic functions (Table 3.26). It follows that cohesion strength increases and adhesion strength is reduced for the maximum packing density of the surface layer structure, regardless of the surfactant molecules' conformation, at $333 \pm 0.1\,\mathrm{K}$; this is the most highly compressed KEP-2 monolayer. Growth of adhesion strength is observed for surfactant concentrations at which the surface layer structure is characterized for the given temperature by constancy of the surface enthalpy and entropy values. If these results are compared with the data for swelling and morphology of these polyurethanes, it is obvious that the increase in polymer cohesion strength is observed and the formation of more ordered and homogeneous microstructure occurs at surfactant mass fractions of 1.5×10^{-4} and 7.5×10^{-4}.

Physical and mechanical characteristics of polyurethane nets based on PPG and ODA mixtures at $343 \pm 1\,\mathrm{K}$ were studied—that is, the properties of polymers formed in the conditions of macrolevel heterogeneity were investigated. As has been shown [167], the separation of the system into phases occurs much earlier than the emergence of microheterogeneous structures connected with network formations when the formed polyurethanes are incompatible. This process is accompanied by stick-slip increase of surface tension in the system of several m/Nm. Consequently, the appearance of microheterogeneous structure and surfactant redistribution due to the appearance of new boundary occurs competitively at two boundaries. In the system considered (Table 3.29), the same cohesion and adhesion strength values are observed as in the system without surfactant, with minima in the concentration dependences of surface enthalpy and entropy ($C_{surf} = 8.5 \times 10^4$ mass fraction) characterizing the ordered layer state [10,11]. For all other KEP-2 concentrations in the incompatible system, adhesion strength falls and cohesion strength rises to 24 MPa (a 2.5-fold increase in the initial value). Ultimate strength at normal fracture and tensile strength are characterized by concentration dependences with two extrema. Maximum growth of cohesion strength in these systems at low surfactant contents is observed for a KEP-2 concentration of 1.7×10^4 mass fraction when the enthalphy and entropy of the surfacant surface layer in PPG are temperature invariant and characterized by positive values [11]. The largest tensile strength increase, as is obvious from Table 3.29, is observed at such surfactant concentrations that its presence ensures independent behavior of ODA and PPG oligoglycols in the mixture. Table 3.29 shows the same regularity observed in investigation of polyurethanes based on compatible oligoglycols, namely, adhesion and cohesion strength vary inversely in the presence of KEP-2. There is no essential change in the polyurethanes' elasticity with surfactant present (Table 3.29).

The data obtained in cohesion characteristics studies of polyurethane systems based on PPG and ODA mixtures are well correlated with the results of specific interactions studies. As in the case of polyurethane systems based on compatible oligomer pairs, growth in the number of self-associations, formation of more ordered microstructure, and increase of polyurethane cohesion strength are observed for the same surfactant contents.

TABLE 3.29 Physical and Mechanical Properties of Polyurethane Nets Based on ODA and PPG Incompatible Oligoether Mixtures, Trimethylolpropane, and Toluylene Diisocyanate at 343 ± 1 K

KEP-2 mass fraction (±3%)	Tensile strength (MPa) (±10%)	Ultimate strength under normal fracture (MPa) (±10%)	Percent elongation (%) (±10%)
0	10.0	12.0	660
1.7×10^{-4}	16.0	9.0	700
3.3×10^{-4}	13.0	9.0	630
5.5×10^{-4}	12.0	9.0	650
8.5×10^{-4}	9.0	13.0	850
1.1×10^{-3}	19.6	11.0	700
1.65×10^{-3}	24.0	10.0	670

Physical and mechanical properties of polyurethane networks based on another incompatible oligoglycol pair, ODA and PTMG at their equimolar ratio, are presented in Table 3.30. As the table shows, polyurethane specimens were synthesized with various KEP-2 concentrations in the oligomer mixture, including those for which extrema were observed for the thermodynamic functions. This concentration is 3.5×10^{-5} surfactant mass fraction for ODA, and $(1.8\text{–}1.9) \times 10^{-4}$ mass fractions of KEP-2 for PTMG. Analysis shows that KEP-2 presence generally leads to twofold worsening of cohesion strength, reduction of percent elongation, and insensitivity to surfactant of the ultimate strength value at normal fracture. A small (20%) adhesion strength reduction and a large (3-fold) decrease in percent elongation occur at a KEP-2 mass fraction of 1.9×10^{-4}. The lowest cohesion and adhesion strength values are observed for surfactant concentration of 1.5×10^{-3} mass fraction. At this surfactant concentration the solution becomes turbid and KEP-2 and oligomer incompatibility may be observed visually. Hence, the cohesion and adhesion characteristics of polyurethane nets do not depend on the structure of KEP-2 adsorption layers in the initial oligomers of the ODA and PTMG mixture. There is an insignificant reduction of adhesion strength, and cohesion properties fall to a significant degree only at the surfactant concentration corresponding to the maximally packed surface layer.

Data from specific interaction studies in the polyurethane system based on ODA and PTMG incompatible oligomer mixtures show correlation between the change in cohesion strength and the number of hydrogen bonds formed due to self-associated urethane groups. A twofold decrease of the number of hydrogen bonds formed in self-associations in the presence of 1.9×10^{-4} surfactant mass fraction results in a twofold reduction in cohesion strength.

TABLE 3.30 Physical and Mechanical Properties of Polyurethane Nets Based on ODA and PTMG Incompatible Oligoether Mixtures, Trimethylolpropane, and Toluylene Diisocyanate at 343 ± 1 K

KEP-2 mass fraction (±3%)	Tensile strength (MPa) (±10%)	Ultimate strength under normal fracture (MPa) (±10%)	Percent elongation (%) (±10%)
0	20.0	19.0	1100
3.5×10^{-5}	10.0	18.0	800
7×10^{-5}	10.0	18.0	800
1.9×10^{-4}	8.0	15.0	400
5×10^{-4}	10.0	18.0	800
1×10^{-3}	10.6	18.0	800
1.5×10^{-3}	7.0	14.0	800

The investigation showed that the physical and mechanical concentration dependences of polyurethanes formed in the presence of surfactant are distinguished by extremal character. The changes in physical and mechanical properties in compatible polyurethane systems are connected with the concentration dependences of the entropy and enthalpy of the surfactant surface layers in the initial oligomer mixture. When polyurethanes are formed in conditions of incompatibility, the change in physical and mechanical characteristics is determined by the particular concentration dependences of the surface thermodynamic functions of KEP-2 solutions in one of the components of the oligomer mixture, namely, polyoxypropyleneglycol. The relationship between the change in physical and mechanical characteristics and the structure of the KEP-2 adsorption (surface) layer is not seen for the initial oligomers when there is no PPG in the mixture. Hence, the existence of a relationship between the thermodynamic functions characterizing surface layers in the initial reaction mixture and the properties and structure of cured polyurethanes depends on the possibility of specific interaction between the initial oligomers (or one of the initial oligomers) and a surfactant.

3.4 Organo-Mineral Composites

Organic and inorganic polymer mixtures were chosen as organo-mineral composites (OMC). The organic polymer used is the product of MDI (crude dihpenylmethane diisocyanate) curing, and the inorganic polymer is the product of waterglass curing (waterglass is liquid glass, i.e., a solution in water of SiO_2).

Polyisocyanates can initiate tripolymerization reactions in the presence of a catalyst, leading to the formation of strong triisocyanurate cycles [67]. Alkalis, acetates, alcoholates of alkali metals, tertiary amines, and many other substances are efficient catalysts of diisocyanate trimerization [50, 55, 57]. However, many of these catalysts proved to be of low effciency for trimerization of industrial polyisocyanates containing large quantities of impurities. Consequently, two catalysts—an alkaline medium formed by sodium waterglass and a tertiary amine [2,4,6-tri(dimethylaminomethyl)phenol]—were used for the polyisocyanate trimerization process.

The organic component (polyisocyanate) and the inorganic component of the OMC (sodium waterglass) constitute a poorly combined two-phase system. The use of interphase catalysts in heterophase systems is well known [103, 106]. The search for an inter-phase catalyst for our system suggested MGF-9 oligoetheracrylate for this purpose. Oxyethyleneglycol fragments can adopt the crown configuration in the MGF-9 chain structure and the presence of sodium silicate ions together with MGF-9 in our system allows formation of complexes between waterglass sodium cations and oxyethyleneglycol fragments of MGF-9 oligoetheracrylate similar to crown type:

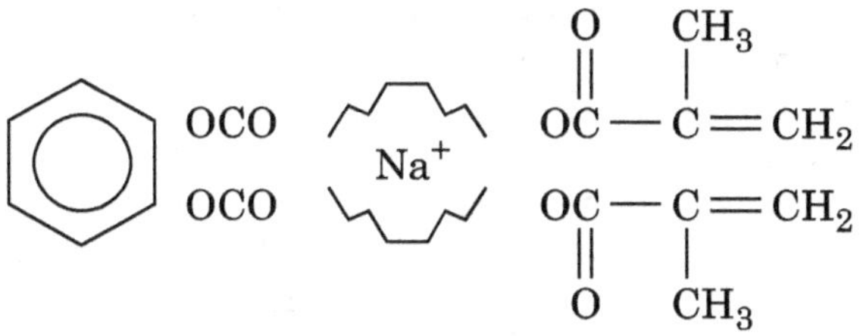

$$
\begin{array}{cccc}
 & & & \overset{\displaystyle O}{\|}\quad \overset{\displaystyle CH_3}{|} \\
\bigcirc & OCO & \frown & OC-C=CH_2 \\
 & & Na^+ & \\
 & OCO & \smile & OC-C=CH_2 \\
 & & & \underset{\displaystyle O}{\|}\quad \underset{\displaystyle CH_3}{|}
\end{array}
$$

Crown-complexes and their open-chain analogs, of which MGF-9 may considered one, are known to act as catalysts carrying ions in two-phase processes and to be efficient catalysts of isocyanate addition polymerization [103, 106].

3.4.1 Consumption of polyisocyanate isocyanate groups in OMC formation processes

A variety of reactive fragments (isocyanate, hydroxyl and ester groups, unsaturated bonds, aromatic rings) are known to facilitate and support parallel and serial competing reactions in the system under investigation: polyisocyanate–waterglass–MGF-9–UP 606/2. The impurities in the commercial products on which OMC development is based further complicate studies of this system. Consequently, investigations of OMC formation were performed on model systems.

The simplest representative of aromatic isocyanates, phenylisocyanate (PHI), served as the polyisocyanate (PIC) model. The choice of phenylisocyanate as the model PIC is connected with the absence from its absorption spectrum of the 1440–1420 cm^{-1} band that could impede spectral interpretation with triisocyanurates in the end products.

The kinetics of NCO group consumption at 295 K was investigated by IR spectroscopy in the following model systems:

1. PHI–benzene
2. PHI–benzene–MGF-9
3. PHI–benzene–waterglass
4. PHI–benzene–waterglass–MGF-9
5. PHI–benzene–waterglass–triethyleneglycol

PHI concentration was 0.1 mol/l, with adduct equimolar NCO ratio.

The content of NCO groups in the model systems remained constant over a period of 24 h; thus no interaction occurred at the ingredient ratios and under the conditions used in the study. The constancy of NCO groups concentration in system No. 2 demonstrates the absence of free triethyleneglycol in MGF-9 resin and of interactions through double bonds. In benzene medium, PHI shows no interaction with the waterglass in sample No. 3, as the system investigated is incompatible. Within 1 h a white-colored deposit had formed in the model system No. 4. Apparently, there was precipitation of sodium phthalate and sodium methacrylate, insoluble in benzene, as the products of MGF-9 oligoetheracrylate hydrolysis in the alkaline medium. In the spectrum an NH band appeared at 1520 cm^{-1}, with a peak at 1220 cm^{-1} characterizing urethane formation. In this case improvement of the compatibility of the mixture components is observed with introduction of MGF-9.

The possibility of MGF-9 ester alkaline hydrolysis was investigated in detail and it was established that both in caustic soda and sodium waterglass media hydrolysis occurs through MGF-9 ester bonds, forming sodium phthalate, sodium methacrylate, and triethyleneglycol. MGF-9 hydrolysis is demonstrated by absorption bands in the regions of 1365–1415 and 1555–1585 cm^{-1} caused by symmetric and asymmetric stretching vibrations of NaOOC, phthalate, and methacrylate groups, as well as by reduction of the 945, 1265, and 1710 cm^{-1} band half-widths, confirming the formation of substances with crystalline structure.

System No. 5 displayed monotonic change of NCO group content with diurethane formation, shown by the appearance of NH bands at 1530 cm^{-1}, amide III bands at 1220 cm^{-1}, and CO absorption bands at 1735 cm^{-1}.

NCO group concentration falls more rapidly for the system with triethyleneglycol than for that with MGF-9. This indicates that incomplete MGF-9 hydrolysis takes place and less TEG is formed as a result than in system No. 5.

Triisocyanurates appear in the end products, while oligoetheracrylate is present with fivefold excess of isocyanate groups in the phenylisocyanate–waterglass mixture, as confirmed by the bands at 760, 1420, and 1715 cm^{-1}. The bands at 760 and 1420 cm^{-1} correspond to the deformation vibrations of triisocyanurate ring bonds and the 1715 cm^{-1} band corresponds to C=O group stretching vibrations in the isocyanate trimer [165]. Loose ion pairs formed by sodium silicate in combination with oxyethylene glycol fragments of MGF-9 resin [104] are responsible for isocyanate transformation up to the cyclotrimer.

3.4.2 Influence of "silica modulus" on OMC end product composition

We characterize the sodium waterglass used by the "silica modulus"—the ratio of SiO_2 to NaOH used in their production, characterizing the degree of transformation of SiO_2 to Na_2SiO_3. The effects of silica modulus on the composition of organo-mineral composite end products were studied using model systems. 2,4-Toluylene diisocyanate was taken as the polyisocyanate model. Sodium waterglasses of various silica modulus in the model systems are generated by using aqueous NaOH solutions of various concentrations.

The end products of the following reaction systems were identified by IR spectroscopy:

$$\text{Water} + \text{2,4-TDI} \tag{1}$$

$$\text{NaOH (5.8\%)} + \text{2,4-TDI} \tag{2}$$

$$\text{NaOH (11.9\%)} + \text{2,4-TDI} \tag{3}$$

$$\text{NaOH (21.8\%)} + \text{2,4-TDI} \tag{4}$$

$$\text{NaOH (37.2\%)} + \text{2,4-TDI} \tag{5}$$

$$\text{NaOH (42\%)} + \text{2,4-TDI} \tag{6}$$

Reactions were run at 295 K at 1:1 and 1:3 ratios of NCO:OH reactive groups. Reaction product spectra were taken on a JR-75 device in the range of 400–4000 cm^{-1} at 295 K with specimens in the form of KBr disks.

It was found that the position and intensity of the IR absorption bands of the products of (1)–(3) were approximately the same. Increase in the initial alkali concentration (to more than 20%) produces changes in the spectra of reaction products of (4)–(6) at 600–1750, 3000–3500, and 2275 cm^{-1} bands, the intensity of which is noticeably reduced. This indicates that the composition of the products of reaction of 2,4-TDI with alkalis differs for low (up to 20%) and high (more than 20%) concentration of alkali.

It is known [67] that the reactions below will proceed provided a strong base is available with isocyanate interaction with water:

$$\text{RNCO} + \text{NaOH} + H_2O \longrightarrow \underset{\text{Carbamate}}{\text{RNHCOONa}} \tag{a}$$

$$\text{RNHCOONa} + \text{NaOH} + H_2O \longrightarrow \underset{\text{Amine}}{RNH_2} + Na_2CO_3 \tag{b}$$

$$RNH_2 + RNCO \longrightarrow \underset{\text{Disubstituted urea}}{RNHCONHR} \qquad (c)$$

IR spectra of the reaction products of (1)–(3) display clear-cut absorption bands at 1550, 1620, 1640, and 3300 cm^{-1}. Comparison of the spectra of diphenylurea and the (1)–(3) reaction products together with published values allowed distinguishing bands to be assigned, since the bands at 1550 and 3300 cm^{-1} are due to deformation and stretching vibrations of the urea NH group and the 1640 cm^{-1} band may be explained by stretching vibrations of the urea group C=O participating in hydrogen bonding [165]. A symmetric band at 1620 cm^{-1} characterizes NH bond deformation vibrations in the amine NH_2 group [166]. Hence, the main products of reactions (1)–(3) are amine [reaction (b) above)] and disubstituted urea [reaction (c)].

The relative band intensities in the spectra of the reaction end products (4)–(6) change as alkali concentration is increased: the bands at 1550 and 1640 cm^{-1} develop shoulders; the absorption intensity varies around 1620 cm^{-1}; and bands appear at 1320, 1580, 1715, and 1920 cm^{-1}. The 1320 and 1580 cm^{-1} absorption bands are explained by symmetric and asymmetric stretching vibrations of the carbamate NaOOC group [164] [reaction (a)] and the 1920 cm^{-1} band is typical for absorption by sodium salts. Consequently, in 2,4-TDI interactions with high-concentration NaOH solutions, mainly reactions (a) and (b) occur: that is, carbamates, amines, and sodium carbonate are predominantly formed and the contribution of the disubstituted urea [reaction (c)] is insignificant.

For the ultimate identification of compositions (2), (5), and (6), the reaction products were washed with double-distilled water and the washed product was dried (product I) and the filtrate was evaporated under vacuum (product II). As may be expected, the amount of products soluble in water increases with increasing alkali concentration. Absorption bands at 760, 1420, 1550, 1640, 1690, 1715, and 3345 cm^{-1} are observed in the product I spectrum. The 760 and 1420 cm^{-1} bands correspond to deformation vibrations of triisocyanurate ring bonds and the 1715 cm^{-1} band to C=O group stretching vibrations in the isocyanate trimer [167]. It may be concluded that isocyanate trimer is formed in 2,4-TDI reactions with high concentrations (over 20%) of alkali. Salts of basic nature, whose production increases with alkali concentration, probably act as catalysts of isocyanate trimerization. This accords with results in [67, 168] showing that strong bases and basic salts catalyze isocyanate cyclotrimerization processes.

The bands at 1550, 1640, 1690, and 3345 cm^{-1} are attributable to absorption of N—H and C=O bonds of disubstituted ureas. In this case appearance of bands with maxima at 1690 and 3345 cm^{-1} demonstrates hydrogen-bond breakage and formation of free C=O (1690 cm^{-1}) and NH groups (3345 cm^{-1}) [169]; this does not conflict with the opinion of the authors of [165] concerning breakdown of the urea cyclic self-associations in strong base media.

Absorption bands at 1320, 1580, and 1920 cm^{-1} are observed in the spectra of product II. The attribution of the absorption bands made above suggests that sodium carbamates, sodium carbonate, and partially amines are formed on contact with water. To demonstrate the presence of amine in the filtrate, the following experiment was performed: 2,4-TDI was added to the product II; product III resulting from this reaction was washed with CCl_4 and dried, and the IR spectrum was obtained. In the product III spectrum the 1620 cm^{-1} absorption band disappeared; the 1550, 1640, and 3300 cm^{-1} bands appeared; and the 1320 and 1580 cm^{-1} bands remained the same. It is concluded that for isocyanate interaction in strong bases (concentration of more than 20%), the carbamates formed in reaction (a) decompose more slowly [reaction (b)] than the amine–isocyanate [reaction (c)].

IR spectral analysis shows that isocyanate trimerization with aqueous alkali solutions takes place independently of the formation of disubstituted urea, amine, carbamates, and sodium carbonate. The qualitative output of these reaction products varies depending on the NaOH concentration and the ratio of NCO and OH groups.

Figure 3.29 depicts individual band intensities of the end products calculated by comparison with internal standard (1600 cm^{-1}) versus NaOH concentration. The quantity of NCO groups failing to react (2275 cm^{-1} band) in the end products of reactions (2)–(6) is sharply reduced at NaOH concentrations above 20%, and the output of disubstituted urea (1640 cm^{-1} band) shows almost no modification with NaOH concentration change. Sodium carbamate (1580 cm^{-1} band), amine (1620 cm^{-1} band), and isocyanate trimer (1715 cm^{-1} band) notably increase. The qualitative ratio of these compounds in reactions (2)–(6) products is probably determined by hydration in the NaOH–H_2O system.

In accordance with [170], 'free' water (3400 cm^{-1}), "bound" water (in the hydroxyl ion solvation sphere, 2800 cm^{-1}), and OH-hydrated ions (2500–3200 cm^{-1}, broad absorption) are found in aqueous alkali solutions from 3% up to 42% concentration. Judging from the reduction in all the end products of the 2800 cm^{-1} band and the disappearance of the 3400 cm^{-1} band, formation of amines and disubstituted urea proceeds with both "bound" and "free" water. The OH ion absorption remains (wide bands at about 2100 and 3100 cm^{-1}): i.e., OH ions are

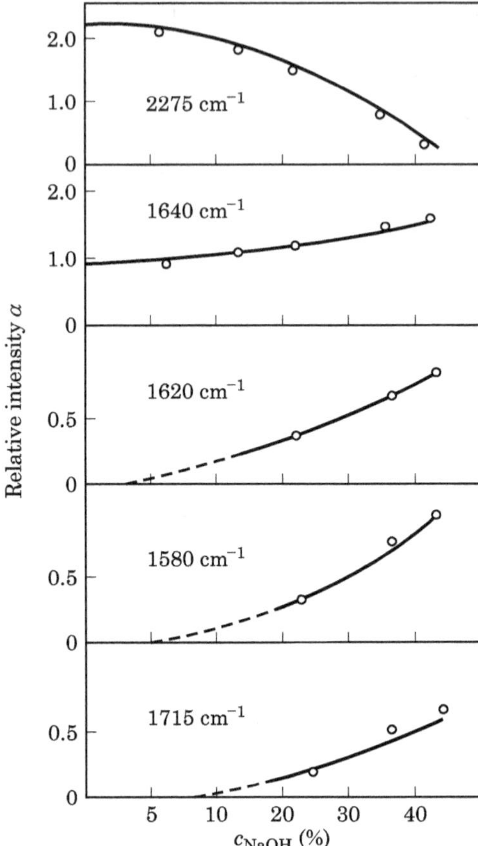

Figure 3.29 Relative intensity $\alpha = D_i/D_{cr}$ of absorption bands in IR spectra of the reaction products depending on NaOH concentration (%).

not consumed in isocyanate reactions with alkali but act as the catalysts of the isocyanate trimerization reaction together with sodium carbamates [67, 168].

Change in the ratio of NCO and OH groups leads to redistribution of the end products. When the NCO/OH ratio is 1:3 rather than 1:1, output of isocyanate trimer and amines increases, whereas the disubstituted urea production remains practically the same. It may be concluded that disubstituted urea formation is determined by availability of "free" water molecules, while alkali concentration and NCO/OH group ratio influence the reactions of isocyanate cyclotrimerization and amine and sodium carbamate formation significantly.

Studies of isocyanate interactions with alkalis of various concentration are of scientific and practical interest as they permit prediction of the chemical composition of the end products and hence the properties of compositions including isocyanates and sodium waterglasses with various values of silica modulus.

3.4.3 The role of MGF-9 oligoetheracrylate in the OMC formation process

It was noted earlier that introduction of MGF-9 oligoetheracrylate into organo-mineral compositions together with an ionic compound of sodium silicate creates the conditions for complexation. Flexible oxyethyleneglycol fragments of MGF-9 resin are, evidently, able to form "crown"-configured cavities and trap sodium cations from waterglass in them. The ability of MGF-9 oligoetheracrylate for complex formation is explained by the presence of negatively charged oxygen donor atoms in its chain structure. The magnitude of the negative charge is to a first approximation determined by the electronegativity of the donor atom. Oxygen is first in the electronegativity table and therefore complexing agents containing oxygen can form strong and stable complexes with the cations of alkali and rare-earth metals as well as with small molecules [105, 108]. A priori the possibility of electrostatic interaction of negative charges located on the oxygen donor atoms with positively charged cations in a macrocycle cavity may be supposed.

The most important characteristic of the cation from the point of view of complex formation possibility is its outer electronic structure. The sodium cation has an electron configuration permitting donor–acceptor interaction, as its unpaired electron is located in the 3*s* subshell. The principle of maximum correspondence of the macrocycle cavity to the metal ion [108, 109] should be observed for the formation of strong complexes of metal ions with macrocyclic ligands. Evidence that the requirements for complex formation were met, and that complexes are formed in OMC, was obtained from electric conductivity data [171].

The electrical conductivity of individual components (Table 3.31) and of OMC mixture (Table 3.32) was investigated. Conductivity measurement was performed on a series of waterglasses with various silica moduli prepared by introducing various quantities of NaOH and holding for one day to allow equilibrium to be reached [174].

In the most general case the waterglass conductivity is defined by mobility of three ion types, namely, alkali metal cations, silicic acid anions of various degrees of polycondensation, and hydroxyl ions formed by hydrolysis [172]. The contribution of hydroxyl ions to the conductivity can be neglected, as the degree of hydrolysis in sodium waterglasses is low. Silica anions in waterglass solutions undergo association, leading to considerable reduction of their mobility.

TABLE 3.31 Characteristics of Sodium Waterglasses

Silica modulus (±0.01)	Density (g/cm^3) (±0.01)	Water mass fraction (%)	Resistance (ohm) (±2.5)
3.00	1.48	53.9	2.95×10^4
2.44	1.47	57.6	2.90×10^4
2.15	1.46	56.7	2.75×10^4
1.87	1.46	57.1	2.75×10^4
1.62	1.44	58.6	2.60×10^4
1.42	1.41	48.9	2.50×10^4

Their role in the conductivity is also not large. The conductivity of waterglass is determined mainly by the motion of alkali cations and depends on the silicate anion structure and on the solution's viscosity.

It is known [128, 172] that three-dimensional polymerization of the silica backbone of sodium waterglasses gives way to layered polymerization when the silica modulus of the waterglass is near 2.5, and with increase of alkali metal cation content it proceeds by chain growth. Chain silica anions of low-modulus waterglass, due to their dense interlacing, do not leave the free space required for cation motion. Anions in high-modulus glasses are branched and form three-dimensional networks through the "holes" of which alkali metal cations can pass. This is the case for concentrated waterglasses.

Silica anions of low-modulus glasses form chain fragments at medium concentrations of the waterglass. Electrical conductivity measured in these glasses is higher than in high-modulus glasses (Table 3.31).

MGF-9 complexing agent at 10 vol% per 45 vol% of the waterglass was introduced into waterglasses with various silica moduli. The resistance of thoroughly mixed waterglass–MGF-9 mixtures (with 0.5 acid number) was measured during 10 days. Measurements showed that the resistance did not change during the first 8 h. At one day and within a further 10 days increase in the resistance was observed, demonstrating reduction of conductivity (Table 3.32). At that point viscosity increase became noticeable and resistance measurements were terminated.

The reduction of electrical conductivity with time can serve as one line of evidence for complexation, as the highly mobile Na^+ ion of the waterglass is localized by MGF-9 oxyethyleneglycol fragments that create pseudo-crown complexes. The same pattern is observed with introduction of dibenzo-18-crown-6 crown-ester into waterglasses with various silica moduli.

The initial stage of the complex formation process is extended in time. This is caused by slow destruction of the hydration spheres

TABLE 3.32 Variations in Resistance of Mixtures of Waterglasses with Different Silica Modulus and MGF-9 over Time[a]

Silica modulus	Resistance (ohm) (±2.5%) Time (days)							
	0	1	2	4	5	6	8	10
3.0	2.90×10^4	3.4×10^4	4.2×10^4	8.1×10^4	2.1×10^5	2.8×10^6	4.9×10^6	7.3×10^6
2.44	2.90×10^4	3.4×10^4	3.9×10^4	7.8×10^4	2.4×10^5	0.8×10^6	2.7×10^6	7.1×10^6
2.15	2.75×10^4	2.95×10^4	3.4×10^4	7.1×10^4	8.7×10^5	1.0×10^6	1.7×10^6	2.9×10^6
1.87	2.75×10^4	2.95×10^4	3.6×10^4	6.9×10^4	1.1×10^5	0.5×10^6	1.1×10^6	3.1×10^6
1.62	2.60×10^4	3.05×10^4	3.2×10^4	8.1×10^4	0.9×10^5	0.7×10^6	2.7×10^6	3.9×10^6
1.42	2.50×10^4	4.05×10^4	4.2×10^4	9.0×10^4	0.3×10^6	0.9×10^6	2.0×10^6	4.3×10^6

[a] Investigated mixture composition was waterglass 45 vol%, MGF-9 10 vol%.

around sodium ions in aqueous solutions and prolonged approach to equilibrium [109]. Triethanolamine, methanol, and other additives are often used to speed up the destruction of the solvation spheres. However, in our case these additives proved to be inapplicable as they cause structuring of the glass.

Sodium cations binding into crown-complexes with the help of MGF-9 ethylene glycol fragments should result in increase of the silica modulus [149]. In practice, the silica modulus of the waterglass–MGF-9 mixture increases noticeably after mixing of the reagents, and thereafter grows insignificantly (Table 3.33). The increase in modulus is further evidence of complex formation.

The efficiency of catalytic systems based on alkali metals and polyoxyethylene glycols depends on the molecular weight of the latter. The dependence of salt binding constant on polymer molecular weight has been discussed in a number of works. For example, the equilibrium constant increased 20 times [109] at the transition from polyethylene glycol with polymerization degree of 300 to the polymer with 2000 links. The explanation of this phenomenon was found in statistical factors. The electronic influence of the end groups in long chains does not play a considerable role in the binding process. Usually the binding constant first increases, and then for MW = 10 000–20 000 remains practically constant. It was noted that polyethylene glycol of MW = 200 can not retain K^+ ions.

Let us consider the influence of catalytic activity of ethylene glycol homologues on the rate of organo-mineral composite formation. Polyoxyethylene glycols (POEG) were taken in equal mole content with the MGF-9 content in actual OMC, where the waterglass/polyisocyanate/MGF-9 ratio is 9:9:2.

As Table 3.34 shows, the gel formation time and composite curing time are reduced with increase of oxyethylene glycol chain length. The first homologues of the POEG series do not possess enough chain flexibility for enfolding of Na^+ cations and complex formation, and therefore do not exert catalytic influence on the system.

The results obtained are well correlated with the data from an investigation [50] of the catalytic activity of the glycol dimethyl ether homo-

TABLE 3.33 Variations of Waterglass–MGF-9 Mixture Silica Modulus with Time

MGF-9/ waterglass ratio	Initial silica modulus	Modulus after mixing at given time				
		0 h	1 h	3 h	6 h	24 h
1:9	1.69	2.53	2.59	2.64	2.75	2.75
1:9	2.85	3.06	3.06	3.07	3.12	3.44
2:9	2.85	2.90	3.04	3.25	3.25	3.45

TABLE 3.34 Characteristics of OMC Modifications Cured with POEG of Various Molecular Weight

OMC modification	Gel formation time (min)	Curing time (h)
OMC	520	32
OMC with EG	480	27
OMC with DEG	120	14.3
OMC with TEG	22	11.5
OMC with POEG, MW = 400	15	3.3
OMC with MGF-9	25	10.3

logues with potassium acetate in cyclotrimerization reactions. These authors showed that lower homologues of glycol dimethyl ethers influence potassium acetate catalytic activity less in the investigated conditions than do their higher molecular analogue oligooxyethylene. The efficiency of glycol dimethyl ethers increases with transition from ethylene glycol dimethyl ether to di- and tetraethylene glycol dimethyl ether. These conclusions were reached on the basis of calorimetric titration.

The authors of [175] established that high flexibility of POEG macromolecules also provides more strength of the complexes in comparison with polypropylene glycol. In 96% ethyl alcohol solution the alkali metal cations each coordinate almost six elementary macromolecules.

Complex formation in aqueous solutions of alkaline salts is extended in time due to slow breaking of hydration spheres and prolonged approach to equilibrium [109]. Results in Table 3.35 show that with increase of waterglass/MGF-9 component hold-up time, the compression strength of the cured composite grows monotonically. This is probably explained by breaking of sodium cation hydration spheres and attainment of equilibrium in the period of preliminary hold-up, which further promotes increase in strength and activity of the complexes formed with the participation of sodium cations and MGF-9 resin oxyethylene glycol fragments.

TABLE 3.35 Dependence of OMC Strength Characteristics on Waterglass–MGF-9 Component Hold-up Time

	Waterglass–MGF-9 component hold-up time (h)				
	0	8	24	72	240
Compression strength at 3 days (MPa)	17.3	19.8	21.0	24.3	29.8
Compression strength at 10 days (MPa)	25.1	27.3	29.9	39.8	52.4

The important factor in heterophase systems is the intensity of mixing of the ingredients [106]. Time and intensity of mixing were not varied in these investigations. The systems investigated were mixed using a mechanical mixer for 3 min at 600 rev/min.

All the results of the organo-mineral composition investigations correlated well with published data, giving sufficiently conclusive evidence of complex proceeding by "crown" type processes.

Complexes of macrocyclic esters and their open-chain analogues with metal ions possess unique properties, and interest in these compounds is continuously growing [103, 109]. The catalytic action of crown-ethers and their analogues as catalysts of phase transfer consists in activation of reagents due to the increase of ion pair dissociation with the formation of nonsolvated reactive anions owing to the transformation of contact ion pairs into solvation-separated ones.

It was considered previously that the most effective phase transfer catalysts are quaternary ammonium bases. However, preliminary experiments with crown-ethers had already shown that these compounds are more powerful phase transfer catalysts than quaternary ammonium bases and are more selective [106, 177]. This is explained by differences in the mechanism of catalytic action. The mechanisms of reaction acceleration in two-phase systems with crown-ethers are as yet little studied, but simple examination of salt extraction with crown-ethers shows that the salt in the aqueous phase (both anion and cation) passes into the organic layer, whereas only anions paired with the "onium" cation pass from the aqueous into the organic phase during extraction with "onium" salts. This considerable difference in the mechanism of action of the two groups of ion-carrying catalysts is the basis for the prospective use of crown-ethers and their analogs instead of quaternary ammonium bases in many fields.

According to the earlier discussion, in the organo-mineral compositions under investigation with MGF-9 interphase transfer catalyst, anions of silicic acid of various degrees of polycondensation, hydroxyl anions, and sodium cations of waterglass will pass from the aqueous phase into the organic as a result of electrostatic interaction with electron-donating oxygen atoms of the MGF-9 resin. Therefore, introduction of MGF-9 oligoester leads to superposition of OMC organic and inorganic components. Due to the action of ion-carrying catalysts, chemical transformations taking phase in the formation of the composition are transferred into the organic phase, with possible reaction also at the phase boundary. Many cases are known [106, 107, 177] when addition of polyether causes dissolution of salts and related compounds in solvents in which they are normally practically insoluble. For example, addition of crown-6 leads to dissolution of such crystalline substances as $KMnO_4$, tert-C_4H_9OK, or K_2PdCl_4 in aromatic hydrocarbons. Crown-ethers and their open-chain analogues are effective catalysts of isocyanate trimerization [103, 114, 168].

Among the anions formed in aqueous alkaline solutions as a result of complexation, the hydroxyl ion is of particular interest. In [47, 67] hydroxyl ion is indicated by the authors as the catalyst of linear homopolymerization and cyclic trimerization of isocyanate. Iwakura et al. showed that output of isocyanate linear polymer grows with the increasing alkaline nature of the aqueous salt solution [312].

Investigations of isocyanate interaction with aqueous NaOH solutions [178] conducted by us established that triisocyanurates are formed with increase of sodium hydroxide concentration above 20% and the output of isocyanate cyclic trimer grows monotonically with the increase of sodium hydroxide concentration above 20%. There are three ion types in the aqueous alkali solutions, namely, Na^+, H_3O^+, and OH^-, but it is reasonable to regard hydroxyl ion as the initiating species in isocyanate polymerization. Hydroxyl ion in the aqueous alkali solutions is enclosed in a hydration sphere, but despite this it displays the initiating action in isocyanate addition polymerization reactions.

The result of MGF-9 pseudo-crown complexation with waterglass sodium cations is a desolvated active hydroxyl ion whose initiating ability is considerably higher in the organo-mineral compositions investigated. Participation of phthalic, methacrylic, and carbamic acids, and silicon-oxygen anions forms “loose” ion pairs with sodium cations whose activity in the cyclotrimerization process is more than twofold higher [50, 94, 103].

3.4.4 Influence of hydroxyl anion on the processes occurring in the inorganic component during OMC formation

Hydroxyl anion action is not limited by the organic component of the composite investigated. According to many authors [128, 132, 179], hydroxyl ion is the catalyst of the waterglass polycondensation process. The essence of hydroxyl ion catalytic action is reduced to proton removal from nonionized molecules of silicic acid. Further polycondensation process can be considered as the result of nucleophilic substitution by ion formed from the hydroxyl group in nonionized molecules of silicic acid. The removed hydroxyl ion, being the catalyst of the waterglass polycondensation reaction, should again neutralize one proton of orthosilicic acid, producing silicate ion, and so on. The presented mechanism of silicic acid polycondensation is based on the ability of silicon compounds to undergo complexation [132].

$$\dashrightarrow (HO)_3{\equiv}SiO^- + SiOH \dashrightarrow [(OH)_3SiO....\overset{(OH)_3}{\overset{|||}{Si}}\quad OH]^- \dashrightarrow$$

$$\dashrightarrow [(OH)_3{\equiv}\, Si - O - \overset{(OH)_3}{\overset{|||}{Si}} - OH]^- \dashrightarrow$$

$$\dashrightarrow [(OH)_3{\equiv}\, Si - O - \overset{(OH)_3}{\overset{|||}{Si}}....OH]^- \dashrightarrow$$

$$\dashrightarrow (OH)_3{\equiv} Si - O - Si \equiv (OH)_3 + OH^-$$

In accordance with the existing notions about silicic acid polycondensation process based on numerous experimental studies, there is seen a clear dependence of influence of the medium alkalinity (silica modulus) on the polymer-homologue composition of silicate anions. This influence of waterglass silica modulus on polymer-homologue composition may be explained on the assumption of the existence of two forms of silicic acid hydrates [136]: $Si(OH)_4$ and $SiO(OH)_2$. For the sodium waterglasses stable in alkali media it is more reasonable to assume a hydrate $SiO(OH)_2$ existing in the form of the corresponding sodium salt, HO—SiOONa. This structure of monomer alkali silicate explains the availability of only the following silicic acid monomers and dimers in the waterglasses with modulus $M < 2$:

```
       O              O           O
       ||             ||          ||
       Si — OH        Si — O — Si
      /              /            \
  NaO            NaO               ONa
```

Dimer formation results in disappearance of OH^- functional groups, which prevents silicate dimer enlargement. This structure of low-modulus waterglasses is confirmed by numerous experimental data [128, 130, 131, 172].

Free metasilicic acid is contained in waterglasses with $M > 2$ and thus formation of linear polymers becomes possible:

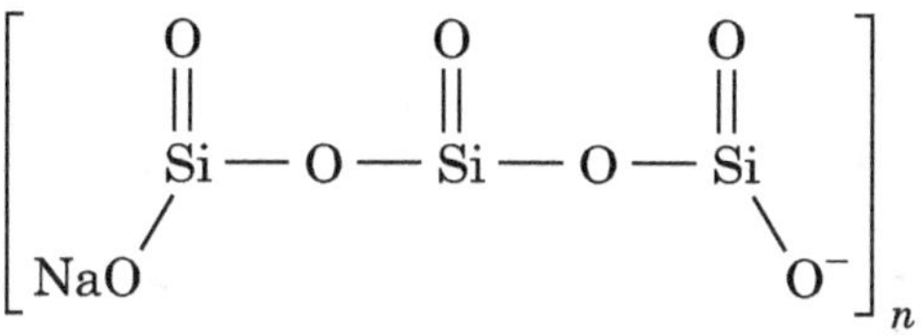

$$\left[\begin{array}{ccccc} & O & & O & & O & \\ & \| & & \| & & \| & \\ & Si & -O- & Si & -O- & Si & \\ / & & & & & & \backslash \\ NaO & & & & & & O^- \end{array}\right]_n$$

A branched structure [128, 172] is supposed for sodium silicates characterized by silica modulus equal to 3 and more:

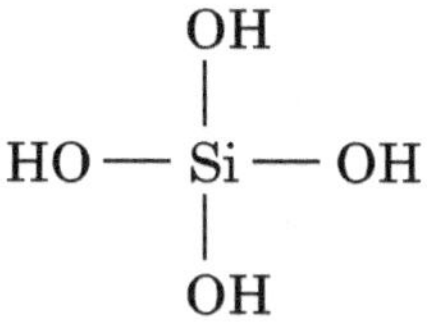

$$\begin{array}{ccc} & OH & \\ & | & \\ HO- & Si & -OH \\ & | & \\ & OH & \end{array}$$

This hydrate, being tetrafunctional, will form spatial polymers as a result of polycondensation.

It thus may be concluded that introduction of sodium waterglass with various silica moduli into organo-mineral compositions leads to multiple influences on the composite formation process. First, change of the medium alkalinity alters the ratios of disubstituted ureas, amines, triisocyanurates, sodium carbamates, and carbonates in the cured OMC end products [178]. These changes are probably closely connected with catalysis of interactions occurring in the OMC organic component, with pseudo-crown complexes, and with the formed sodium cations and oxyalkylene fragments of the MGF-9 resin. Second, the composition of silicon-oxygen anions depends on waterglass silica modulus.

Composite structure investigations by optical microscopy have shown that increase in waterglass silica modulus leads to spherical particles enlarging due to the larger size of silicon-oxygen anions.

Considerable changes in OMC organic and inorganic components, depending on the waterglass silica modulus, affect the physical, mechanical, chemical, and technological characteristics of the composites investigated.

3.4.5 Strength characteristics of organo-mineral composites

Chemical studies of the cured organo-mineral composition end products by IR spectroscopy showed that the ratio of triisocyanurates, disubstituted ureas, amines, sodium carbamates, carbonates, phthalates, and methacrylates, as well as the composition of the silicic acid polymer-homologues, changes with the variation of the initial characteristics of the waterglass (silica modulus). It is important to trace how those changes influence the compression strength of the composite and its deformability indices.

Compression strength tests of the organo-mineral composition were performed on cylindrical specimens of 16 mm diameter and 40 mm height, prepared at 295 K. The results showed that the nature of the strength dependence on the waterglass silicate modulus changes with aging (Fig. 3.30). At 1 day, for compositions curing with silica modulus increasing from 1.35 to 2.35, the compression strength (σ_p) of the specimens grows twofold, and is then sharply reduced. This dependence is also exhibited with specimen hold-up in air for up to 24 days. With more extended aging of the composites and with increase of waterglass silica modulus, the rate of achievement of optimal strength under compression is significantly reduced.

In the aging process of cured compositions, their compression strength noticeably increases, more so for compositions with lower waterglass silica modulus. For example, σ_p of cured compositions with $M = 1.35–2.35$ after 15 days of aging increases 4–5 times, whereas the compositions with $M = 2.35–3.0$ show only 1.8–2 times growth of this index. The compression strength is reduced to a minimum with introduction of waterglass with $M = 3.2$.

The influence of the waterglass silica modulus on physical and mechanical properties of organo-mineral composites is explained mainly by change in the composition of the products of reaction of

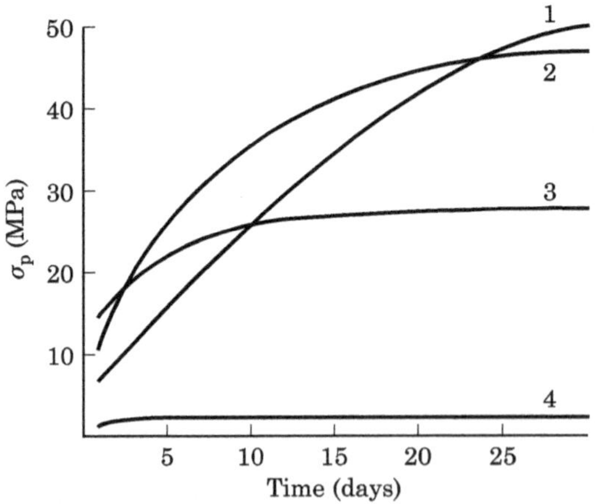

Figure 3.30 Compression strength (σ_p) versus time for OMC containing waterglass with various values of silica modulus M: (1) $M = 1.35$; (2) $M = 1.64$; (3) $M = 2.35$; (4) $M = 3.2$.

the polymer organic component. Silica modulus increase leads to reduction of sodium cation concentration. In the composition investigated, sodium ions participate in several processes:

- Complex formation of "crown" type with oxyethylene fragments of MGF-9 resin as the catalysts of ion phase transfer and isocyanate cyclotrimerization.
- The formation of sodium carbonate, a structure-loosening agent [182].
- The formation of sodium mono- and disilicates in low-modulus glasses and of linear polymers in waterglasses of average concentration.

Change of sodium ion content will cause variation of contribution of the above factors in OMC formation. Thus, composite materials including low-modulus glasses are distinguished by increased Na^+ concentration as compared with composites with high-modulus glasses and, consequently, interphase catalysis, increased content of triisocyanurates, formation of little sodium carbonate amount, and primary formation of polymer-homologue silicic acid of low degree of polymerization ($n = 1$–2). Consequently, composites with low-modulus glasses have high physical and mechanical indices. Low strength properties in the material with waterglass having $M \geq 3.2$ (Fig. 3.30, curve 4) are apparently connected with the fact that the contribution of interphase catalysis and triisocyanurates is less, due to the low concentration of sodium cations. As a result, a material with a poorly ordered, defective structure that does not provide strength is obtained.

Compression strength increase in compositions with low-modulus waterglass is observed over 30 days. The rate of achieving maximum strength increases with silica modulus, but absolute values of this index are reduced considerably. For example, the maximum strength of cured composites with $M = 3.2$ is achieved in 2 days but is only 1 MPa, whereas for the composites with $M = 1.35$ the maximum strength is reached over 30 days but reaches 49 MPa.

This character of specimens' σ_p dependence on time can be connected with prolonged chemical or physical structuring or with the purely physical process of moisture loss by the specimens. The kinetics of moisture loss by the specimens under hold-up in air in normal conditions was investigated to test the last assumption. Moisture loss was assessed by the change over time in specimen mass relative to initial specimen mass. It was established that the rate of moisture loss during hold-up in the air gradually reduces, but even at 30 days it remains important and the maximum moisture loss by specimens was 11.9%.

The rates of change of moisture loss kinetic curves and those of compression strength against time differ noticeably for compositions with the same silica modulus over the same duration of aging, despite their qualitative correspondence. In addition, loss of moisture producing a plasticizing action in the cured composition should reduce the deformability (ε_p) of specimens in the aging process, but experiment indicated that both σ_p and ε_p increase with time. Consequently, moisture loss from the cured specimens is not the main reason for their strength change with time.

Along with the analysis of the kinetics of moisture loss from OMC specimens, the kinetics of NCO groups consumption was also investigated by IR spectroscopy. The results indicated that consumption of polyisocyanate NCO groups occurs monotonically during 30 days, as shown by reduction of the 2275 cm^{-1} band intensity. Thus, composite chemical structuring is extended in time and continues when the material is in the solid state. Parallel running of the physical structuring process is possible.

Triisocyanurates were detected in the composition of cured OMC products by IR spectroscopy [104, 178]. The isocyanate cyclotrimerization reaction is often used for obtaining branched network structures [50, 92, 93] possessing unique properties. Sodium waterglasses are also able to form inorganic polymer network materials, according to numerous publications [128, 130, 131]. This is an obvious precondition for obtaining interpenetrating networks (IPN) or semi-IPN in the systems under investigation. Detailed studies are needed to identify the opportunities for forming inorganic networks in OMC.

Gel-sol analysis was used to obtain further structure information. Cured OMC specimens were studied in toluene and dimethyl formamide. For washout, five samples were loaded in parallel into Soxhlet apparatus. An average 50% washout was obtained for OMC end products, the remainder appearing to be powderlike toluene and dimethyl formamide insoluble substances.

The theoretical and practical significance of polymer dynamic/mechanical properties is well appreciated [180, 183]. The dynamic elastic modulus and mechanical losses are the most sensitive indicators for all forms of polymer molecular mobility, especially for the glasslike state. The dynamic elastic modulus is the primary index for polymer deformation properties. Apart from their purely theoretical interest, understanding of the mechanisms of molecular motion within polymers and of mechanical losses may have significant practical importance in properly defining other polymer mechanical properties.

Dynamic mechanical spectroscopy using a device described in [157] at 100 Hz frequency and over a 243–533 K temperature interval was used to investigate the viscous-elastic behavior of OMC. Cured composite blocks obtained from industrial sodium waterglass, D-grade polyisocyanate, and MGF-9 oligoether acrylate by serial input of components and thorough mixing were used to prepare specimens for further study. Waterglass featuring a variety of silica modulus (1.42 to 3.0) was used to make specimens.

Elastic modulus temperature dependences obtained for all specimens thus studied (see Fig. 3.31) have the typical appearance of polymer composites based on linear binders. A sharp fall of elastic modulus for all composites studied above the glass-transition temperature and the absence of a viscoelasticity plateau indicate that the composite polymer binder is not a network polymer.

From the OMC viscous-elastic behavior and gel-sol analyses, it is clear that no continuous network is formed in systems thus investigated. Apparently, the OMC dispersion medium studied here constitutes the organic component interaction products, displaying the basic

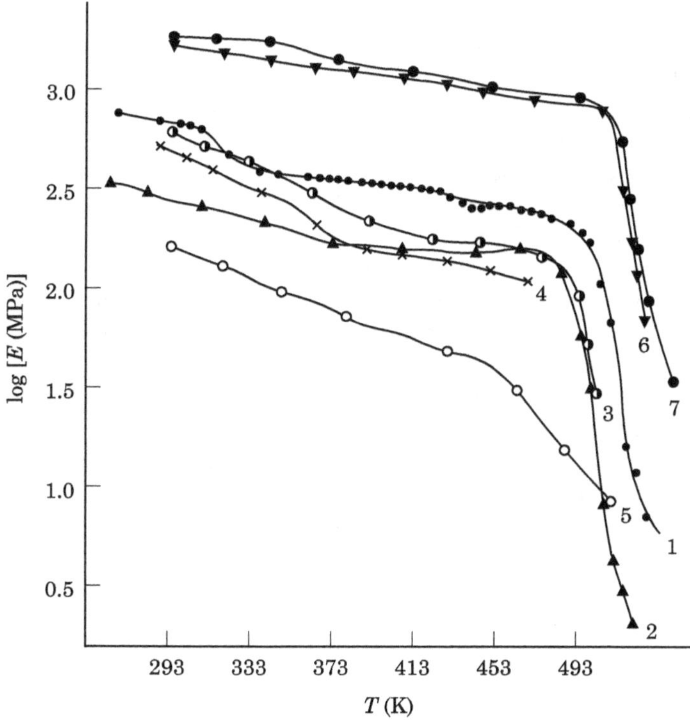

Figure 3.31 Elastic modulus ($\log E$) temperature dependence on OMC structure (% of vol.): (1–3) WG-45 of $M = 3.0$, 2.15, and 1.42, respectively, and MGF-9; (4, 5) WG-45 of $M = 3.0$ and 1.42, respectively, without MGF-9; (6, 7) WG-30 and -20 of $M = 2.15$ and MGF-9.

polymer index of glass transition, and the OMC inorganic polycondensation products represent the dispersed phase.

The temperature dependence of elastic modulus for curves 7, 6, and 2 in Fig. 3.31 indicates that for 20% and 30% contents of waterglass the elasticity moduli are practically the same, and a rapid drop of elasticity module for 45 vol% of waterglass is observed, which is probably explained by polymer matrix transition to the boundary layer state. Enhanced elastic modulus might be expected for a composite containing 10 vol% of waterglass, but a fragile material displaying 0.2 MPa compression strength was obtained in this case, and specimens could not be made for OMC viscous-elastic studies. When the waterglass content is low, sharp modifications of the polymer matrix occur that are apparently connected with insufficient formation of crown-complexes by waterglass sodium cations and the MGF-9 oxyethyleneglycol fragments that are responsible for isocyanate addition polymerization catalysis as well as being the interphase catalyst. The main interest of the systems studied here lies in the fact that no phase inversion is observed even for 60–80 vol% of waterglass content and bulk material is formed as from large waterglass contents. The shift of the glass-transition temperature of the polymer component to lower temperatures (523–507 K) is noted when sodium silicate (Fig. 3.31, curves 7, 6, and 2) is increased, again due to polymer transition into the boundary layer state.

The influence of waterglass silica modulus on the viscous-elastic behavior of the composite has to be considered. Although there is no firm consensus concerning the nature of sodium waterglass, a number of researchers agree that its structure and properties depend upon the silica modulus. We earlier described the influence of silica modulus on the polymer matrix content of the composites studied. Silica modulus modification leads to correlative changes in the end products of cured OMC triisocyanurates, amines, disubstituted ureas, sodium carbonate, phthalate, carbamate, and methacrylate. The physical and chemical properties of composites with dispersed fillers are defined by the dispersion medium [183]. Therefore, waterglass silica modulus may be expected to influence all of the OMC viscous-elastic properties as well.

Figure 3.31 shows temperature dependences of viscous-elastic functions of composites that include 45 vol% waterglass with various silica modulus. Comparing curves 4 and 5 ($M = 3$ and $M = 1.42$), note that increase of silica modulus leads to increase in the elastic modulus of the composite.

More complicated OMC viscous-elastic behavior may be observed when MGF-9 modifier is introduced (see Fig. 3.31, curves 1–3). First, increase of elastic modulus is seen for all composites regardless of silica modulus when MGF-9 modifying additive is introduced; at $M = 1.42$ the increase is most marked (see curves 5 and 3), and is less apparent at $M = 3.0$ (see curves 4 and 1).

The decisive influence appears to be due to crown-complexes formed by waterglass sodium cations and MGF-9 fragments. The catalytic activity of this system in isocyanate cyclotrimerization reactions for interphase catalysis conditions is determined by the concentrations of sodium cations and MGF-9. For a given MGF-9 content, sodium ion concentration is higher in compositions comprising low-modulus glasses than in those with high modulus, and the influence of crown-complexes is thus more apparent. In composites featuring low-modulus glasses a shift occurs toward greater formation of isocyanurate structures, with no carbon dioxide liberation. As a result, sodium carbonate, which acts as a loosening agent of the composite permolecular structure [182], is present in significantly lower quantities and can only be formed from CO_2 liberated by reaction of polyisocyanate isocyanate groups with waterglass water.

These are not the only limiting influences exerted by the silica modulus on the viscous-elastic properties of OMC. An influence on the structure of the polycondensation products has also been established [128–132]. Thus, silica modulus induces significant changes of the morphology and structure of both composite material phases, and the viscous-elastic behavior observed supports this.

When mechanical loss temperature dependences are considered (see Fig. 3.32) variable levels of "background" loss in the composite glass state (273–453 K) are noted. High background loss ($\tan\delta = 0.1$–0.12) in specimens 4 and 5 is apparently caused by absence of MGF-9 modifier, and demonstrates the increased molecular mobility in these structures.

Studies of the viscous-elastic behavior of OMC with various waterglass silica moduli and MGF-9 contents have revealed the considerable influence of these factors on composite viscous-elastic properties. Evidently, study of the viscous-elastic properties of OMC is an effective method for prediction of physical, mechanical, and operating characteristics of the composites investigated.

Composite structure was studied using an MBS-9 optical microscope and an EM-100C electron microscope. Specimens for morphological investigation were prepared from cured composite blocks and initial components by thorough grinding and polishing of their surfaces. Specimens for electron-microscopic investigation were obtained by a metal–carbon replica method. Specimens for study had fixed composition of organic component (45 vol% polyisocyanate + 10 vol% MGF-9 oligoetheracrylate + 0.02 vol% UP 606/2), and the inorganic content (waterglass with $M = 2.15$) was varied from 0 to 60 vol%.

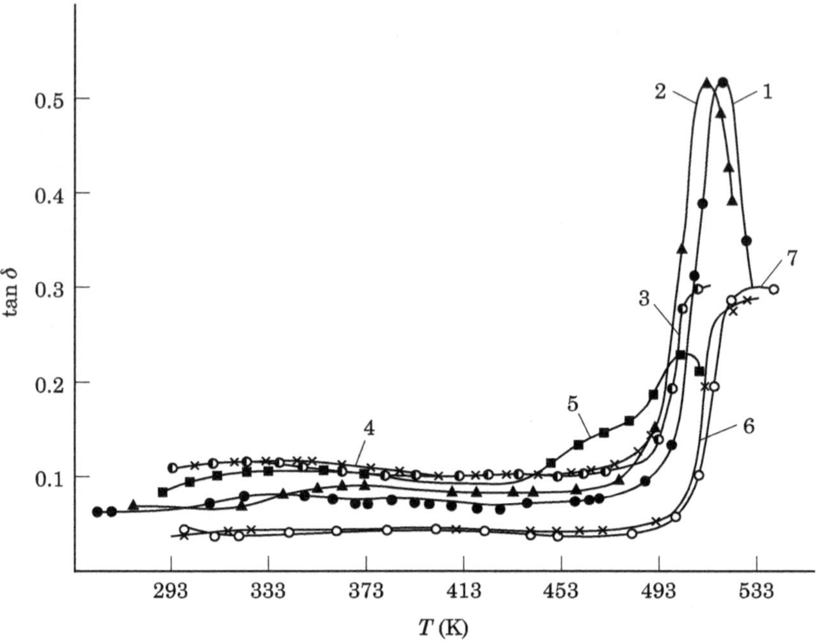

Figure 3.32 Mechanical losses (tan δ) temperature dependence on OMC structure (% of vol.): (1–3) WG-45 of $M = 3.0$, 2.15, and 1.42, respectively, and MGF-9. (4, 5) WG-45 of $M = 3.0$ and 1.42, respectively, without MGF-9; (6, 7) WG-30 and -20 of $M = 2.15$ and MGF-9

Optical microscopy showed that the composite material formed is a heterophase system consisting of a continuous polymer matrix and a dispersed phase comprising spherical particles. Increasing the waterglass content resulted in increased size of the spherical particles. Polycondensation products of the glass form the dispersed phase. The fact that no phase inversion is observed up to a waterglass content of 60 vol% is of particular interest.

The omission of MGF-9 leads to a sharp change in the composite structure: no spherical particles are seen. The composite structure becomes heterogeneous, with structural elements of irregular shape and a wide scatter of sizes (10–170 μm).

The dispersion medium in the composites studied includes triisocyanurates, disubstituted ureas, and amines observed in the cured OMC end products by IR spectroscopy [104, 178]. Sodium carbonate, carbamate, phthalate, and methacrylate also found in the end products probably form intermediate layers that cannot be detected by optical microscopy.

The volume of a spherical particle is related to its surface area by

$$V = \tfrac{1}{3}RS \tag{3.29}$$

where V = volume (cm^3), R = particle radius (cm), and S = surface area (cm^2). For compositions containing n_1 particles with concentration C_1 and n_2 particles with concentration C_2, the following ratios hold:

$$\frac{C_1 V}{n_1} = \tfrac{1}{3} R_1 S_1 \tag{3.30}$$

$$\frac{C_2 V}{n_2} = \tfrac{1}{3} R_2 S_2 \tag{3.31}$$

where V = composition volume (cm^3), C_1, C_2 = waterglass volumetric fractions, and n_1, n_2 = numbers of particles in the composition.

Dividing Equation (3.30) by Equation (3.31) we obtain

$$\frac{C_1}{C_2}\frac{n_1}{n_2} = \frac{R_1}{R_2}\frac{S_1}{S_2} \tag{3.32}$$

From the experimental data,

$$\frac{C_1}{C_2} = \frac{R_1}{R_2} \tag{3.33}$$

so that

$$\frac{n_2}{n_1} = \frac{S_1}{S_2}$$

or

$$n_1 S_1 = n_2 S_2 \tag{3.34}$$

The product $W = nS$ is the magnitude of the interphase surface.

Thus, from the above and the fact that the dispersed phase particle size depends linearly on the waterglass content, we conclude that the interphase boundary area is constant and does not depend on the volumetric content of sodium waterglass in the composite.

Optical microscopy was used to study the change of composite structure with silica modulus. It was found that increase of the silica modulus leads to structural changes similar to those caused by increase of the content of waterglass, namely, enlargment of the spherical particles of waterglass polycondensation products. Composites with low-modulus glasses ($M = 1.42$) are distinguished by 8–20 μm spherical structural elements. With $M = 2.6$, a size of 20–40 μm is typical for the structural elements. Spherical particles of 40–100 μm are observed in composites containing high-modulus glasses. The change in the particle size is connected, among other factors, with the structure of the silicion-oxygen anions, which depends on the silica modulus.

Optical microscopy allows assessment of dispersed phase morphology but provides little information about the dispersion medium. In contrast, electron microscopy gives more comprehensive information on the morphology of the continuous phase.

Electron micrographs show that the matrix is of globular structure, containing mainly structural elements of two specific sizes, which depend monotonically on the waterglass content in the composite. At 15 vol% sodium silicate content, the structural elements are 0.4 and 0.18 μm; at 25 vol% the corresponding values are 0.1 and 0.04 μm. One particular size, namely, 0.07 μm, becomes typical at less than 35 vol% waterglass. Large structural formations in specimens with 15 vol% waterglass make up approximately one-third of the total volume, and large particles make up to 50% of the volume in specimens with 25 vol% waterglass. Thus change of waterglass content in the composites results in changes in the structural elements of both dispersed phase and dispersion medium.

Comparing specimens differing in chemical composition, the structure of the composite containing polyisocyanate, sodium waterglass, MGF-9 interphase catalyst, and UP 606/2 polymerization catalyst is identical to that of the composite consisting of polyisocyanate and UP 606/2 only. Similar distribution and size of the structural elements are observed.

It is known that compounds with mobile oxyalkylene chains similar to those available in MGF-9 are able to form crown-complexes with the alkali metal cations in waterglass and to act as interphase catalysts in heterophase systems [103–109]. According to the supposed mechanism of their action [106], they act at the phase boundary. Therefore, in the systems investigated, MGF-9 enters the waterglass–polyisocyanate boundary, leading to the formation of waterglass particles of spherical shape and to the independence of the phase boundary area on the waterglass content in the composite. It can be assumed that MGF-9 forms gel-like films of dual action at the waterglass boundary: these films prevent phase inversion and also localize dispersed particles of waterglass polycondensation products, preventing their enlargement.

Chapter

4

Internal Stresses in Adhesive-Bonded Joints and Ways of Decreasing Them

4.1 Effect of Internal Stresses on Properties of Adhesive-Bonded Joints

The internal stresses in adhesive-bonded joints arise for two reasons. In the course of setting of the adhesive, its volume decreases due to volatilization of solvents, polymerization or physical structurization. As a result of the adhesion interaction of the adhesive and the substrate, the film can contract only in thickness, which is why stresses that appear in it are parallel to the surface. The film extends while contraction stresses appear in the substrates. Rapid growth of stresses, which tend to reduce the length of the film, begins from the moment the polymer loses yield.

The second component of the internal stresses is thermal stress caused by differences of the coefficients of linear thermal expansion of the adhesive and the substrate. They appear in the course of heating or cooling of the adhesive-bonded joint. The mechanism of the internal stresses occurring in adhesive-bonded joints does not generally differ from that in coatings, but because there are two solid surfaces the magnitude of the stresses in the first case appears to be substantially greater.

In some cases the adhesive film cannot contract in thickness during curing—for example, in the case of "tube-in tube" adhesive-bonded joints. Then the tangential internal stresses are added to the normal ones. The total stresses in this case increase substantially. The occurrence of internal stresses during formation of the polymeric film on the solid surface results in decrease of the adhesion strength because the stresses are directed oppositely to the forces of the adhesion bond and tend to pull the film off the substrate. The stronger the bond of the polymer with the surface, the greater is the effect of the internal stresses because the strong interaction of the polymer chains with the substrate prevents the relaxation processes on the interphase boundary. In this connection there is a fairly clear dependence of the internal stresses on the adhesion strength [158–160]. Hence, because the internal stresses act oppositely to the forces of adhesion, this dependence is of complex character as the increase of the internal stresses is accompanied by decrease of the adhesion strength.

The internal stresses can be equated to a long-period acting load [161]. They can cause spontaneous peel-off of the polymer and fracture of the adhesive-bonded joint [160–163], even at 15–50% of the instantaneous breaking stress. In this case there is a linear dependence in semi-log coordinates between the lifetime of the joint and the internal stresses [164]. An analytical dependence has been proposed that relates the strength of the adhesive-bonded joint to the internal stresses in the adhesive layer [160].

The value of the internal stresses in the adhesive does not depend significantly on the layer thickness [160]; with the increase of layer thickness the tangential stresses grow linearly [160, 165]. But at a certain film thickness the internal stresses begin to decrease and reach a minimum value. This decrease was found to be induced by the disturbed contact between the film and the substrate.

The internal stresses in adhesive-bonded joints are not distributed uniformly, being concentrated on the edges and especially at corners. The edge effect is more noticeable the higher the ratio of the length of bonding adhesive in contact with air to the contact surface. The absence of proportionality between the failure stress and the area of the adhesive-bonded joint is explained to a great extent by the influence of edge effects upon the adhesion strength. Stresses can concentrate on the solid surface [166, 167]. The stresses have been shown [168] to reach 30 MPa on the boundary between fiber and phenol–formaldehyde resin. Such overloads cause failure of the binder, occurring especially at low temperatures and under cyclic loads and heat and mechanical impacts. In this case the crack grows from the boundary of the polymer contact with the solid surface.

The presence of stresses in the adhesive layer alters the polymer's physical and chemical properties. The glass-transition temperature depends on the internal stresses that appear in the polymer. For example, the glass-transition temperature is known to decrease with increase in internal stresses in films based on polystyrene [169] and other polymers. Note that restriction of the macromolecular mobility due to adsorption interaction with the solid surface increases the glass-transition temperature to greater extent than it is decreased by the internal stresses [170]. This is why in practice one observes increase of the glass-transition temperature of filled polymers.

Effects due to changes in the internal stresses in the binder seem to explain the changes in the temperatures of softening and glass-transition with various quantities of binder in reinforced plastics [171, 172]. This was confirmed by studies of internal stresses in specimens which use nonwoven polyamide fiber fabric cemented with polystyrene [171].

Increase of internal stresses in the polymer causes increased vapor permeability of the polymer and corrosion failure [173], and the formation of cracks that accelerate oxidation; the abrasion resistance decreases [174], and the thermal and physical properties change.

This brief review of the literature shows that the effect of internal stresses on the properties of adhesive-bonded joints is considerable. The creation of adhesives with predetermined properties, providing high strength and joint stability, requires that the level of the internal stresses be minimized: efficient and practically acceptable ways of reducing internal stresses are necessary. Structural adhesives and coatings that do not create undue internal stresses on the boundary with the substrate can be applied in novel ways in engineering, such as cementing uneven surfaces, when the adhesive layer may need to be quite thick, or applying thick reinforced coatings onto metal surfaces; these coatings themselves can serve as structural elements and can be used to strengthen elements of structures.

4.2 Determination of Internal Stresses in Adhesive-Bonded Joints

Methods for determining the internal stresses in coatings have been developed. In our experience the photoelasticity method is suitable for transparent materials, the tensometric method is subject to substantial error at high temperatures. The internal stresses in coatings have also been studied using the Polani instrument with a rigid ring. The quadrupole resonance method, which requires microwave spectroscopically active substances, and IR spectroscopy, which allows determination of the concentration of overstressed bonds, have been proposed for study of the internal stresses in filled polymers. At present the most widely used method is the cantilever method [160] because of its experimental simplicity and because the calculations can be performed with the need to determine only one value—deflection of the cantilever end. All of these methods determine only averaged values of the internal stresses.

Given the requirement for knowledge of how to specify and preserve the properties of adhesive-bonded joints and for efficient methods of reducing the internal stresses, let us consider the distribution of internal stresses throughout the cross-section of a nontransparent adhesive interlayer both perpendicular and parallel to the cemented surfaces.

4.2.1 Thermal stresses in adhesive-bonded joints

Thermal internal stresses in polymeric paint coatings and adhesives have been less studied than shrinkage stresses, although they can considerably exceed shrinkage stresses [168]. Let us consider a method for determining the thermal internal stresses in an adhesive-bonded joint that permits investigation of the stress distribution in the adhesive, in the substrates, and at the interface. The specimen used comprised a three-layer plate of length l much greater than the width b. The interlayer of the plate consists of adhesive, while the top and bottom layers may be of metals with various linear thermal expansion coefficients (Fig. 4.1).

4.2.1.1 Calculation of stresses.

The stress in each of the layers of the plate is determined by the equation

$$\sigma_i = E_i\left(\varepsilon^0 + \frac{y_i}{R^*} - \alpha_i\,\Delta t\right) \tag{4.1}$$

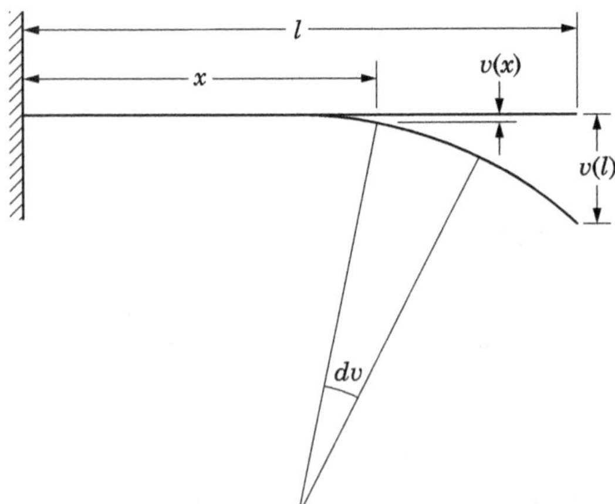

Figure 4.1 Design of the cemented plate and the definition of displacements.

where y_i are displacements, with the conditions that $-h_2 \leq y \leq 0$ (in the interlayer) and $-(h_2 + h_3) \leq y_3 \leq -h_2$ (in the bottom layer), where E_i is the modulus of elasticity of the material of the respective layer; α_i is the linear thermal expansion coefficient of the material; h_1, h_2, h_3 are the layer thicknesses; ε^0 is the relative elongation of the layer, assumed to be zero; and $1/R^*$ and Δt are the change of the layer curvature and temperature, respectively.

$1/R^*$ is given by

$$\frac{1}{R^*} = \frac{1}{R} - \frac{1}{R_0} \tag{4.2}$$

where $1/R$ is the curvature that corresponds to the final equilibrium state of the plate after cementing, and $1/R_0$ is the initial curvature of the plate.

To determine the values ε^0 and $1/R^*$ let us specify the conditions that reflect the fact that the normal effort and bending moment in the cross-section are equal to zero. The condition under which the normal force is equal to zero is

$$\int_0^{h_1} E_1\left(\varepsilon^0 + \frac{y_1}{R^*} - \alpha_1\,\Delta t\right) b\alpha y_1 + \int_{-h_2}^{0} E_2\left(\varepsilon^0 + \frac{y_2}{R^*} - \alpha_2\,\Delta t\right) b\alpha y_2$$

$$+ \int_{-(h_2+h_3)}^{-h_2} E_3\left(\varepsilon^0 + \frac{{y_3}^3}{R^*} - \alpha_3\,\Delta t\right) b\alpha y_3 = 0 \tag{4.3}$$

The condition under which the bending moment is equal to zero is

$$\int_0^{h_1} E_1\left(\varepsilon^0 + \frac{y_1}{R^*} - \alpha_1\,\Delta t\right) b y_1 \alpha y_1 + \int_{-h_2}^{0} E_2\left(\varepsilon^0 + \frac{y_2}{R^*} - \alpha_2\,\Delta t\right) b y_2 \alpha y_2$$

$$+ \int_{-(h_2+h_3)}^{-h_2} E_3\left(\varepsilon^0 + \frac{{y_3}^3}{R^*} - \alpha_3\,\Delta t\right) b y_3 \alpha y_3 = 0 \tag{4.4}$$

After integration, Equations (4.3) and (4.4) take the forms

$$A\varepsilon^0 + B\frac{1}{2R^*} = C \tag{4.5}$$

$$\frac{B}{2}\varepsilon^0 + D\frac{1}{3R^*} = F \tag{4.6}$$

where

$$A = E_1h_1 + E_2h_2 + E_3h_3$$

$$B = E_1h_1^2 - E_2h_2^2 - E_3h_3^2 - 2E_3h_2h_3$$

$$C = E_1\alpha_1\,\Delta th_1 + E_2\alpha_2\,\Delta th_2 + E_2\alpha_3\,\Delta th_1$$

$$D = E_1h_1^3 + E_2h_2^3 + E_3(h_3^3 + 3h_2h_3^2 + 3h_2^2h_3)$$

$$F = E_1\alpha_1\Delta t\frac{h_1^2}{2} - E_2\alpha_2\,\Delta t\frac{h_2^2}{2} - E_3\alpha_3\Delta t^2\frac{h_2h_3 + h_3}{2}$$

From Equation (4.5),

$$\varepsilon^0 = \frac{C}{A} - \frac{1}{2R^*}\frac{B}{A} \tag{4.7}$$

Substituting Equation (4.6) into (4.7), we obtain the following equation for $1/R^*$:

$$\frac{1}{R^*} = \frac{E_1E_2(\alpha_1 - \alpha_2)L_1^3 + E_1E_3(\alpha_1 - \alpha_3)L_2^3 + E_2E_3(\alpha_2 - \alpha_3)L_3^3}{E_1^2h_1^4 + E_2^2h_2^4 + E_3^2h_3^4 + E_1E_2H_1^4 + E_1E_3H_2^4 + E_2E_3h_3^4} \tag{4.8}$$

where the following designations are introduced:

$$L_1^3 = h_1h_2(h_1 + h_2) \qquad L_2^3 = h_1h_3(h_1 + 2h_2 + h_3)$$

$$L_3^3 = h_2h_3(h_2 + h_3) \qquad H_1^4 = 2h_1h_2(2h_2^2 + 3h_1h_2 + 2h_1^2)$$

$$H_2^4 = 2h_1h_3(2h_3^2 + 6h_2h_3 + 6h_2^2 + 3h_1^2 + 3h_1h_3 + 6h_1h_2)$$

$$H_3^4 = 2h_2h_3(2h_3^2 + 3h_2h_3 + 2h_2^2)$$

Using Equations (4.7) and (4.8), we can find the value of stresses in the respective layers according to Equation (4.1).

4.2.1.2 Determining displacements. Under the condition $1/R_0 \to 0$, the differential equation of displacements can be presented as

$$\frac{v''}{[1 + (v')^2]^{3/2}} = \frac{1}{R} \tag{4.9}$$

where $1/R$ is determined experimentally; v is the displacement measured along the normal to the undeformed surface of the plate. If one edge of the plate is fixed, the other free edge is displaced by a certain value (l).

Multiplying both sides of Equation (4.9) by dv we obtain

$$\frac{v'' \, dv}{[1 + (v')^2]^{3/2}} = \frac{dv}{R} \tag{4.10}$$

After integrating the l.h.s. of Equation (4.10) with respect to v' and the r.h.s. with respect to v, we obtain

$$-\frac{1}{\sqrt{1 + (v')^2}} = \frac{v}{R} + C \tag{4.11}$$

The arbitrary constant C is determined by the boundary conditions:

$$x = 0 \qquad v = 0 \qquad \alpha = 0 \tag{4.12}$$

In this case

$$v' = \tan\alpha \qquad \cos\alpha = \frac{1}{\sqrt{1 + (v')^2}} \tag{4.13}$$

Equation (4.11) then takes the form

$$-\cos\alpha = \frac{1}{Rv} + C \tag{4.14}$$

Using conditions (4.12) we find

$$C = -1$$

Thus, the equation for the displacement is

$$v = R(1 - \cos\alpha) \tag{4.15}$$

where $\alpha = 1/R$.

To determine the modulus of elasticity of the adhesive layer taking account of temperature, solving Equation (4.8) for E_2 we find

$$E_2{}^2 + E_2(E_1 a E_3 d) + E_1{}^2(h_1/h_2)^4 + E_3{}^2(h_3/h_2)^4 + E_1 E_3 C = 0 \tag{4.16}$$

where

$$a = \frac{H_1{}^4}{h_2{}^4} - \frac{(\alpha_1 - \alpha_2)L_1{}^3}{h_4{}^2(1/R^*)} 6\Delta t$$

$$C = \frac{H_2{}^4}{h_2{}^4} - \frac{(\alpha_1 - \alpha_3)L_2{}^3}{h_4{}^2(1/R^*)} 6\Delta t$$

$$d = \frac{H_3{}^4}{h_2{}^4} - \frac{(\alpha_1 - \alpha_2)L_3{}^3}{h_4{}^2(1/R^*)} 6\Delta t$$

In Equation (4.16) the value $1/R^*$ is determined experimentally according to Equation (4.15).

Equation (4.16) gives the following expression for E_2:

$$E_2 = -\frac{E_1 a + E_3 d}{2} \pm \sqrt{\left(\frac{E_1 a + E_3 d}{2}\right)^2 - E_1\left(\frac{h_1}{h_2}\right)^4 + E_3{}^2\left(\frac{h_3}{h_2}\right)^4 + E_1 E_3 C} \qquad (4.17)$$

Taking Sprut-5M adhesive as an example, consider the modulus of elasticity of the polymer in the adhesive layer in terms of the distance to the substrate.

The adhesive layer thickness was controlled within the limits 5×10^{-3} to 3×10^{-2} cm, and the cemented materials were 1Kh18N10T steel and aluminum. The thickness of the substrates h, their moduli E, and the linear thermal expansion coefficients α were

$$h_1 = 7 \times 10^{-3}, \quad h_3 = 5 \times 10^{-3}$$

$$E_1 = 2 \times 10^5 \text{ MPa}, \quad E_3 = 7.7 \times 10^5 \text{ MPa}$$

$$\alpha_1 = 0.12 \times 10^{-4}, \quad \alpha_3 = 0.24 \times 10^{-4}$$

The results obtained are presented in Fig. 4.2 (curve 2). Beginning with some value, the modulus of elasticity increases abruptly on approaching the substrate [175, 176]. The general shape of curve 2 (Fig. 4.2), which represents the form of the distribution of the polymer modulus of elasticity through the thickness of the adhesive layer, is close to parabolic.

Now consider the dynamic method of determining the modulus of elasticity E_c of the polymer in the adhesive layer. To determine E_0 of the adhesive directly in the joint, we set up an equation for the rigidity parameter D of the three-layer plate for bending relative to the neutral surface:

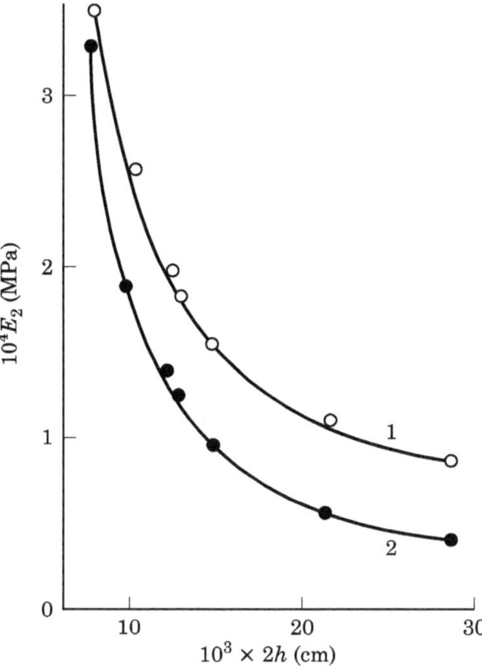

Figure 4.2 Dependence of the polymer elastic modulus in the adhesive layer on the distance to the substrate. The modulus is determined by the dynamic (1) and static (2) methods.

$$D = D_1 + D_2 + D_3 - B_1 {H_1}^2 + B_2 {H_2}^2 \tag{4.18}$$

where

$$D_1 = \frac{E_1 {\sigma_1}^3}{12(1 - {\upsilon_1}^2)} = \frac{B_1 {\sigma_1}^2}{12} \qquad D_2 = \frac{E_2({h_1}^2 + {h_2}^3)}{3(1 - {\upsilon_0}^2)}$$

$$D_3 = \frac{E_3 {\sigma_2}^3}{12(1 - {\upsilon_2}^2)} = \frac{B_3 {\sigma_2}^2}{12}$$

$$H_1 = h_1 + \frac{\sigma_1}{2} \qquad H_2 = h_2 + \frac{\sigma_2}{2}$$

The polymer rigidity parameter D_c can be expressed as

$$D_c = B_2 \frac{h^2}{3} + B_3 \left[\frac{B_1(h + \sigma_1/2) - B_3(h + \sigma_2/2)}{B_2 + B^*} \right] \tag{4.19}$$

The rigidity parameter D is determined experimentally by finding the frequency of vibrations of the model under consideration.

Using Equation (4.19), the relationship (4.18) can be converted into an equation to determine E_c:

$$D_3 = B_2\frac{h^2}{3} + B_2{}^2\left[B_1\left(\frac{\sigma_1{}^2}{4} + \sigma_1 h + h^2\right)\right.$$

$$\left. + B_3\left(\frac{\sigma_2{}^2}{4} + \sigma_2 + h^2\right) - D + 2B^*\frac{h^2}{3}\right]$$

$$+ B_2\left\{B_1B_3\left(\frac{\sigma_1{}^2}{2} + \frac{\sigma_2{}^2}{2} + \sigma_1\sigma_2 + 8h^2 + 4\sigma_1 h\right)\right.$$

$$\left. - 2D^*B^*\frac{h^2}{3} + \left[B_1\left(h + \frac{\sigma_1}{2}\right) - B_3\left(h + \frac{\sigma_2}{2}\right)\right]^2\right\}$$

$$+ \tfrac{1}{2}B_1B_3B^*\left(\frac{\sigma_1{}^2}{2} + \frac{\sigma_2{}^2}{2} + \sigma_1\sigma_2 + 8h^2 + 4\sigma_1 h + 4\sigma_2 h\right) - D^*B^{*2} = 0$$

$$D_c = D - D_1 - D_3 \tag{4.20}$$

The cubic equation (4.20) allows determination of the adhesive modulus of elasticity E_2 in a plate made of three layers. For this purpose it is necessary to find the frequency of the natural vibrations of the three-layer plate under investigation with one edge fixed and the other free.

The transcendental equation to determine the frequency of the natural vibrations has the following shape in this case:

$$Chkl\cos kl + 1 = 0$$

$$k^4 = \frac{\mu\omega^2}{D}; \qquad \mu = \frac{\gamma_1}{g}\sigma_1 + \frac{\gamma_c}{g}2h + \frac{\gamma_2}{g}\sigma_2 \tag{4.21}$$

where ω is the natural frequency of vibrations, γ is the specific weight, and μ is the weight per unit of surface.

The first root of Equation (4.21) is $kl = 1.875$. Then the equation for the rigidity parameter D for bending of the three-layer plate has the form

$$D = \frac{\omega^2 l^4}{1.875}\left(\frac{\gamma_1}{g}\delta_1 + \frac{\gamma_c}{g}2h + \frac{\gamma_2}{g}\delta_2\right) \tag{4.22}$$

Having determined D experimentally and applying Equation (4.20), we find the value of the dynamic modulus of elasticity of the adhesive layer in the joint (see Fig. 4.2, curve 1). The figure presents satisfactory coincidence of the value and functional form of the modulus of elasticity determined by the two methods. The value of the modulus of elasticity of the polymer in the adhesive interlayer shows good agreement with the data presented in Chapter 1 in its dependence on the distance to the substrate.

4.2.1.3 Determining thermal internal stresses. Experimentally, the stress with a particular temperature change is determined by the deflection of the free end of the cemented three-layer plate fixed as a cantilever. The deflection of the plate end was determined by means of a projection optical system. Figure 4.3 displays the distribution of the thermal internal stresses in a Sprut-5M adhesive-bonded joint for temperature change from 20 to 42°C. Plates made of 1Kh18N10T steel 0.07 mm thick and aluminum foil 0.02 mm thick were used as substrates. It is evident that the internal stresses are characterized by maximum values on the interfaces.

To check the method of determining internal stresses, resistance tensometers were cemented to the external surfaces of the substrates. From a series of experiments it was found that the values of the stress on the external surfaces of the plate calculated by the quoted formulas and determined by means of the resistance tensometers differ by not more than 9%.

4.2.2 Shrinkage internal stresses in adhesive-bonded joints

There are established methods for evaluating the shrinkage internal stresses in the adhesive-bonded joint using cantilevered three-layer beams [160, 175], but in practice these methods are frequently too complicated.

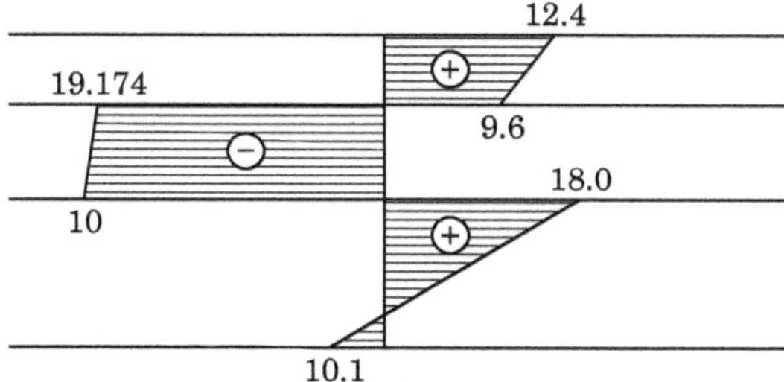

Figure 4.3 Internal stresses (MPa) in the adhesive-bonded joint occurring on temperature elevation from 20°C to 43°C.

A paper of Myshko and Garf [175] was the first to attempt to create a method of calculating the distribution of the shrinkage internal stresses along an adhesive-bonded joint cross-section for a cantilevered specimen made of two plates of different flexural rigidity with an adhesive layer between them—i.e., to generalize the equations from [160] for adhesive-bonded joints.

Consider the effect of internal stresses in two cases: between rigid substrates (only the adhesive layer deforms) and between flexible substrates (combined deformation). The pattern of the tension–contraction forces P_0, P_1, P_2 of the substrates and the adhesive layer and their bending moments M_0, M_1, M_2 caused by shrinkage is represented by a system of four equations:

$$P_1 - P_0 - P_2 = 0 \tag{4.23}$$

$$M_1 + M_2 - M_0 + P_0\left(\frac{h_0}{2} + h_1 + \frac{h_2}{2}\right) - P_1\frac{(h_1 - h_2)}{2} = 0 \tag{4.24}$$

$$\varepsilon_{\text{full}} = \frac{P_1}{E_1F_1} + \frac{M_1}{EW_1} = \frac{P_0}{E_0F_0} + \frac{M_0}{E_0W_0} \tag{4.25}$$

$$\varepsilon_{\text{full}} = \frac{P_1}{E_1F_1} - \frac{M_1}{E_1W_1} = \frac{P_2}{E_2F_2} + \frac{M_2}{E_2W_2} \tag{4.26}$$

where F_0, F_1, F_2 are the areas of the cross-section; W_0, W_1, W_2 are the moments of resistance; h_0, h_1, h_2 is the thickness; E_0, E_1, E_2 are the moduli of elasticity of the top and bottom substrates and of the adhesive layer; and $\varepsilon_{\text{full}}$ is the relative shrinkage of the adhesive layer without substrates (free shrinkage).

Analyzing equations (4.25) and (4.26), it is easy to note that the authors proceed from the equality of the relative deformations of the adhesive layer and the substrates. Such an approach raises some doubts insofar as the bond between the adhesive and the substrate is not absolutely rigid, as a result of which Equation (4.27) has no definite solutions for real systems:

$$\varepsilon_{\text{full}} = \frac{2}{l^2}\frac{AD - BC}{A - C}f \tag{4.27}$$

where f is the sag value on the beam end; l is the length of the beam; and A, B, C, D are parameters that depend on the thickness and the modulus of elasticity of the substrates and the adhesive layer.

Actually, if

$$A = \left(\frac{1}{E_1F_1} + \frac{1}{E_0F_0}\right)\frac{h_1 + h_2}{h_0 + h_1} + \frac{1}{E_0F_1} \tag{4.28}$$

$$C = \left(\frac{h_1 + h_2}{h_0 + h_1} + 1\right)\frac{1}{E_1F_1} + \frac{1}{E_2F_2} \tag{4.29}$$

with $A = C$, we have

$$\frac{1}{E_0F_0}\frac{h_1 + h_2}{h_0 + h_1} = \frac{1}{E_2F_2} \tag{4.30}$$

or

$$\frac{E_2}{E_0} = \frac{h_0}{h_2}\frac{h_0 + h_1}{h_1 + h_2} \tag{4.31}$$

With the steel ($E = 2 \times 10^5$ MPa) and aluminum ($E = 0.7 \times 10^5$ MPa) substrates and at $h_1 = h_0$, Equation (4.30) is reduced to the quadratic equation

$$2(h_1/h_2)^2 - 0.35h_1/h_2 - 0.35 = 0 \tag{4.32}$$

the roots of which produce thickness relationships of the adhesive layer and the substrates, reducing Equation (4.27) to indeterminacy.

Thus, the approach proposed in [98] and [177] is not rigorous, which restricts its application. In addition, this method cannot be applied to determine the distribution of the internal stresses through the cross-section of the adhesive interlayer.

The authors of [178] attempted to remove these shortcomings by proposing a method for calculating the distribution of stresses in the adhesive layer and the substrates as well as for determining the internal stresses in the adhesive layer and the stresses that originate in each substrate in accordance with the Dugamel–Neimann hypothesis. The position of the neutral surface was determined by means of the equations obtained for three-layer plates. Hence, the proposed method has a number of unjustified assumptions that do not allow its application in practice. For example, the relaxation processes in the polymerizing adhesive are not taken into account in determining the modulus of elasticity of the polymer in the adhesive interlayer by tensometering the free surface of one of the substrates.

The method provides too high values of the modulus of elasticity. Let us analyze the formula to calculate the adhesive elasticity proposed by the authors [178]. Applying the designations adopted in [178], the proposed equation can be presented as

$$B_1 = \frac{2}{h_1{}^2}\{B_0(h_0{}^2 + \tfrac{3}{2}h_0h_1) - B_2(2h_2{}^2 + 3h_1{}^2 + \tfrac{9}{2}h_1h_2 + 3h_0h_1 + 3h_0h_2)$$

$$+ 3\varepsilon_{0ex}\rho[(h_2 + h_1) - B_0(h_0 + h_1)]\} \qquad (4.33)$$

where ε_{0ex} is the relative deformation of the substrate free surface determined experimentally; and B_0 and B_2 are the reduced rigidity of the top and bottom substrates. The modulus of elasticity of the adhesive layer E_1 can be expressed through the reduced rigidity B_1 in the following way:

$$E_1 = \frac{B_1(1 - \nu_1{}^2)}{h_1} \qquad (4.34)$$

where ν_1 is the Poisson coefficient of the adhesive layer.

Let us analyze Equation (4.33) for a certain adhesive-bonded joint with $h_0 = 5 \times 10^{-5}$ m, $h_2 = 10 \times 10^{-5}$ m, $h_1 = 10 \times 10^{-5}$ m, $B_0 = 3.846 \times 10^2$ N/m (aluminum foil), $B_2 = 21.978 \times 10^2$ N/m (steel). The expression takes the form

$$B_1 = 11.45\varepsilon_{0ex}\rho \times 10^6 - 541.758 \times 10^3 \qquad (4.35)$$

The authors of [178] determine the value ε_{0ex} by tensometering the free surface of the top substrate.

Figure 4.4 presents a portion of the length of the adhesive-bonded joint. BB' is the deformation of the free surface of the top substrate and corresponds to ε_{0ex}; AA' is the layer elongation, assumed zero, and corresponds to ε_1; AO is the radius of curvature of the adhesive interlayer deformed by shrinkage. As the figure displays,

$$\varepsilon_{0ex} = \varepsilon_0 + CB' \qquad (4.36)$$

From the similarity of OAA' and $A'CB'$ we obtain

$$CB' = \frac{h_0\varepsilon_1}{\rho} \qquad (4.37)$$

i.e.,

$$\varepsilon_{0ex}\rho = \varepsilon_1\rho + h_0 \qquad (4.38)$$

After substituting the values of ρ and ε_1 obtained experimentally by deflection of the cantilever-fixed specimen, we find $\varepsilon_{0ex}\rho = 6.303 \times 10^{-3}$ cm, $E_1 = 40.562 \times 10^5$ MPa. The value obtained for the modulus of elasticity of the polymer in the adhesive layer exceeds that of steel, which clearly indicates the inadequacy of the method.

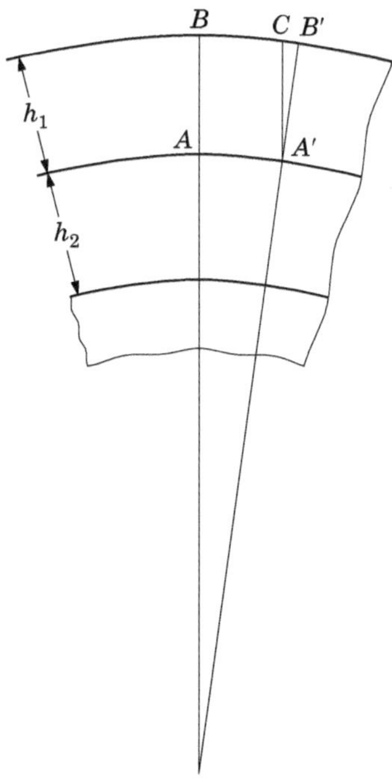

Figure 4.4 Element of the adhesive-bonded joint.

Let us consider a method of determining internal stresses that lacks the above shortcomings by applying it to an adhesive-bonded joint represented by a three-layer plate (substrate–adhesive–substrate). The adhesive layer shrinkage causes deformation of the whole plate. The resistance forces appearing in this case can be represented as a system of normal forces and bending moments, which act layer by layer throughout the plate, and stresses can be represented as

$$\sigma_0 = E_0\left(\varepsilon_1 + z\frac{1}{\rho}\right) \tag{4.39}$$

$$\sigma_1 = E_1\left[\varepsilon_1 + z\frac{1}{\rho} + (1+\nu_c)\Delta\right] \tag{4.40}$$

$$\sigma_2 = E_2\left(\varepsilon_1 + z\frac{1}{\rho}\right) \tag{4.41}$$

where $\sigma_0, \sigma_1, \sigma_2$ are the stresses in the top substrate, adhesive layer, and bottom substrate, respectively; $1/\rho$ and ε_1 are the curvature and linear deformation of the coordinate surface; Δ is the elastic shrinkage of the adhesive layer; and z is the current coordinate representing the distance from the coordinate surface.

Further calculations substantially simplify the representation of the forces and bending moments as integrals of stresses through the cross-sectional area of the respective layers:

$$P_0 = \int^{z} \sigma_0 b\,dz \tag{4.42}$$

$$P_1 = \int^{z} \sigma_1 b\,dz \tag{4.43}$$

$$P_2 = \int^{z} \sigma_2 b\,dz \tag{4.44}$$

$$M_0 = \int^{z} \sigma_0 bz\,dz \tag{4.45}$$

$$M_1 = \int^{z} \sigma_1 bz\,dz \tag{4.46}$$

$$M_2 = \int^{z} \sigma_2 bz\,dz \tag{4.47}$$

To determine the values of ε_1 and Δ, let us set up a system of equations that reflect the fact that the sums of the bending moments and of the forces across the specimen cross-section are equal to zero:

$$\begin{aligned} \sum_{i=1,3} P_i &= 0 \\ \sum_{i=1,3} M_i &= 0 \end{aligned} \tag{4.48}$$

The values of ε_1 and Δ are found after the respective conversions:

$$\Delta = \frac{1}{C_\rho} \frac{B_0{}^2 h_0{}^2 + B_2{}^2 h_2{}^2 + B_1{}^2 h_1{}^2 + 2B_0 B_1 H_1{}^2 + 2B_2 B_1 H_2{}^2 + 2B_0 B_1 H_1{}^2}{B_1[B_0(h_0 + h_1) - B_2(h_2 + h_1)]} \tag{4.49}$$

$$\varepsilon = \frac{1}{3\rho}\frac{B_0L_1{}^2 + B_2L_2{}^2 + B_1h_1{}^2/2}{B_0(h_0 + h_1) - B_2(h_2 + h_1)} \tag{4.50}$$

where

$$M_1{}^2 = 2h_1{}^2 + 2h_0{}^2 + 3h_0h_1$$

$$H_2 = 2h_1{}^2 + 2h_2{}^2 + 3h_2h_1$$

$$H_3{}^2 = 2h_0{}^2 + 2h_2{}^2 + 6h_1{}^2 + 6h_0h_1 + 6h_2h_1 + 3h_0h_1$$

$$L_1{}^2 = 2h_0 + \frac{3}{2}h_1{}^2 + 3h_1h_0 - 3h_1h_0 - 3h_2h_0$$

$$L_2{}^2 = 2h_2{}^2 + \frac{3}{2}h_1{}^2 + 3h_2h_1 + 3h_1h_0 + 3h_2h_0$$

The coordinate h_0 determines the position of the neutral surface assumed as a coordinate:

$$h_1' = \frac{B_0(h_1 + h_0) - B_2(h_1 + h_2)}{2(B_0 + B_1 + B_2)} \tag{4.51}$$

Let us consider how to determine experimentally the shrinkage internal stresses. An optical system similar to that described in Section 4.2.1 was used to determine the displacement of the free end of the cantilever-fixed adhesive-bonded plate. Plates made of aluminum foil 5×10^{-5} m thick ($E = 0.7 \times 10^5$ MPa) and of steel 10×10^{-5} m thick ($E = 2 \times 10^5$ MPa), both with areas of $1 \times 10\,\text{cm}^2$, served as substrates. Sprut-5M adhesive was used to cement the plates. The shrinkage internal stresses were determined at 293 K 2 days after the specimens were prepared.

To determine the modulus of elasticity of the adhesive layer, the temperature of the specimens was elevated to 314 K at a rate of 0.1 K/min. The values of the modulus of elasticity of the adhesive as $E_1 \times 10$ MPa for two specimens at temperatures in the above range, calculated by the formula (4.17), are tabulated below.

	10E (MPa)					
Specimen thickness, h (10^{-5} m)	294 K	298 K	302 K	306 K	310 K	314 K
13	0.086	0.086	0.085	0.085	0.085	0.084
6	0.089	0.089	0.089	0.088	0.088	0.087

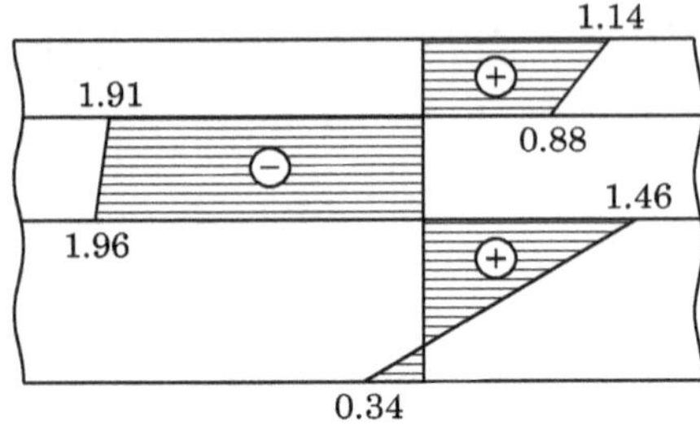

Figure 4.5 Shrinkage internal stresses (MPa) in the adhesive-bonded joint.

It is evident that with temperature elevation from 293 to 314 K the modulus of elasticity of the Sprut-5M adhesive changes insignificantly, demonstrating the applicability of the method of temperature stresses.

Figure 4.5 displays the calculated distribution of the shrinkge stresses over the adhesive-bonded plate cross-section. At the adhesive–metal interface a concentration of stresses is observed that can exceed twice those in the bulk adhesive.

4.2.3 Calculation of internal stresses by the lattice cell method

Due to complexity and limited capacity of the analytical methods, recourse is more frequently had to numerical modeling methods, above all the finite elements method (FEM), to determine the stress–strain state.

In models based on FEM, either the composite material is considered as a homogeneous body whose physical and mechanical characteristics correspond to the averaged characteristics of the system, or account is taken of the structural features of the composite material. For this purpose one introduces the physical and mechanical characteristics of the system components into the designed model and can thus study their mutual effects. For all its advantages, FEM also has its shortcomings. The most serious is that the method is based on the assumption of infinitesimal deformations, which is why when calculating the stressed state of articles made of high-elasticity materials by means of FEM one applies multistage iteration procedures using the finite deformations obtained as a result of a number of solutions with small deformations [179]. In addition, the limitations on the shape of the finite element (triangular elements must be close to equiangular) make the modeling of the fine adhesive interlayer difficult because of limited computer memory.

Discretization of the elastic medium is achieved by dividing it into separate elements along the coordinate lines of the selected coordinate system. Each discrete element of the initial medium based on an electromechanical equivalent is replaced by a cell of an electrical network involving electric components with concentrated parameters. The latter are determined by similarity between Hooke's law written for separate elastic elements of the initial medium and Ohm's law written for the lattice cells.

The lattice cells integrate into a single system, which models the initial medium. The integration is performed by means of the Kron conversion [181]. According to this, the variety of electrical systems that can be obtained as a result of various combinations of a set of the electrical elements is considered a consequence of the transformation of the coordinated system. This means that a concrete method of connecting the electrical components determines the concrete system of generalized coordinates that are the electrical variables.

Kron showed that processes that occur in similar electric circuits are described by tensor equations. The network parameters and electrical variables are the physical objects that are the components of these equations.

Such an approach makes it possible to find components of a physical object in a new system of coordinates from its components in a known system. This can be done by means of a formal procedure involving application of special matrices. The analysis of any electrical circuit goes through the following stages:

1. Formulation of an equation of the circuit state in the coordinate system that is characterized by minimal analytical difficulty.
2. Setting up of the conversion matrix that connects the new and old systems of coordinates.
3. Application of this matrix to determine the electrical variables and parameters in the new system of coordinates.

A set of separate elements that are not interconnected is a convenient choice as an initial system that produces no analytical difficulties. Kron called similar cross-cut systems "primitive."

The equation of the primitive network can be obtained by means of the Kron postulate of generalization [181], which helps to find the type of system equation. The postulate is based on the fact that integration of elements into a single system is not accompanied by addition of new physical phenomena or objects that would not have been in the initial element.

The equation of the lattice cell state is defined by Ohm's law in the symbolic form

$$\{I\}_{ci} = [y]_{ci}\{E\}_{ci} \tag{4.52}$$

Here $\{I\}_{ci}$ is the current that affects the ith cell; $\{E\}_{ci}$ is the potential difference that occurs on the cross-link couples; and $[y]_{ci}$ is the matrix of cell conductivity.

In the primitive, i.e. cross-cut, network consisting of n lattice cells, there are three physical objects: $\{I\}_{\mathrm{p}}$ represents the points that come in and out of the cells; $\{E\}_{\mathrm{p}}$ represents the potential differences that occur on the cross-link couples; and $[y]_{\mathrm{p}}$ represents the intrinsic and mutual conductivity of individual cells.

Consequently, from Kron's Postulate I on generalization, the equation of the primitive system will also have the form of Ohm's law:

$$\{I\}_{\mathrm{p}} = [y]_{\mathrm{p}}\{E\}_{\mathrm{p}} \tag{4.53}$$

The same equation expressed in the form of block matrices that correspond to each lattice cell of the system, takes the form

$$\begin{Bmatrix} \{I\}_{\mathrm{c1}} \\ \{I\}_{\mathrm{c2}} \\ \cdots \\ \{I\}_{\mathrm{c}n} \end{Bmatrix}_{\mathrm{p}} = \begin{bmatrix} [y]_{\mathrm{c1}} & 0 & \cdots & 0 \\ 0 & [y]_{\mathrm{c2}} & \cdots & 0 \\ \cdots & \cdots & \cdots & \cdots \\ 0 & 0 & \cdots & [y]_{\mathrm{c}n} \end{bmatrix}_{\mathrm{p}} \begin{Bmatrix} \{E\}_{\mathrm{c1}} \\ \{E\}_{\mathrm{c2}} \\ \cdots \\ \{E\}_{\mathrm{c}n} \end{Bmatrix}_{\mathrm{p}} \tag{4.54}$$

As was noted earlier, the transition from the equations of the primitive system to the equations of the integrated system can be treated as a transformation of the system of coordinates. Values indicated by subscript p belong to the primitive system and those indicated by subscript b belong to the integrated system within the system of coordinates of which there are the objects $\{I\}_{\mathrm{b}}, \{E\}_{\mathrm{b}}, [y]_{\mathrm{b}}$.

The state equation of the integrated system has the form

$$\{I\}_{\mathrm{b}} = [y]_{\mathrm{b}}\{E\}_{\mathrm{b}} \tag{4.55}$$

According to [181], the transition from the primitive variables to the integrated ones is performed by means of the transformation matrix $[A]$. As a result we have

$$\{E\}_{\mathrm{p}} = [A]\{E\}_{\mathrm{b}} \tag{4.56}$$

The currents of the integrated system are defined by

$$\{I\}_{\mathrm{b}} = [A]^{\mathrm{t}}\{I\}_{\mathrm{p}} \tag{4.57}$$

and the conductivity is defined by

$$[y]_{b} = [A]^{t}[y]_{p}[A] \tag{4.58}$$

Expressions (4.56)–(4.58) and the state equation (4.55) allow definition of the unknown variables of the integrated circuit.

The key provision of the Kron approach is the application of the rearrangement matrix $[A]$. The distinguishing feature of this matrix is that it is constructed using the specific regularities typical of circuits, Kirchhoff's law in particular. Application of the Kron approach to the theory of elasticity is defined by the possibility of forming the rearrangement matrix $[A]$. For this purpose it is necessary to introduce a network model of an elastic body. Such a model can be based on the electromechanical analogy. The method of formation of the network model and the matrix $[A]$ are described in [180].

The ability to form the matrix $[A]$ as well as the existence of analogues $\{I\}_{ci} \sim \{F\}_{ci}$, $\{E\}_{ci} \sim \{\Delta\}_{ci}$, $[y]_{ci} \sim [k]_{ci}$, where $\{F\}_{ci}$ is the vector of forces, $\{\Delta\}_{ci}$ is the vector of absolute deformations, and $[k]_{ci}$ is the matrix of rigidity of the ith elastic element, permit Equations (4.56)–(4.58) to be applied for solution of problems of elasticity theory.

To define the stress–strain state of the elastic body by the method of lattice cells, the following actions should be performed. To form a primitive system, an elastic body should be cross-cut along the coordinate lines of the selected system of coordinates into a certain number of discrete elements that are not interconnected. Based on Hooke's law, each element is described by the relationship

$$\{F\}_{ci} = [k]_{ci}\{\Delta\}_{ci} \tag{4.59}$$

The force vector of the primitive system is defined by the forces of each of the elements in the following way:

$$\{F\}_{p} = \begin{Bmatrix} \{F\}_{c1} \\ \{F\}_{c2} \\ \cdots \\ \{F\}_{cn} \end{Bmatrix}_{p} \tag{4.60}$$

The rigidity matrix of the primitive system represents a set of blocks, which are the rigidity matrices of separate elements:

$$[k]_{p} = \begin{bmatrix} [k]_{c1} & 0 & \cdots & 0 \\ 0 & [k]_{c2} & \cdots & 0 \\ 0 & 0 & \cdots & [k]_{cn} \end{bmatrix}_{p} \tag{4.61}$$

To obtain the matrix equation that describes the stressed state of a body, the parameters and the variables of the primitive system should be converted to the respective values of the solid system by means of the matrix $[A]$. To form the matrix $[A]$ the elastic body should be replaced by an equivalent electrical circuit.

With the known matrix $[A]$, the known forces and rigidities of the primitive system can be converted to the respective variables and parameters of the solid body:

$$\{F\}_{\mathrm{b}} = [A]^{\mathrm{t}}\{F\}_{\mathrm{p}} \tag{4.62}$$

$$[k]_{\mathrm{b}} = [A]^{\mathrm{t}}[k]_{\mathrm{p}}[A] \tag{4.63}$$

The equation of state of the solid body takes the form

$$\{F\}_{\mathrm{b}} = [k]_{\mathrm{b}}\{\Delta\}_{\mathrm{b}} \tag{4.64}$$

Unknown here is the vector $\{\Delta\}_{\mathrm{b}}$, which is defined by the relationship

$$\{\Delta\}_{\mathrm{b}} = [k]_{\mathrm{b}}^{-1}\{F\}_{\mathrm{b}} \tag{4.65}$$

The solution obtained cannot be applied directly to define the distribution of deformations over the body insofar as it is necessary to know absolute deformations of each elastic element into which the body is notionally cross-cut. These data are obtained by the conversion

$$\{\Delta\}_{\mathrm{p}} = [A]\{\Delta\}_{\mathrm{b}} \tag{4.66}$$

Now, substituting the found solution into Equation (4.59), one can find stresses of each of the elements.

The sequence of defining the stress–strain state by the lattice cell method is the following:

1. The body is cut into separate elastic elements along the coordinate lines of the selected system of coordinates.
2. The matrix of rigidities and the vector of specified forces are formulated.
3. The matrices of rigidity and of force of each elastic element are integrated into the respective matrix and force vector of the cross-cut system.
4. Using the electrical–mechanical analogy, the body consisting of separate elastic elements is replaced by the equivalent electrical circuit, which is the basic tool for formation of the Kron conversion matrix.

5. By means of the Kron conversion matrix, the rigidity matrix and the force vector of the cross-cut system are converted to the respective characteristics of the integrated system, which serve for formulation of the system state equation.
6. After this equation is solved, the vector of absolute deformation is determined in the integrated variables.
7. By applying the Kron matrix and the obtained solution in the integrated variables, we find absolute deformations for each elastic element.
8. The stresses are calculated by the defined deformations and the known rigidities of each element.

The thermal-stressed state of a polymer layer in the shape of a rectangle that is characterized by arbitrary internal structure was determined with the help of a computer. The configuration of the structure is represented with precision of up to the size of an individual element. Its components may differ in physical characteristics due to combining a number of these components into the number of elements into which the rectangle is cross-cut.

The effect of the relative sizes of the elements on the results of the solution is of interest. The analysis showed that with sides of the rectangular element in the ratio 1:100, the solution is not distorted. With a ratio of 1:200, possible distortion should be taken into consideration.

The rectangular region should be cross-cut along the coordinate lines. Within the limits of the allowable ratios of the element sides, the coordinate lines can be arbitrarily thickened, but narrow elements must be placed right after the broad ones. These two conditions are important when studying the stress-state of thin interlayers: for example, polymers in adhesive-bonded joints.

Within the limits of the given rectangular region, the transition from structure to structure is fairly fast and simple, because of the small amount of initial data. The data include four data files and two parameters. The modulus of elasticity, the Poisson coefficient, and the coefficient of linear expansion are pointed to in the EM data array for each of the media or material components coded by a number. The M_{at} data file points to the coordinates and the number of the material of that element with which the medium properties begin to change. Parameters N_j and N_i are the numbers of elements into which the sides of the rectangle are cross-cut along the x and y axes. The N_{jx} and N_{iy} data files enumerate the distances between the coordinate lines along the x and y axes along which the body is cross-cut into elements.

4.2.3.1 Structure of the program. The primary relationships of the lattice cell method, (4.62)–(4.66), are related to the matrix rearrangements and to the solutions of the matrix equations. All these standard operations are included in the software of modern computers. Application of the transformation matrix $[A]$ and the matrix of rigidities of the primitive system $[k]_p$ (the first is sparse, consisting of zeros and units; the second is a narrow ribbon-shaped matrix) would consume intolerable amounts of computer resources, memory in particular. For this reason, the primitive variables and the system parameters were not transformed in the form of the matrix equations (4.62) and (4.63) but by a software application.

In addition to determining stresses in the adhesive layer and overstress at the adhesive–substrate interface, the lattice method permits calculation of the stressed state of any polymer layer of interest because of the possibility of arbitrary densing of the cross-cutting network of the region under study into elements along the coordinate lines.

Figure 4.6 presents the distribution of internal stresses in the epoxy rubber coating determined by the lattice cell method. A feature of the calculation is the large ratio of the length to the width of the specimen, with a substantial difference of the elastic characteristics that characterize the system:

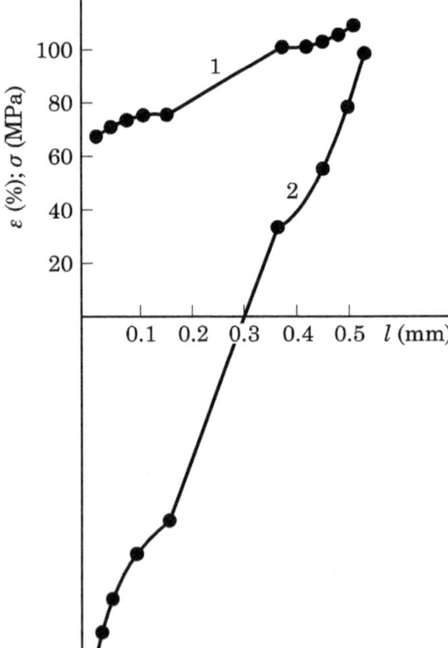

Figure 4.6 Distribution of deformations (1) and stresses (2) in the two-layer specimen along the cross-section of the polymeric coating with division into cells and subsequent densing.

$E_{\mathrm{coat}} = 9 \times 10^2\,\mathrm{MPa}$, $\mu = 0.122$; linear thermal expansion coefficient (LTEC) $= 30 \times 10^{-6}$

$E_{\mathrm{basic}} = 2 \times 10^5\,\mathrm{MPa}$, $\mu = 0.3$; LTEC $= 12.6 \times 10^{-6}$

The temperature range is 30 K. One layer of elements modeled the base; the coating was divided into five layers of similar width. As Fig. 4.6 displays, the maximal value of the internal stresses in the epoxy rubber coating is observed at the interface with metal.

4.2.4 Edge internal stresses in adhesive-bonded joints

The determination of edge stresses in adhesive-bonded joints is based on their property of counteracting the bonding forces between the adhesive and the substrate—in other words, of decreasing the adhesion strength.

The pin-shaped specimens used to determine edge stresses are shown in Fig. 4.7. They consist of substrates 1 and 3, adhesive layer 2, and pin 4 lapped to a conical hole in the substrate 3. An axial force P or a torsion moment M_{edge} that cause normal and tangential failure stresses can be applied to the pin. The internal stresses, the pattern of

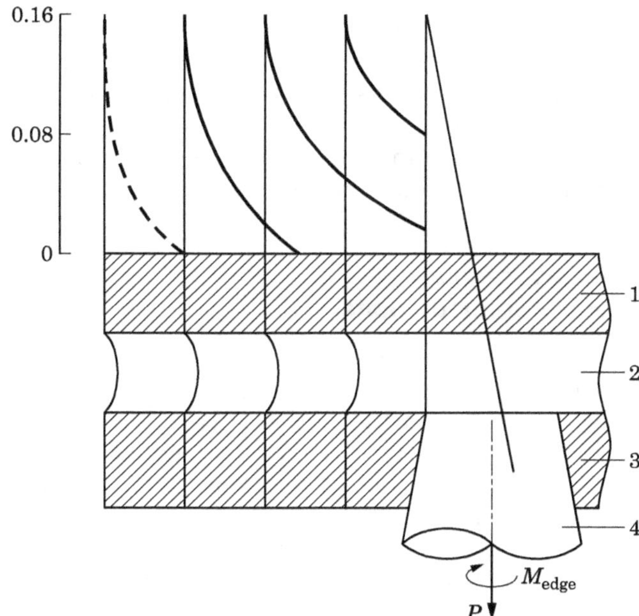

Figure 4.7 Patterns of edge stresses in the adhesive-bonded joint.

which is also presented in Fig. 4.7, occur in the adhesive in the course of curing. A 40% solution of PBMA in MMA was used as adhesive, which was cured by the redox system of benzoyl peroxide (b.p.) (1%)–dimethylaniline (DMA) (0.5%) at 20°C. The peak stresses on the edges of the adhesive-bonded joint are the issue of attention. The edges are brought closer to the hole in a step-by-step way by decreasing the size of the top substrate. At each step the pin is pulled out, with measurement of the failure force P or the torsion moment M_{edge}, and the failure stresses are calculated. The pattern of their dependence on the distance from the edge was plotted. With observed σ_0 or $\tau_{\text{ad}0}$ adopted as initial values, i.e., when the zone of influence of the edge internal stresses is substantially far from the pulled-out pin and does not affect its pull-out force, the coefficient of stress concentration in the edges is calculated by the formula

$$k_{is} = \frac{\sigma_{a,b,c,\ldots,i}}{\sigma_0} \tag{4.67}$$

or

$$k_{is} = \frac{\tau_{a,b,c,\ldots,i}}{\tau_0} \tag{4.68}$$

where $\sigma_{a,b,c,\ldots,i}$ and $\tau_{a,b,c,\ldots,i}$ are the normal and tangential failure stresses at the edge positions $a, b, c, \ldots, i$.

As Fig. 4.7 displays, the concentration of stresses on the edges of the adhesive-bonded joint in this case reached 1.12, although a comparatively elastic adhesive was used. For the adhesive based on the PN 609-21-RK polyester resin, the overstresses on the edges exceed 2.1; for epoxy adhesives they are 2–10.

This method for studying the edge internal stresses can be used for coatings as well, but in this case there is no second substrate.

4.3 Method of Decreasing Internal Stresses in Adhesive-Bonded Joints

All the practically acceptable methods of decreasing the internal stresses in adhesive-bonded joints can be divided into three groups: decrease of the adhesive shrinkage in the course of cure; increase of the relaxation rate of the internal stresses in the adhesive; and decrease of the difference between the coefficients of linear thermal expansion of the adhesive and the substrate.

Widely used methods for decreasing the internal stresses relate to increasing their rate of relaxation, for example, by decreasing the modulus of elasticity when adding plasticizers to adhesives or when they are combined with elastomers. The residual stresses were found to begin to relax noticeably only after the elastic modulus of the polymer becomes less than 2.4×10^3 MPa [182]. The decrease of polymer rigidity commonly results in decrease of its heat resistance and stress–strain behavior stability, and decrease of strength limits under bending, contraction, and shear [182]. The mechanism of the decrease of internal stresses when plasticizers are added to the polymeric adhesive is related to the decrease of the intermolecular interaction due to adsorption along the interfaces of the structural elements [183]. In this case a large amount of plasticizer is added, which enhances the decrease of strength and adhesion of the polymer. During operation of the adhesive-bonded joint, the plasticizer is redistributed and removed from the system, resulting in loss of elasticity and an increase of the internal stresses.

A more convenient method is not to decrease the elastic modulus of the whole adhesive but to form an elastic finish coat [182] not less than 30 μm thick [183] on the adhesive–substrate interface. With lower elastic modulus, the mobility of structural elements decreases. Application of an elastic sublayer decreased the internal stresses during the formation of coatings made of unsaturated polyester resins [185], epoxides [186], and polymer solutions [187]. The use of elastic finish coats found comparatively wide application for paint coatings but seems less promising for adhesive-bonded joints because of decrease of the adhesion strength due to the low cohesion strength of the finish coat itself and because of the labor requirements in producing the adhesive-bonded joints.

Some papers [183, 188, 189] have stated that internal stresses can be decreased and the coating lifetime can be increased by means of additives that create an ordered crosslinked structure in solutions and dispersions as well as in coatings based on them. The formation of this structure promotes the rapidity of the relaxation processes [190]. The mechanism of such a process [191] is related to transition of the system into the gelatinous state due to the formation of local physical or chemical links between structural elements, which shortens the relaxation time [192]. In oligomeric systems the thixotropic structure can be formed before polymerization begins by introducing polyfunctional additives that interact with oligomer active groups, resulting in formation of hydrogen bonds and physical links [183]. The rate of relaxation of the internal stresses in polymers can also be increased by adding other substances such as fillers or alloying additives [193].

High adhesion strength with simultaneous minimization of the internal stresses can be obtained in the formation of adhesive-bonded joints under conditions in which the rates of internal stress relaxation and polymer formation are compatible. This approach to decreasing the internal stresses is acceptable when their rate of relaxation is comparatively high.

The thermal stresses in the layers and films of the polymer can be decreased by heat treatment due to the increase in the rate of their relaxation [194].

Let us consider in more detail methods of decreasing the internal stresses that are most acceptable for application in adhesive-bonded joints. These are based on the increase of the relaxation rate at the time of formation of the adhesive layer, on decrease of the adhesive shrinkage, and on the decrease of overstresses at the adhesive–substrate interface.

4.3.1 Effect of surfactant on internal stresses in adhesive-bonded joints

Addition of the IS agent OP-10 into adhesive based on ED-20 epoxy resin results in abrupt decrease of the internal stresses in the adhesive layer (Fig. 4.8a). With the RS substance L-19, which is close to OP-10 in composition, the internal stresses again decrease, but much less so. Similar dependences are observed when the internal stresses are studied in adhesive-bonded joints based on acrylate adhesives (Fig. 4.8b).

As Fig. 4.8 shows, RS and IS agents have a noticeable effect on the kinetics of both the growth and relaxation of stresses. In the case under consideration, the failure stress of the adhesive-bonded joints and the internal stresses do not correlate (Fig. 4.8a). Thus, addition of RS substances to the adhesive results in increase of the adhesion strength and in decrease of the internal stresses. IS substances decrease both the adhesion strength and the internal stresses. A similar dependence is observed for the acrylate adhesives. The adhesive containing ATG and fluorinated alcohol has maximal adhesion strength and minimal internal stresses.

Let us consider these results. IS substances separating as an individual phase during the oligomer polymerization concentrate in the interstructural regions of the polymer and close to the solid surface, thus increasing the rate of stress relaxation. The adhesive interlayer can "slide" along the boundary polymolecular layer of surfactant relative to the substrates, with a drop in the internal stresses.

RS agents do not form weak interface layers and increase the adhesion interaction of the adhesive with the substrate. It might be supposed that addition of RS substances to the adhesive results in increase of the internal stresses, while the actual picture is quite the opposite. Molecules of RS substances, low-molecular weight sub-

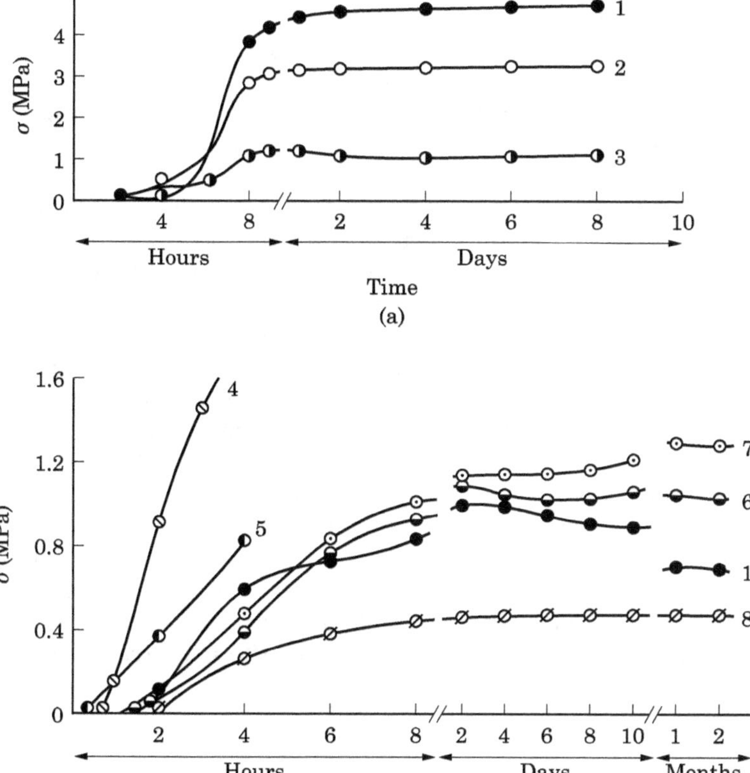

Figure 4.8 Growth and relaxation of the internal stresses in the adhesive interlayer based on epoxy adhesive (a) and on 40% solution of PBMA in MMA (b): (1) without additives. With additives: (2) 3% L-19; (3) 3% OP-10; (4) 2% triallylisocyanurate; (5) 30% ATG and 0.01% triethylenediamine; (6) 10% ATG; (7) 30% ATG; (8) 30% ATG and 2% fluoropentyl alcohol.

stances, are adsorbed on the metal substrate, ensuring a higher level of packing compared with that without RS additive (this is confirmed by the decrease of the interphase surface tension), and thus serve as an intermediate zone between the adhesive and the substrate. In addition, by increasing affinity of the substrate for the polymers, RS agents help to make the molecular properties of the adhesive and the substrate become more similar, which results in decrease of overstress on the interface.

The addition of surfactant into the polymers can cause change of the macromolecular conformation and decrease in the sizes of their aggregates [155]. The packing density of the polymer structural elements on the substrate surface must increase, which also will result in the decrease of overstress on the interface. Actually, as shown in [195], the decrease in the size of polymer structural elements results not only in increase of the adhesion strength but also in decrease of the internal stresses. Thus, even without adsorbing on the polymer–substrate interface, surfactant can decrease the overstress in this zone.

When added to the adhesive, surfactant that can form hydrogen bonds with the polymer macromolecules can promote ordering of its structure [196], which increases the rate of relaxation of the internal stresses. Hence, with addition of RS surfactants this rate remains insignificant (Fig. 4.8a). Moreover, the internal stresses at the initial stage of formation of the joint in adhesives containing RS surfactant are higher than those in adhesives without the additive, whereas the formation of thixotropic structure most abruptly decreases the internal stresses at the initial stage of polymer formation [183].

Thus, when an IS substance is added to the adhesive, the decrease of the internal stresses occurs due to increase of the rate of the stress relaxation and due to "sliding" of the adhesive layer relative to substrates; the most probable mechanism of action of RS substances lies in the increase of the packing density of adhesive molecules on the substrate surfaces and the related decrease of the overstress on the interface.

4.3.2 Controlling internal stresses in adhesive-bonded joints by taking account of the separation in time of the formation of linear and crosslinked polymers

For there to be substantial separation in time of the processes of linear polymer formation and of intermolecular crosslink formation, they would have to proceed by different mechanisms. For the relaxation of the internal stresses that occur at the stage of formation of the linear polymer, the polymer must possess sufficient pliability under creep. Because internal stresses may also appear at the stage of formation of the intermolecular crosslinks, the concentration and rate of formation of crosslinks must be minimal.

Let us consider the above method for decreasing the internal stresses in acrylic adhesives based on 40% solution of PBMA in MMA. The adhesive was cured in the course of copolymerization of MMA and BMA under effect of the b.p.–DMA redox system at room temperature. The cured adhesive has comparatively low elastic modulus of 6.2×10^2 MPa. The polymer has a high rate of relaxation of the internal stresses, which does not exceed 0.5 MPa in the adhesive interlayer.

The internal stresses in the adhesive under consideration can decrease for another reason. As pointed out in Chapter 1, the substrate surface can inhibit the polymerization of MMA and BMA, which plasticizes the adhesive layer on the interface and facilitates the increase of the rate of relaxation of the internal stresses for some time after the adhesive in bulk has cured. In fact, noticeable growth of the internal stresses in the joint begins to be observed only 10–15 min after the adhesive has cured in bulk. The relationship between the residual stresses in a polymer and the differences in rates of curing in bulk and on the surface was discussed in [197].

The linear structure of the macromolecules of the polymers comprising adhesive components and the adhesive's capacity to creep do not actually allow application of the adhesive in liquid solvent media and their use for structural purposes because the long-term strength of the adhesive will be low despite there being internal stresses.

ATG crosslinking was added to increase the strength and resistance of the adhesive to solvents. ATG copolymerizes with MMA and BMA during curing of the adhesive and a copolymer of linear structure results with side-branches containing isocyanate groups. Under the influence of the moisture that diffuses into the adhesive interlayer, the isocyanate groups react one with another and form urea bridges.

$$\left[-CH_2-\underset{COOCH_3}{\overset{CH_3}{\underset{|}{\overset{|}{C}}}}-\right]-CH_2-\underset{CH_2OOCNH(C_6H_3)(CH_3)NCO}{\underset{|}{CH}}-\left[CH_2-\underset{COOC_4H_9}{\overset{CH_3}{\underset{|}{\overset{|}{C}}}}-\right]-$$

$$+\,H_2O \rightarrow \left[-CH_2-\underset{COOCH_3}{\overset{CH_3}{\underset{|}{\overset{|}{C}}}}-\right]-CH_2-\overset{CH_2OOCNH(C_6H_3)(CH_3)NCO}{\overset{|}{CH}}-\left[CH_2-\underset{COOC_4H_9}{\overset{CH_3}{\underset{|}{\overset{|}{C}}}}-\right]-$$

$$\left[-CH_2-\underset{COOCH_3}{\overset{CH_3}{\underset{|}{\overset{|}{C}}}}-\right]-CH_2-\underset{CH_2OOCNH(C_6H_3)(CH_3)NH}{\underset{|}{CH}}-\left[CH_2-\underset{COOC_4H_9}{\overset{CH_3}{\underset{|}{\overset{|}{C}}}}-\right]-$$
$$\hspace{6em} \searrow C=O \swarrow$$
$$\left[-CH_2-\underset{COOCH_3}{\overset{CH_3}{\underset{|}{\overset{|}{C}}}}-\right]-CH_2-\overset{CH_2OOCNH(C_6H_3)(CH_3)NH}{\overset{|}{CH}}-\left[CH_2-\underset{COOC_4H_9}{\overset{CH_3}{\underset{|}{\overset{|}{C}}}}-\right]-$$

TABLE 4.1 Alteration of Acrylic Adhesive Properties with Time

	Time after cementing (days)						
Index	0	1	3	5	10	30	60
Relative concentration of isocyanate groups (%)	100	98	80	71	53	28	22
Gel fraction concentration (%)		0	3	6	16	20	21
Adhesive elastic modulus, E (10^{-2} MPa)		0.5	8.2	9.4	11	12	12.1
Adhesion strength (MPa)		14.6	17.9	11.2	29.3	36.1	38

Table 4.1 presents data on the alteration with time of the isocyanate groups in the adhesive, the amount of gel fraction, the elastic modulus, and the adhesion strength. The adhesive layer was 0.3 mm thick; the ATG concentration in the adhesive was 10%. The concentration of the isocyanate groups was determined by differential IR spectroscopy of the band of valence oscillations at 2280 cm^{-1}. The adhesive film for investigation was obtained by separating it from the foil substrates. Acetone was used as a solvent for determining the concentration of gel fractions. As Table 4.1 shows, increase of the gel fraction concentration, of the elastic modulus, and of the adhesion strength accompanies consumption of the isocyanate groups in the adhesive.

Let us consider the effect of crosslink formation on the internal stresses in the adhesive layer. After the adhesive is cured (in one hour after cementing) the internal stresses are low both in the adhesive with ATG and in that without ATG (see Fig. 4.8b). Crosslink formation in the adhesive with ATG does not cause increase of the internal stresses if ATG concentration is 10%; the internal stresses increase at 30% concentration of ATG.

The internal stresses in the adhesive can be decreased using surfactant before formation of the intermolecular crosslinks. Addition of 2% of fluorinated alcohol into the adhesive containing 30% of ATG decreased the internal stresses almost 3-fold.

Thus, it is the separation in time of the formation processes of the linear polymer and the intermolecular crosslinks that produces adhesive-bonded joints with low internal stresses. In the adhesive containing 20% of gel fraction but in which the crosslinks were formed using triallyl isocyanurate along with the polymer formation, the internal stresses were substantially higher than in the adhesive containing ATG only (see Fig. 4.8b, curve 4). Increase of the crosslink formation rate by adding triethylenediamine, as catalyst of the reaction of the isocyanate groups with water, to the adhesive simultaneously with addition of ATG also increased the internal stresses (see Fig. 4.8b, curve 5).

4.3.3 Decrease of internal stresses in adhesive-bonded joints using adhesives based on interpenetrating networks

Obtaining interpenetrating networks (IPNs) is in principle a new method of blending nonmelting and insoluble three-dimensional polymers. IPNs are known to be characterized by a number of thermodynamic and physical-mechanical peculiarities. Application of IPNs based on crosslinked polymers of various chemical compositions should produce adhesives with a wide range of properties.

The components of IPNs are thermodynamically incompatible, and a transition region of two phases of these components is formed in the system. All the properties are determined by the existence and characteristics of this region. In particular, the density of macromolecular packing in IPNs is less than in the individual networks, which results in decrease of shrinkage during IPN formation. The mobility of the chain segments between the crosslinks of the network is higher than in the individual networks, which can be explained by the formation of a loose transition region in the systems of such a type; the extent of the looseness depends on the ratio of components. The increase of mobility of the chain segments may result in increase of the rate of relaxation of the internal stresses in polymers.

Thus, the application of IPNs as adhesives allows control of such properties as shrinkage and relaxation processes, which directly affect the internal stresses inside the adhesive layer. New ways of decreasing the internal stresses in the adhesives are based on differences in the curing rates of two individual networks that comprise the IPN. We consider adhesives consisting of a mixture of polyester and polyurethane. The IPN resulted from simultaneous formation of two networks. For this purpose a styrene solution of maleate–phthalate oligodiethylene glycol (PE resin) was blended with MDI, which is the product of reaction of 1 mole of oligodiethylene glycol adipate with 2 moles of TDI, and this was cured with 2% MEKP(O) and 1% CN (cobalt naphthenate). The end hydroxyl and carboxyl groups of the PE resin were prebonded by phenyl isocyanate to prevent chemical reaction between the individual networks. The resin was cured at 20°C for 40 min. MDI was polymerized by the reaction of the isocyanate groups with atmospheric moisture that diffused into the adhesive layer (the polymerization ran for 90–100 days) and was monitored by IR spectroscopy of the band of valence oscillations of the isocyanate groups in the region of 2200 cm^{-1}. The ratio of polyester to MDI was assumed to be 1:0.8.

In connection with the low rate of MDI polymerization one can consider two stages of the state of the adhesive in the layer. At the first

TABLE 4.2 Change of Elastic Modulus of IPN Adhesive Layer with Time

	Modulus of elasticity (MPa)	
Time after adhesive cure (days)	Torsion shear	Tension shear
3	3.8	3.6
10	0.5×10^2	0.6×10^2
100	2.1×10^2	4.9×10^2
200	2.2×10^2	4.9×10^2

stage, after the PE resin is polymerized, the adhesive comprises a polyester plasticized by MDI; at the second stage, the IPN is formed.

The properties of the system change in the course of IPN formation: both the modulus of elasticity (Table 4.2) and the adhesion strength (Fig. 4.9) increase. The adhesion strength of IPN exceeds that with polyester or polyurethane adhesive. One reason is the low level of the internal stresses in the adhesive layer of IPN compared with polyester (Fig. 4.10). The decrease of the internal stresses in this case is explained by the increased rate of relaxation of the internal stresses in the adhesive layer due to the presence of a substantial amount of liquid MDI. Because polymerization of MDI runs for a long time, the internal stresses manage to relax completely; even 100 days after the adhesive polymerization, the stresses do not exceed 1–1.5 MPa.

A second reason for the decrease in internal stresses in IPN-based adhesives is their low shrinkage. To separate these two factors, the rate of MDI curing was increased by adding 2,4-butylene glycol to the adhesive. As Fig. 4.9 displays, the internal stresses increased somewhat but nevertheless remained much lower than those of the pure polyester adhesive. The shrinkage of the adhesive layer, which causes the internal stresses, was determined by Equation (4.49). For the IPN-based adhesive (with added 2,4-butylene glycol) it appeared on average to be 2.6 times less than that of the adhesive based on polyester.

4.3.4 Methods of decreasing edge internal stresses in adhesive-bonded joints

The methods of controlling the edge stresses are based on adding substances that produce different mechanisms of adhesive cure in the bulk and at the interface with air. This results in alteration of the physical-mechanical properties of the adhesive depending on the distance to the edge—i.e., decrease of the adhesive modulus of elasticity at the edge and increase of the rate of relaxation of the internal stresses.

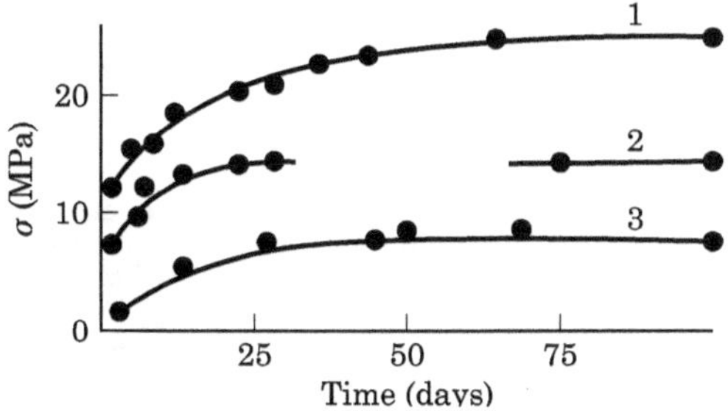

Figure 4.9 Change of the adhesion strength of the adhesive-bonded joints with time: (1) adhesive based on IPN; (2) polyester adhesive; (3) polyurethane adhesive.

Let us consider methods of decreasing the edge stresses of a system based on unsaturated compounds. For this purpose the capacity of oxygen to inhibit the process of radical polymerization of unsaturated compounds is exploited. In addition to other factors, the location of the layer under study determines the oxygen concentration in the system.

At the stage of inhibition of radical polymerization by the oxygen, the following reactions occur [200]:

Development of oxidation chain

$$R + O_2 \xrightarrow{k_1} RO_2 \qquad RO_2 + M \xrightarrow{k_2} R$$

Growth of polymer chain

$$R + M \xrightarrow{k_3} R$$

Termination of chain

$$RO_2 + RO_2 \rightarrow \text{nonreactive products}$$

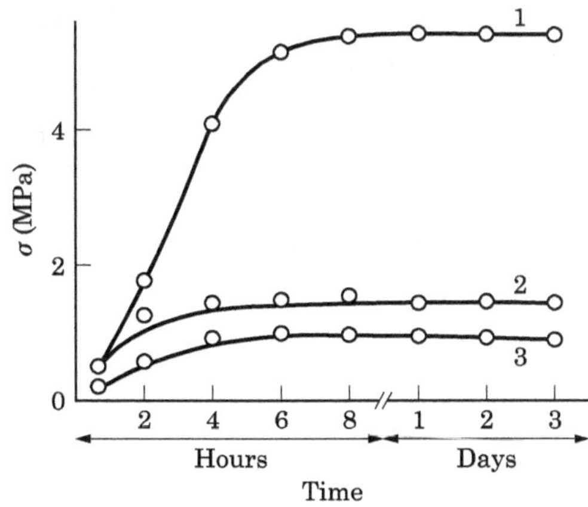

Figure 4.10 Kinetics of internal stress elevation and relaxation: (1) polyester adhesive; (2) IPN-based adhesive with added 2,4-butylene glycol; (3) the same with additives.

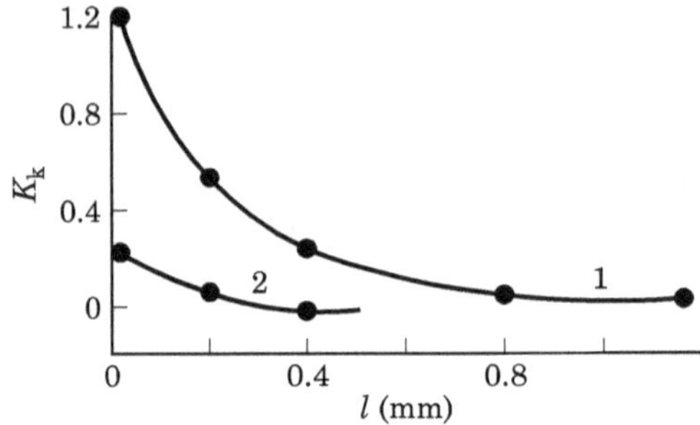

Figure 4.11 Effect of the crosslinking/plasticizing agent on the internal stress concentration (K_k is the concentration coefficient) at the edges of the adhesive-bonded joints

Given that, for example, for MMA $k_3/k_1 = 3 \times 10^{-5}$, practically all the polymer radicals react with oxygen but not with a monomer, although the concentration of the latter is 10^4 times higher than the oxygen concentration (oxygen solubility in MMA is 10^{-3} mol/l when saturated with air). As the reactivity of RO_2 relative to monomer is far less than that of R, the conversion of R into RO_2 is similar to chain termination. The monomer that has not reacted volatilizes from the adhesive layer.

Let us add to an adhesive a substance containing two or more double bonds of allyl type, which is not able to undergo the reaction of radical homopolymerization. In the bulk of the adhesive this substance copolymerizing with a monomer serves as a crosslink of the polymer chains. Due to the inhibiting effect of the oxygen and to the volatilization of the monomer, it cannot copolymerize at the edge and is in the initial state in the adhesive, exerting a plasticizing effect and increasing the rate of relaxation of the internal stresses. Thus, depending on the distance to the boundary between the adhesive and the air, the same substance may act as a crosslink or a plasticizer.

As Fig. 4.11 shows, the addition of 10% of $CH_2{=}CH{-}CH_2{-}O{-}CO{-}NH_2)C_6H_{12}$ to an adhesive results in practically complete disappearance of the edge stresses. A 40% solution of PBMA in MMA was used as adhesive, which was cured at standard temperature under the effect of b.p. and DMA. The modulus of elasticity of the adhesive film is 1.5×10^3 MPa, and that of the 0.2 mm thick polymer on the boundary with air is 22 MPa.

Chapter

5

Cementing and Operation of Adhesive-Bonded Joints in Liquid Media

5.1 Cementing in Liquid Media

The principles considered so far of controlling the properties of adhesives are related to the cementing of dry surfaces. Practical experience reveals the necessity for adhesives to be operated in different liquid media. In fact the use of adhesives is often demanded outdoors where it is almost impossible to remove liquid from the surfaces to be cemented—for example, in the repair of oil pipelines or oil tanks when working in rain or actually under water. The presence of liquids on the substrate prevents the formation of a strong adhesive joint. Because of the possible formation of water adsorption films of substantial thickness on the surfaces to be cemented, even atmospheric humidity affects the adhesion strength in the course of cementing.

The formation and operation of adhesive-bonded joints in liquids is characterized by a number of features, some physical-chemical aspects of which will be considered in the present chapter.

As has been stressed in earlier chapters, wetting of the substrate by an adhesive is one of the principal conditions for formation of a strong adhesive-bonded joint. Without the formation of a close contact between the adhesive molecules and the solid surface, the formation of a strong adhesive-bonded joint is in fact impossible.

When applying adhesive to a solid in a liquid environment, the process of wetting the solid surface with adhesive, i.e., of forcing liquid out, is of primary importance. When the cementing is performed in air, the adhesive-bonded joint is formed even with incomplete wetting. If the adhesive is applied to a solid surface in a medium of another liquid, their simultaneous presence on the surface is observed only for a narrow range of interfacial angle of wetting of the solid by the liquids. In other cases, the solid surface is wetted by only one liquid. Thus, if the adhesive does not preferentially wet the solid surface in a liquid ambience, there will be a layer of liquid between the adhesive and the surface. In this case the strength of cementing is close to zero. It is impossible to increase the degree of filling of the surface defects with the adhesive under these conditions because of the incompressibility of liquids.

Analysis of the theoretical underpinnings indicates [201–206] that to provide for wetting of the solid surface by adhesive it is necessary first to achieve the minimal value of the interphase tension between the adhesive and the solid. Actually, cementing of low-energy surfaces (polymers, for example) under water poses no problem, while cementing of metals is impeded because of the higher affinity of water to metals than of adhesive. At first glance, hydrocarbons do not seem to hinder formation of strong adhesive-bonded joints because the affinity of the majority of the adhesives to metal is higher than that of hydrocarbons. In practice it is very difficult to perform cementing in oils, lubricants, and diesel fuels because they contain impurities with surface-active properties [202] that decrease the interphase tension between oil and metal. Additionally, many oil products have high viscosity, which strongly decreases the rate at which they can be forced from the surface by the adhesive.

The addition of surfactant to the adhesive changes the surface properties of the polymers, and consequently changes the wetting of the solid surface. As Fig. 5.1 shows, the addition of 1% OP-10 or of the fluorinated alcohol $CF_3(CF_2)CH_2OH$ to an adhesive based on a 40% solution of PBMA in MMA decreases the interfacial angle for wetting of the steel surface by the adhesive from 72° to 30–50°.

The liquid in which cementing is performed can affect the process of adhesive polymerization. Thus, when cementing under water with VAK and Sprut-4 adhesives containing MMA as an active solvent, the rate of cure increases to 1.2–1.5 times that when cementing in air. This seems to be explained by the higher content of oxygen, which inhibits MMA polymerization, in the air than in water. When cementing under water, the molecular weight of the polymethyl methacrylate can also increase, which results in change in some of the physical-mechanical characteristics of the adhesive. It is well known that when 1.25% of water is added to a monomer the molecular weight of the polymethyl methacrylate

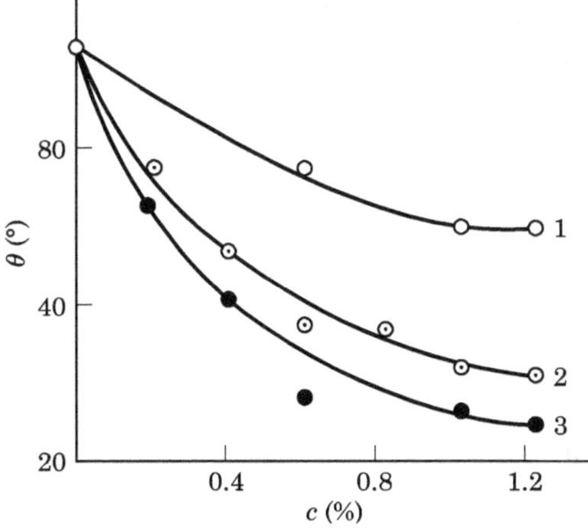

Figure 5.1 Dependence of the interfacial angle of wetting of steel by the cured adhesive under water on the surfactant concentration: (1) without surfactant; (2) OP-10; (3) fluorinated alcohol.

increases almost 2-fold from 3.68×10^6 to 7.13×10^6. This occurs because water is not a solvent for polymethyl methacrylate and, although the polymerization process remains homogeneous with this addition of water, the increase of the coiling probability of the growing macrochains decreases their termination rate [208]. With the increase of the water content in the monomer, the system becomes heterogeneous and adopts a colloid-dispersed state that does not result in further increase of the molecular weight of the polymethyl methacrylate formed.

The water present can chemically react with components of the adhesives. Thus, adhesives containing large quantities of isocyanate groups cannot be used for cementing under water if their time and area of contact with water will be considerable because the carbon dioxide released forms pores and pits in the adhesive (Equation 5.1).

$$2R{-}NCO + H_2O \rightarrow R{-}NH{-}\underset{\underset{\displaystyle O}{\|}}{C}{-}NH{-}R + CO_2 \tag{5.1}$$

The water can also inhibit the polymerization of unsaturated compounds at the boundary between the adhesive and the metal, which decreases the adhesion strength. To obtain a strong adhesive-bonded joint the adhesive also must not dissolve and swell in the liquid medium in which the cementing is performed.

In the case of selective wetting of a solid surface by liquid 1 in a medium of liquid 2, mechanical work has to be performed at the interface to separate a drop of liquid 2 from the surface. Similar processes have been fairly well investigated in studies of the mechanism of detergency. Equation (5.2) determines the specific mechanical work necessary for complete replacement of liquid 2 on the solid surface by liquid 1:

$$A_{\text{liq}} = v_{\text{liq1 liq2}}(1 + \cos\theta) + v_{\text{liq1 liq2}}\left(\frac{S_1 - S_2 - S_3}{S_2}\right) \tag{5.2}$$

Here, S_1, S_2, S_3 are the interfaces of solid–liquid 1, solid–liquid 2, and liquid 1–liquid 2, and v is the rate of separation. When cementing in water and oil products, the adhesive is liquid 1 and the medium is liquid 2. Equation (5.2) indicates that spontaneous wetting (without mechanical work to the interface) is possible when the surface tensions of the adhesive and the liquid are equal, i.e., for $\sigma_{\text{ad liq}} = 0$, although the maximal value of $\sigma_{\text{ad liq}}$ is desirable because it is directly proportional to the adhesion thermodynamic work.

Thus, it is impossible to obtain a strong adhesive-bonded joint when cementing in liquid media without supplying mechanical work to the adhesive–substrate interface, even when thermodynamic conditions are appropriate for selective wetting of the substrate by the adhesive. Practically, the mechanical work can be supplied by mutual displacement of the adhesive and the cemented surfaces, for example, by pumping the adhesive into the gap between the surfaces, by pressing the surfaces together so that the adhesive is expelled from the gap, by ultrasonic treatment of the adhesive layer, and so on.

The influence of this work on the strength of the adhesive-bonded joint is illustrated by Fig. 5.2. Compounds containing surfactant and capable of cementing in water and oils were used as adhesives. The compositions of the adhesives were as follows (parts by mass):

A. 40% solution of PBMA in MMA	100
Product $(CH_3(C_6H_3)(NCO)NHCOOCH_2{=}CH_2$	10
B. PN-1 resin (styrene solution of PDEGMP)	100
Polyurethane prepolymer (product of reaction of 1 mol of polyethylene glycoladipate and 2 mol of TDI)	50
Product $CH_3(C_6H_3)(NCO)NHCOOCH_2{-}CH{=}CH_2$	30
C. NPS609-21S (solvent TGM)	100
Polyurethane prepolymer	50
Product $CH_3(C_6H_3)(NCO)NHCOOCH_2{-}CH{=}CH_2$	30

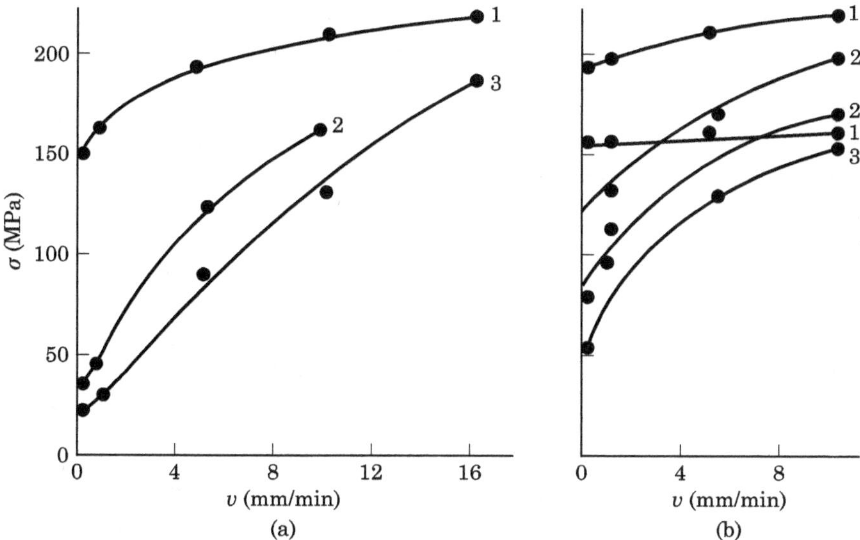

Figure 5.2 Dependence of the failure stress of the adhesive-bonded joint on the feed rate between two surfaces of adhesives C (a), A and B (b) in air $(1, 1')$, in water $(2, 2')$, and in oil (3). Primed numbers refer to adhesive A.

The viscosity of the adhesives was increased to 20 PA · s by adding AM-1 aerosil. The adhesives were cured at room temperature by the redox system b.p.–DMA for adhesive A, and by MEKP(O)–CN for adhesives B and C. The curing time was 0.5–1 h. The specimens for cementing were made of St3 steel in the shape of a dolly. The specimens were tested 24 h after cementing.

Application of mechanical work to the adhesive–metal interface (by adhesive flow through the gap between the dollies at various speeds) had greater effect on the adhesion strength when cementing was performed in the liquids than in the air (see Fig. 5.2). This can be explained by the fact that the work done facilitates the selective wetting of the substrate by the adhesive: the flowing adhesive picks up drops of liquid from the substance surface and removes them.

5.2 Effect of Liquids on the Properties of Adhesive-Bonded Joints

When adhesive-bonded joints are operated in liquids, their adhesion strength frequently decreases. For example, the decrease in strength of steel adhesive-bonded joint held in water at 20°C for 72 h is 22% (21.0) for the VK-3 adhesive, 50% (26.7) for VK-13M, 18% (21.0) for VK-32-200, 16% (16.1) for VK-24M, 5% (27.9) for VK-24-50, and 16% (26.5) for VK-24-100 adhesive (figures in parentheses are the initial strength of the joint in MPa). When treated with cold or hot water for 7–30 days, the strength of the epoxy-based adhesive-bonded joints decreases by 40–50%. Even when the joints are operated in air their strength depends on the air humidity [210].

The water resistance of adhesive-bonded joints also depends on the nature of the materials cemented. The maximal decrease of strength is characteristic for aluminum and seems to occur due to the high hydrophilicity of the porous oxide film on the aluminum surface. The water resistance of steel adhesive-bonded joints increases if the steel is etched.

Impurities in the water have a great effect on the water resistance of the joints. The maximal strength decrease of glass-reinforced plastics is caused by distilled water. The strength of steel joints cemented with adhesive based on the Epoxy-1000 resin decreased by 44% when subjected for 60 days to rainwater, by 21% with river water, and by 18% with sea water. Adhesives based on PN polyester resin form comparatively water-resistant joints with glass-reinforced plastics [211], while steel joints cemented by these adhesives are not water-resistant. When subjected to warm (60°C) water for 400 h and to boiling water for 120 h, the strength of steel joints bonded by polyester adhesives decreased 3–4 times.

One of the principal reasons for failure of the adhesion bonds is a specific adsorption reaction of the medium with the material to be cemented at the boundary with the adhesive. There is an adsorption substitution of adhesive–substrate bonds by medium–substrate bonds. Surface structural defects that are present in each solid are the first to be subjected to adsorption. It is to be expected that the probability of appearance of such defects is higher at an interface of two materials with different properties. The rate of penetration of the medium along the polymer–substrate interface frequently substantially exceeds the rate of diffusion of the medium in pure polymer [212]. Adsorption substitution of the polymer macromolecules by water molecules on the metal surface explains the low water resistance of such adhesive-bonded joints as fluoroplastic–steel or polyethylene–steel [34]. The adhesion strength, which decreases during hold-up of adhesive-bonded joints in water, is frequently reestablished after the joints are dried [213].

It can be assumed that when metals are cemented by nonpolar adhesives there is a layer of water molecules (always present on the metal under normal conditions) between the adhesives and the metal surface. The reversibility of the adhesion strength will be governed by the thickness of the water layer at the polymer–metal interface. It should be noted that the properties of thin water layers on metal surfaces differ from those of bulk water [214]. Thin layers of polar liquids, including water, on hydrophilic surfaces are known to have abnormally high shear elasticity. From this point of view, the decrease of adhesion in liquid water or in humid atmospheres is explained by the increase of the water layer thickness at the polymer–substrate interface. When the properties of the water in the layer approach those of bulk water, the strength of the adhesive-bonded joint drops practically to zero. The decrease of the layer thickness at the interface when the adhesive-bonded joint is dried results in enhancement of the strength of the joint, i.e., in reestablishment of adhesion.

Enhancement of the reliability of adhesive-bonded joints operated in liquid media is a high-priority task. A metal–polymer interlayer can be exploited for this purpose. Substances forming strong bonds with both the polymer and the substrate are often used as interlayers. For example, to enhance the water resistance of some coatings on glass, silanes are applied that form chemical bonds with both glass and coating [215]. To enhance the adhesion strength, as well as water resistance, of polyethylene with metals a monolayer of fatty acids can be applied to the metal surfaces. Mechanical treatment, such as finishing, of metal surfaces with a fatty acid present further increases the liquid-resistance of coatings. The water resistance of polyester compounds can be enhanced by adding surfactant [216]. The water resistance of adhesives based on epoxy and alkyd resins can be enhanced by adding low-viscosity organosilicon liquids. Tars and bitumen are used [217] to modify epoxy resins, whereby increased water resistance is accompanied by decrease of the adhesion strength.

As this brief literature review displays, at present there are no scientifically substantiated and practically applicable methods for enhancing the water resistance of adhesive-bonded joints.

Let us consider the effect of water on the adhesion strength of elastic polyurethane adhesives (for example), which exhibit the lowest water resistance. The adhesives were based on MDI which was synthesized from polyoxypropylene glycol (POPG) or polytetramethylene (PTMG) glycol of molecular weight 1000 or 2000 and TDI or diphenylmethane diisocyanate (DPMDI). In the course of synthesis of MDI, the ratio of isocyanate and hydroxyl groups was assumed to be 2:1, i.e., the MDI looked like:

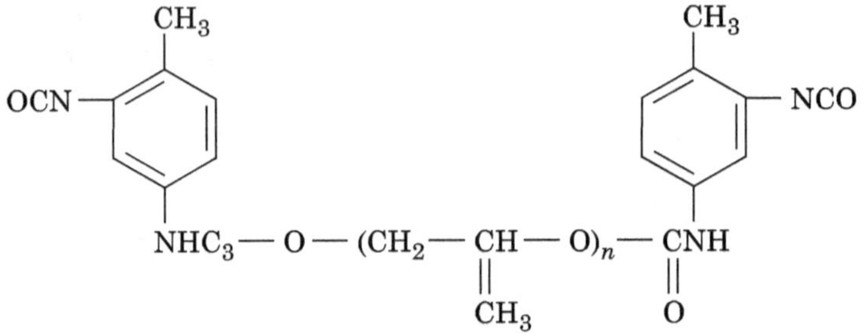

CH_3 CH_3

OCN — NCO

$NHC_3—O—(CH_2—CH—O)_n—CNH$

CH_3 O

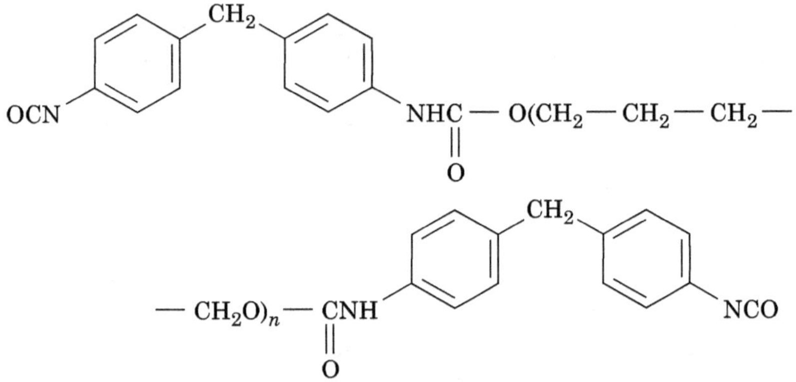

CH_2

OCN

$NHC—O(CH_2—CH_2—CH_2—$

O

CH_2

$—CH_2O)_n—CNH$ NCO

O

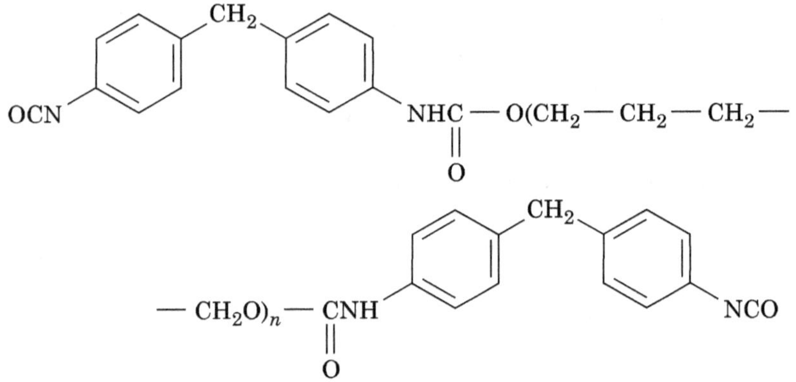

CH_2

OCN

$NHC—O(CH_2—CH_2—CH_2—$

O

CH_2

$—CH_2O)_n—CNH$ NCO

O

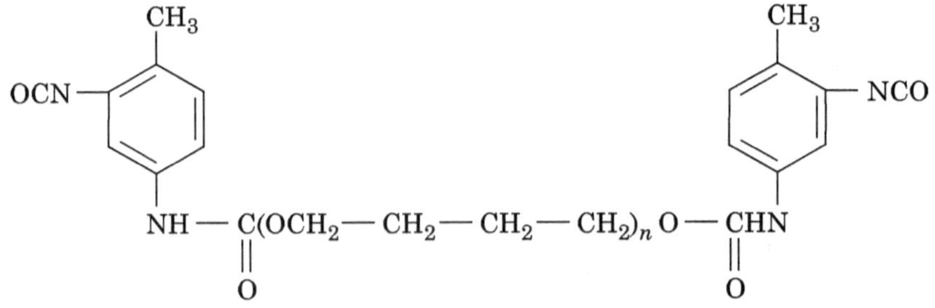

CH_3 CH_3

OCN — NCO

$NH—C(OCH_2—CH_2—CH_2—CH_2)_nO—CHN$

O O

TABLE 5.1 Composition of Adhesives

Diisocyanate	Polyester	Polyester molecular weight	A-200 aerosil
$(C_6H_4NCO)_2CH_2$	$HO[CH_2CH(CH_3)O]_nH$	1000	0
$(C_6H_4NCO)_2CH_2$	$HO[CH_2CH(CH_3)O]_nH$	1000	10
$CH_3C_6H_3(NCO)_2$	$HO[CH_2CH(CH_3)O]_nH$	1000	10
$CH_3C_6H_3(NCO)_2$	$HO[CH_2CH(CH_3)O]_nH$	2000	10
$(C_6H_4NCO)_2CH_2$	$HO[CH_2CH_2CH_2CH_2OI]_nH$	1000	10

The composition of the adhesives is given in Table 5.1.

The adhesives were cured by atmospheric humidity and by moisture on the surfaces to be cemented and on the surface of the filler particles.

The water resistance of the adhesives was determined by measuring the adhesion strength after holding the cemented specimens in water as well as according to a specially developed procedure based on the change of electrical resistance as a result of water penetration. Water resistance was evaluated in the course of water penetration through the coating, along the adhesive–substrate boundary, and through the adhesive layer. The authors of [218] demonstrated that this method can be applied to determining water resistance for coatings.

The test specimens comprised two metal cylinders, one of which had a central hole. One of the surfaces was coated with adhesive to which a platinum electrode was applied. The free end of the electrode, insulated from the body of the specimen, was brought out through the hole. After measurement of the distance between the electrode and the edge of the specimen, the cylinders were cemented and, after cure, were placed in a vessel containing distilled water.

For determination of the rate of water penetration along the adhesive–substrate interface, the specimens were prepared in the same way but the electrode was placed between the substrate and the adhesive, being insulated from the metal by a water-soluble film.

The results of the water resistance study using the electrical resistance change are displayed in Fig. 5.3b. Water penetration along the adhesive–substrate boundary proceeds much faster than through the adhesive layer or coating. It seems to be governed by the selective sorption of water by metal, by defects of the polymer supermolecular structure at the substrate surface, and by the presence of gas bubbles at the interface due to the reaction of isocyanate groups with the adsorbed moisture. The water resistance of adhesive-bonded joints is highly dependent on the origin of the isocyanate and polyester components. MDI synthesized on DPMDI produces more water-resistant

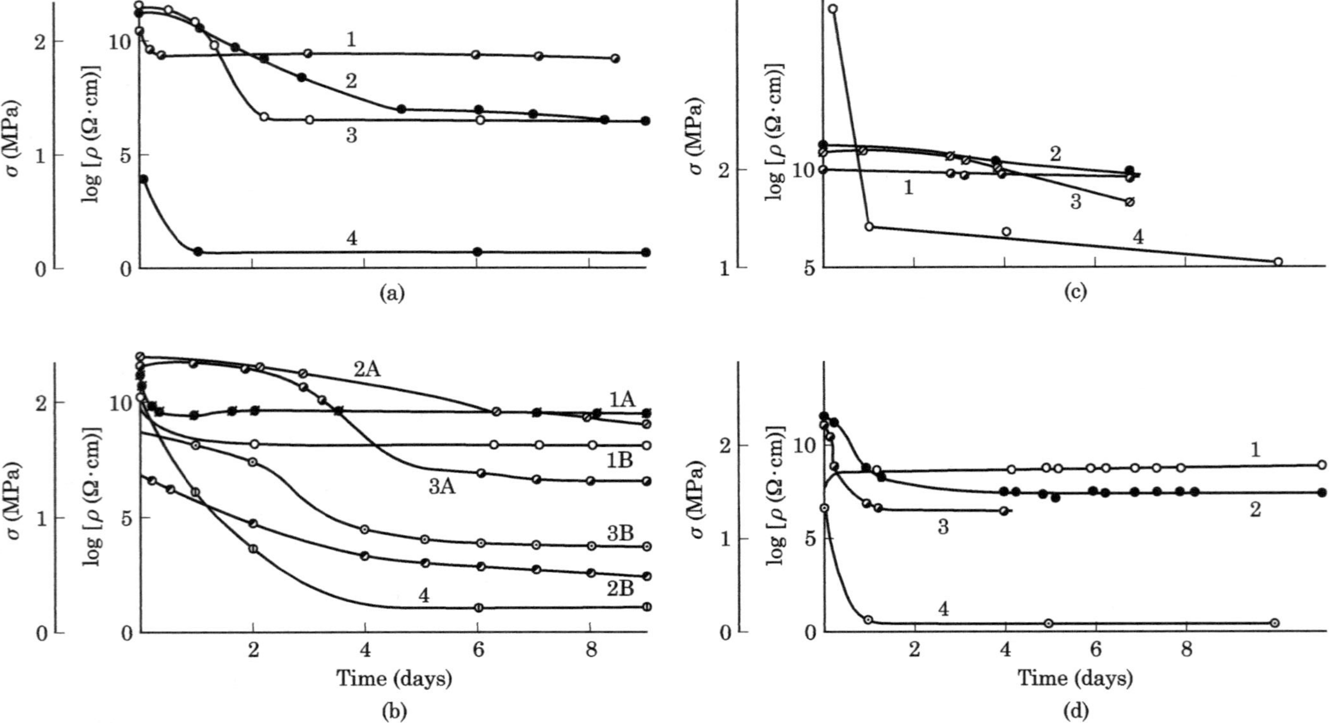

Figure 5.3 Effect of water on the specific electrical resistance and the adhesion strength (steel 3–steel 3) of the adhesive based on macrophenylmethane diisocyanate (a), and on macrodiisocyanates made of POPG-1000 and diphenylmethadiisocyanate (b), POPG-2000 and TDI (c), and POPG-1000 and TDI (d) containing 0.2% catalyst and 10% filler (b, A; c; d), and without filler (b, B): (1) adhesive 0.25 mm thick applied to steel foil; (2) electrode in the bulk adhesive; (3) electrode on the substrate–adhesive boundary; (4) pull-off adhesion strength.

adhesive-bonded joints compared with MDI based on TDI. So, for the first 2–3 days the specific electrical resistance of the adhesive-bonded joint made with MDI based on TDI drops virtually at the moment of contact with water (Fig. 5.3b, d).

The addition of aerosil as a filler of the polyurethane adhesive hinders the rate of water penetration. Along the boundary between unfilled adhesive and steel, water penetrates twice as fast as along the boundary with the filled adhesive; and through the bulk of the unfilled adhesive, water penetrates four times faster than through that of the filled adhesive. The specific electrical resistance of a water-submerged coating made of filled MDI is higher throughout the test range than that of the unfilled coating (Fig. 5.3b).

The enhancement of MDI water resistance with added aerosil is caused both by the additional structurization of the adhesive when the isocyanate groups interact with the active centers of the filler surface, and by the adsorption of the diffusing moisture by aerosil.

The water resistance of polyurethane adhesive depends to a great extent on the molecular weight of the initial polyester (Fig. 5.3c, d). As this increases, the electrical resistance and water resistance of the adhesive decrease. This is displayed most clearly when the water penetrates along the adhesive–steel boundary. Similar ($3 \times 10^6 \, \Omega \cdot \text{cm}$) specific electrical resistance was obtained with the polyurethane adhesive based on POPG-1000 for 4 days, and with the polyurethane adhesive based on POPG-2000 for 1 day.

Increase of the POPG molecular weight noticeably decreases not only the adhesive's water resistance but also the adhesion strength (Fig. 5.3c, d, curves 4), which seems to be related to the decrease of the concentration of urethane groups.

The initial effect of water on the coating can increase its electrical resistance (Fig. 5.3d). This seems to be related to the adhesive post-cure. The water diffuses very slowly through the coating, since the noticeable drop in electrical resistance begins to be observed after some months of submersion of the coating in water.

Replacement of POPG by PTMG of the same molecular weight substantially decreases the rate of water diffusion along the adhesive–steel boundary and along the adhesive, and increases the original adhesion strength and delays its decrease in water (Fig. 5.3a).

Note that in all cases the initial strength is reestablished after the adhesive-bonded joint is dried. Thus, the drop in strength of adhesive-bonded joints operated in water is explained mainly by diffusion of water at the adhesive–substrate interface. Methods that increase water resistance decrease the rate of water diffusion along the interface, although it still takes place. Increase in the rate seems to be caused by both adsorption and diffusion mechanisms of the medium.

Resolution of the effect of the medium into separate components is of great interest, but the problem presents substantial difficulties in that the various factors act simultaneously and the result is not simply the sum of separate components but the resultant of their interaction. In some cases it is advisable to restrict the approach to treatment of the predominant mechanism, which in our opinion is adsorption. In this case the principal means of increasing the water resistance of the joints is the creation of conditions that decrease or eliminate the selective sorption of water onto the substrate surface under the adhesive layer. This condition is decrease of the interphase tension of solid adhesive–substrate.

Figure 5.4 shows the correlation between the water resistance of steel adhesive-bonded joints and the adhesive–mercury interphase tension. Given the different states of aggregation of the steel and the mercury, such a correlation can only be qualitative, but it nevertheless clearly indicates ways of controlling the water resistance of adhesive-bonded joints.

The epoxy, acrylate, polyurethane, and polyester adhesives described earlier in Chapter 2 were used for investigations, and RS agents were used to control the interphase tension. Water resistance was evaluated by the retention of strength of the joints after hold-up in water for 30 days; no noticeable hydrolysis was observed in this period.

For interphase tension below 220–230 mN/m and use of hydrolytically stable adhesives, the joints do not fail along the adhesive–metal interface even after hold-up in water for periods of years. Presumably with this interphase tension the process of substitution of adhesive

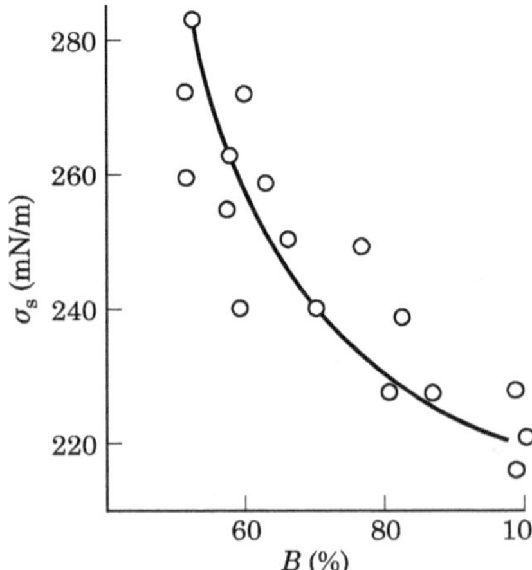

Figure 5.4 Dependence of the water resistance B of adhesive-bonded joints on the adhesive–mercury interphase tension.

macromolecules by the water molecules at the metal surface is energetically disadvantageous, which implies retention of adhesion strength with time. For example, adding two RS agents, $CF_3(CF_2)CHOH$ and $CH_3C_6H_3(NCO)NHCOOCH_2CH{=}CH_2$, to the adhesive based on a 40% solution of PBMA in MMA produced an adhesive–mercury interphase tension less than 230 mN/m. The adhesion strength of metal joints cemented with adhesive does not decrease even after six years in water.

Let us consider a particular example of the effect of RS substances on the water resistance of adhesive-bonded joints. Adhesives based on unsaturated polyester resins, such as PN-1, are distinguished by low water resistance. The influence of water on a steel joint cemented by such an adhesive actually results in some initial increase of the specific electrical resistance along the adhesive–steel interface and then in an abrupt drop (Fig. 5.5). The increase is explained by more complete consumption of the monomer in the system. When ATG is added to the adhesive (which decreases the interphase tension) the specific electrical resistance stabilizes after a drop. The decrease seems to be related to the processes of relaxation of the internal stresses in the adhesive interlayer. The stresses facilitate the diffusion of liquids in polymeric materials, in particular the stress concentration at the polymer–metal interface.

Thus, minimizing the adhesive–substrate interphase tension is a necessary condition for obtaining both strong and liquid-resistant adhesive-bonded joints. But it is not necessarily sufficient, because the strength and the water resistance can be determined by various factors, such as internal stresses, low hydrolytic stability, etc.

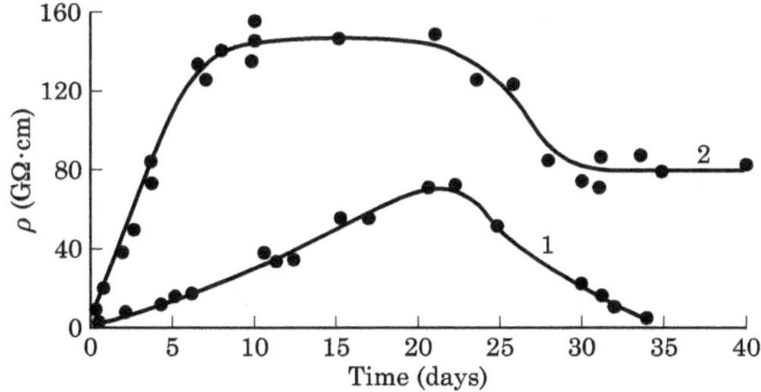

Figure 5.5 Dependence of the specific electrical resistance at the polyester adhesive–steel interface on the holding time of the joint in water: (1) initial adhesive; (2) with addition of 10% ATG.

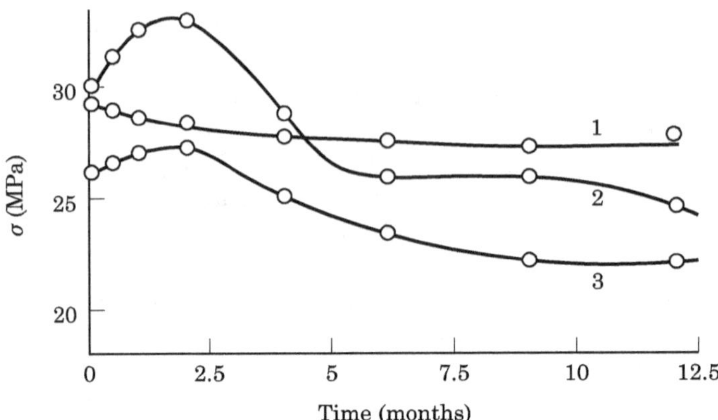

Figure 5.6 Effect of time of holding in water on the adhesion strength of joints based on (1) MMA; (2) PM 609-21M resin; and (3) ethyl methacrylate.

Figure 5.6 shows the dependence of the adhesion strength of steel adhesive-bonded joints on the holding time in liquids. Acrylate (curves 1 and 3) and polyester (curve 2) adhesives were used. In this particular case the polyester adhesive was based on PN 609-21M unsaturated polyester resin, and the acrylate adhesives were based on MMA (1) and ethyl methacrylate (3).

The adhesive-bonded joints were placed into water for 3 months after cementing. It is evident that the adhesion strength changes over time, although the failure of the adhesive-bonded joints is always of cohesive character. Thus, when operating adhesive-bonded joints in liquid media it is necessary to take into account the effect of the media on the polymeric adhesive itself.

The strength characteristics of polymeric materials deformed in liquid media depend on the change of the free surface energy of the system and of the material structure at the tip of the fracture crack, which govern the successive processes of adsorption and diffusion of the medium [219–222]. Penetration of liquid medium into the materials at the crack tip can induce failure of the polymer intermolecular bonds and the formation of weaker polymer–liquid–polymer bonds, facilitating conformational displacements of the polymer chains. This is why the orientation of the polymeric chain segments in micro-bulk at the crack tip is more intensive when the material is deformed in liquid than in air [219]. Consequently, in general the strength of polymeric materials in liquid media will decrease due to adsorption of the medium and to the formation of weaker bonds, but the increase of the rate of orientation at the tip of the fracture crack

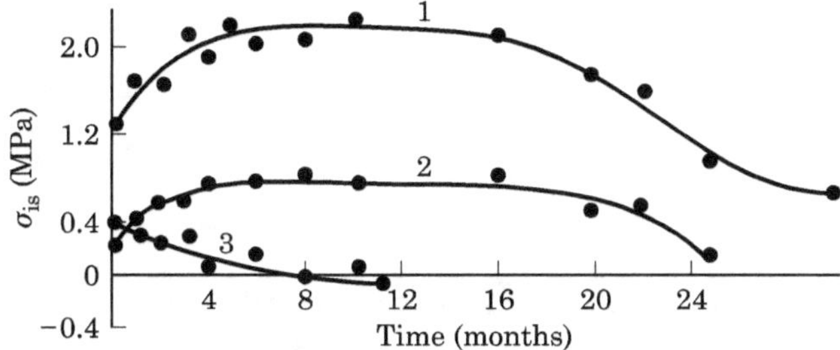

Figure 5.7 Kinetics of internal stress elevation and relaxation in the adhesive interlayer based on 40% solution of PBMA in MMA in various media: (1, 2) water; (3) gasoline; the adhesive interlayer is 0.008, 0.016 and 0.003 mm thick, respectively

can enhance the strength. Thus, the cohesion strength can decrease or increase compared with the strength in air, depending on the origin of the polymer and the medium.

The liquid medium diffusing into the adhesive interlayer plasticizes it and causes a volume change that affects the internal stresses. Figure 5.7 indicates that the rate of relaxation of the internal stresses in adhesive-bonded joints in water or gasoline is substantially increased. The increase of the adhesive layer volume can change the sign of the internal stresses, as in the case of gasoline. Stresses due to humidity, shrinkage, and thermal effects can decrease the strength of the adhesive-bonded joints. Methods of decreasing the humidity stresses and of increasing the hydrolytic stability of adhesives lie in the application of space-linked polymers resistant to polymer hydrolysis, in increasing their hydrophobicity by adding fluorine- or silicon-containing additives, and in adding water adsorbents and hydrophobic fillers.

Chapter

6

Adhesion and Molecular Mobility of Filled Polymers

6.1 Control of Polymer-to-Solid Surface Adhesive Bond Strength by Addition of Fillers

The present chapter deals with the possibility of controlling adhesive bonds from an essentially new point of view, based on the principle of directed change of the interaction of polymer chains in systems containing two fillers of different chemical natures and macrostructures, namely, fibrous reinforcing fillers and finely dispersed ones. Mineral filler with lyophilic or lyophobic surface character, imparted by chemisorbed long-chain aliphatic molecules, was used as the finely dispersed filler. These interactions occur in two types of surface layers formed in polymeric composite materials: polymer on the surface of finely dispersed fillers, and a surface layer of the filled polymer on the surface of the fibrous reinforcing component. The principal feature here is the contribution of organic molecules grafted to the filler surface in changing the polymer properties in the boundary layer, since the modified filler surface of this kind promotes polymer plasticization at the phase boundary under the influence of modifier molecules, resulting in improvement of the physical-mechanical properties of the composite material.

This section gives experimental support for the novel method of improving adhesion of polymer coatings by introduction of dispersed or reinforcing filler, or both, into the polymer.

Polyurethanes were used that were based on trimethylolpropane–toluylene diisocyanate (TMP-TDI) and tetrahydrofuran and propylene oxide copolymer (THF-PO) obtained with various ratios of isocyanate

TABLE 6.1 Influence of the Reinforcing Component on Equilibrium Force of Breakdown from Steel of Polyurethane Coatings with Various Network Densities

NCO:OH group ratio	Equilibrium force of breakdown of the polyurethane coating from steel (N/m)	
	Plain	Reinforced
1.1:1	22	25
1.5:1	32	37
2.0:1	45	52
3.0:1	51	63

and hydroxyl groups; as fillers, amorphous silicon dioxide (aerosil) with specific surface area of 1.75×10^5 m^2/kg (A-175) and aerosil modified by ethylene glycol (ATG), diethylene glycol (ADEG), and butyl alcohol (ABA) were used.

Let us consider the experimental data presented in Table 6.1 on the adhesion strength to steel of polyurethane coatings characterized by various spatial network densities determined by the NCO:OH group ratio. It is obvious that higher network density leads to increase in the coating adhesion. Reinforcement of the coating with cloth based on poly-ε-caproamide fiber containing —CO—NH— functional groups, close to urethane groups, in the main chain results in noticeable increase in adhesion strength (by 10–20%). This effect is probably explained by increase in cohesion strength of the reinforced coating itself, in accordance with a conception of the rate of cohesion strength in adhesion developed earlier [329].

Now let us consider the influence on adhesion of the fillers introduced into the coating. Data presented in Figs. 6.1 and 6.2 show that in all cases the introduction of fillers differing in surface properties influences the adhesion strength. This influence, slight for aerosil A-175, is displayed to a greater extent by ADEG and ABA. At the same time, there is an optimal filler concentration at which adhesion strength reaches a maximum and is increased by 75–95% in comparison with that of unfilled coatings.

The dependence of the adhesion strength of polyurethane coatings on the concentration of unmodified aerosil has an indistinctly marked maximum in the region of 1–2 wt% concentrations. Note in particular the results showing that modified lyophobic (ABA) and lyophilic (ADEG) filler surfaces result in the same maximum breakaway force (Figs. 6.1 and 6.2, curves 1). For polyurethane containing modified filler with lyophobic surface, increase in the filler content beyond the maximum adhesion strength does not cause as sharp a drop as occurs for polyurethane filled with ADEG. These results show good correla-

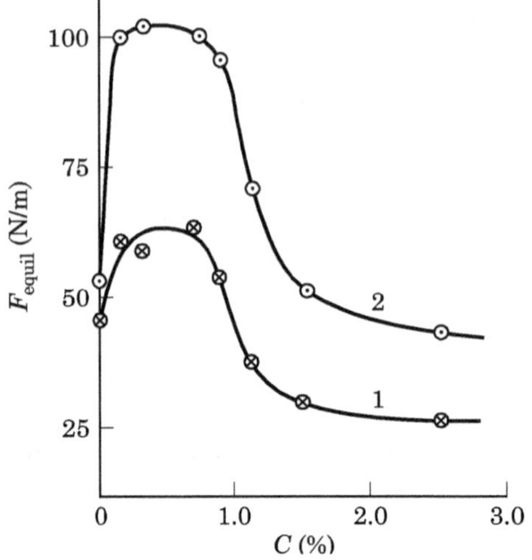

Figure 6.1 Dependence of equilibrium force of breakaway from steel (F_{equil}) of polyurethane films based on TMP-TDI and THF-PO on ADEG content (C): (1) plain film; (2) film reinforced with polyamide cloth.

tion with those obtained earlier from study of the mobility of polymer molecules at the solid boundary that show the same molecular mobility changes in the layers on fillers characterized by various surface energies [334, 336, 339]. This demonstrates that, in the systems investigated, lyophobic filler surfaces interact with the binder similarly to

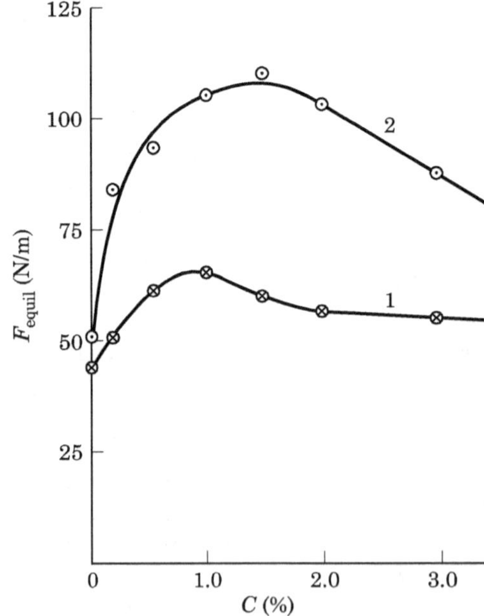

Figure 6.2 Dependence of equilibrium force of breakaway from steel (F_{equil}) of polyurethane films based on TMP-TDI and THF-PO on ABA content (C): (1) plain film; (2) film reinforced with polyamide cloth.

lyophilic, so that lyophobic interactions also contribute to changing the properties of polyurethane surface layers.

Introduction into polymer of both reinforcing fiber and filler produces considerable increase in the coating–substrate adhesive bond (Figs. 6.1 and 6.2, curves 2).

Since the increase in adhesion is considerably greater for reinforced polyurethane compositions filled with modified aerosils, it is of interest to clarify the influence of the nature of the organic modifier molecule. Comparison of the data in Figs. 6.1 and 6.2 shows that the nature of the organic component change influences the filler concentration boundaries. In this case we observed the influence of both reinforcing component and filler on the adhesive bond strength. Thus, for aerosil modified with diethylene glycol high adhesion strength is observed only up to 0.8% filler content; and for coatings containing aerosil modified with butyl alcohol the region of increased bond strength of substrate ranges from 0.2% to 2.5% filler.

In filled polyurethane coatings containing ADEG and ABA we observe a similar character of adhesion change with modified aerosil concentrations (Figs. 6.1 and 6.2, curves 1) that does not depend on the value of the modified filler's surface energy. The only difference lies in the ranges of concentration at which extremal values of adhesion strength are observed: for systems with ADEG at 0.5–0.6% aerosil and for the system with ABA at 0.9–1.0%. This difference can be attributed to different states of aggregation of the particles of the fillers used, as is detailed below. Another feature of this dependence is observed in the system containing unmodified aerosil: the presence on the filler surface of organic modifier molecules that are chemically bound to the surface is also important for the formation of the coating–substrate adhesive bond.

The patterns established should be explained starting from the notion of the generation of two surface layer types in the composite system, namely, polymer on the filler particles and filled polymer on the fibrous reinforcing material [335]. We may take it that existence of two surface types, differing in chemical nature and the potential for chemical interaction of binder components with them in the course of curing, changes the structure of the network formed and will certainly result in microheterogeneity of the polymeric coating. In conformity with Bikerman's concept of the role of weak boundary layers in adhesion [329] and the role of surface energy in the formation of the interphase layer [330], we may assume that the affinity of the surfaces for the components of the rating system determines the migration of low-molecular weight impurities, including low-molecular weight fractions formed during curing, from the binder volume to the filler boundary. As a result, low-molecular weight fractions, bound by the filler, leave the polymer–metal substrate boundary, thus allowing the formation of a less defective boundary layer. This assumption is confirmed by the extremal character of the dependence of adhesion strength on filler content. Evidently, in the region of the maximum the filler surface is saturated with fractions sorbing onto it, and further increase in filler content does not result in growth of adhesion strength. The other possible explanation of the observations is increase in the cohesion strength of the coating itself, which in accordance with theory [329] leads to growth in adhesion strength. However, within this conceptual framework it is difficult to explain the very sharp drop in the 1–3% region of dispersed phase concentrations. Decrease in the internal stresses system plays a clear role in increasing the polymer–substrate adhesive bond strength. Experiment shows that residual internal stresses in the filled polyurethane coating for small amounts of modified filler (0.5% ADEG; 1.0% ABA) are less than those in unfilled coatings (Table 6.2). The internal stresses in the coating grow as the filler content increases. The form of internal stress change can be explained on the assumption that modified filler simultaneously displays effects of restriction of mobility and polymer structural plasticization in the boundary layer due to the presence of long-chain modifier molecules. Structural plasticization of the polymer in the region of high solid phase concentrations improves the mobility of segments at the phase boundary with the filler and increases their packing density. Also, the filled polymer system contracts, increasing internal stresses due to polymer–substrate adhesion contact. Increased internal stresses result in reduced strength of the coating–metal substrate adhesive bond.

The regularities observed in the changes of adhesion strength and internal stresses at low concentration of the filler particles are apparently connected with the breaking of permolecular aggregations in the polymer and formation of more homogeneous structure of the polymer material, producing better adhesion interaction with the metal substrate. This is supported by the results of investigations of the polymer

TABLE 6.2 Residual Internal Stresses in Polyurethane Coatings Based on TMP-TDI and THF-PO (NCO:OH = 2:1) Containing Modified Filler, 48 h after Curing at 353 K

Filler	Filler concentration (mass %)	Residual internal stresses (MPa)
Without filler	–	1.50
ADEG	0.5	0.47
ADEG	2.0	1.80
ABA	1.0	0.55
ABA	2.0	1.35

interphase layer structure at the boundary with the metal substrate.

Undoubtedly, the influence of filler on the polymerization reaction kinetics [336–338] is very important for the polymer/substrate adhesive bond strength and formation of polymer layer structure at the phase boundary with the filling. To eliminate this influence, all subsequent investigations were performed with filled polymers derived from solutions of linear polyurethanes, synthesized in the absence of filler.

The results presented thus demonstrate the possibility of influencing the adhesion interaction by a novel method—introduction of reinforcing and dispersed fillers without resort to modification of the binder or the surfaces being glued or coated.

The experimental data and their explanation necessitate investigation of the dependence of adhesion and of the molecular mobility of polymer filled with modified filler on the linear dimensions of modifier molecules chemisorbed on the surface of the filler.

6.2 Influence of the Molecular Size of the Filler Surface Modifier on the Strength of Adhesive Bonds with Solid Substrates and the Molecular Mobility of the Filled Polyurethane

Studies of the influence of the molecular size of the modifier, chemically bound on the filler surface, on the strength of the polymer–substrate adhesive bond and the relaxation processes were performed on specimens of polyurethane based on ODA, 2,4-TDI, and OPG containing 2.0% filler by mass. Aerosil was used as the filler, to which surface glycols of various molecular masses were bound.

Figure 6.3 presents the temperature dependences of the tangent of dielectric loss angle in the polyurethane specimens investigated, in which two clearly expressed maxima are seen and a third maximum, that is less marked, at low temperature. The maximum of dielectric losses in the 245–247 K region is governed by the mobility of macromolecular chain segments and is related to the dipole-segmental relaxation process; the indistinct maximum at 150–190 K is related to relaxation of main-chain groups and side-chains. The maximum around 373–393 K can be attributed to the mobility of large kinetic aggregates—nodes of the network of physical bonds.

A shift of 18 K toward higher temperature of the tan δ maximum due to dipole-segmental relaxation in the polymer filled with A-175 in comparison with the unfilled polymer indicates restriction of the molecular mobility by the solid surface, in accordance with the results obtained earlier.

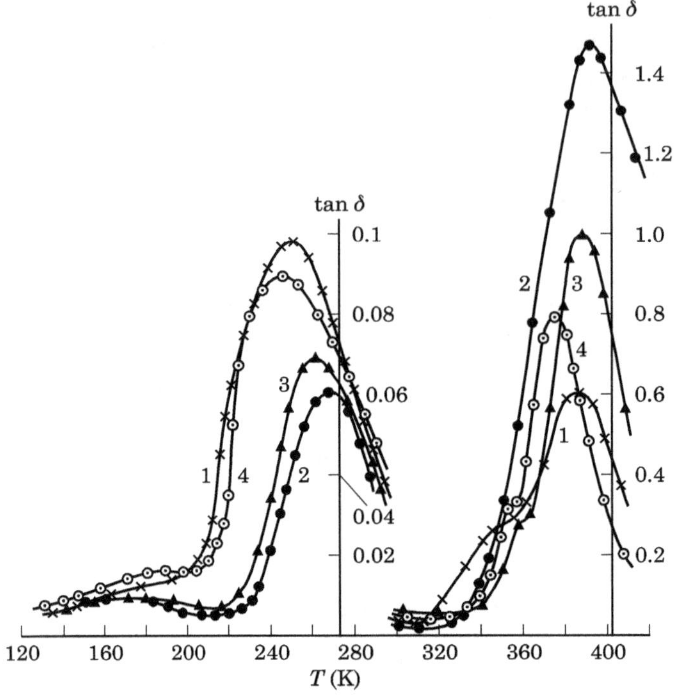

Figure 6.3 Dependence of tan δ on temperature (T) of polyurethanes based on OPG, 2,4-TDI, and ODA at 1 kHz frequency: (1) unfilled polymer; (2) with 2 wt% A-175; (3) with 2 wt% AEG; (4) with 2 wt% ADEG.

Introduction of glycol-modified aerosils into the polyurethane leads to an increase in segment mobility (Fig. 6.3, curves 3, 4) in comparison with A-175. For example, the maximum in the dielectric losses in the polyurethane composition containing AEG shifts 9 K toward higher temperature, and in the polyurethane–ADEG system the maximum corresponding to relaxation of macromolecule chain segments is observed at even lower temperature than that in the unfilled polyurethane. The shift of the tan δ maximum toward lower temperature for this composite material is 5 K relative to unfilled polyurethane and 22 K relative to polyurethane filled with A-175. The dependence of the temperature of the tan δ maximum for relaxation of nodes of the network of physical bond on the size of the glycol molecule on the aerosil surface has similar character.

Consequently, for filling with modified aerosils, two competing processes occur: stiffening of polyurethane macromolecules by the filler surface, and their plasticization by glycol molecules grafted onto filler surface. The double influence of these fillers on the relaxation processes occurring in filled polymers is reflected in the density of macromolecular packing in the boundary layer and in the morphology of the polymer surface layers as a whole.

Comparison of the experimental results of dielectric relaxation investigations with data on the strength of adhesion to steel of polyurethanes filled with A-175, AEG, ADEG, and ATEG show that, with the growth of the molecular mass of the glycol modifying the filler surface, increase in polyurethane chain mobility is accompanied by increase in polymer–metal adhesive bond strength. Thus, in this case with polymer plasticization, there is a relation between adhesion strength and molecular mobility. Hence, modification of the filler by organic molecules chemically bound to its surface changes both polymer interaction with the filler and polymer composite interaction with the substrate.

The regularities revealed in the strength of filled polyurethane–steel adhesive bonds permit the preliminary conclusion that, from the nature of its influence on adhesion strength, increase in the length of modifier molecule grafted to the filler surface cannot be reproduced by increase in the content of modified filler with grafted molecules of lesser length, since larger linear dimensions of the modifier molecule lead to improved relaxation conditions, i.e., improve its own mobility, and consequently promote the creation of a more balanced state of the polymer in the boundary layer and better interaction of the filled binder with substrate.

To evaluate this conclusion, it is necessary to consider the molecular mobility and adhesion strength of the filled polyurethane elastomer according to the concentration of modified filler.

6.3 Molecular Mobility in Filled Polyurethanes and Their Adhesion Properties at Different Filler Concentrations

As already noted, introduction of filler into a polymer matrix causes changes in molecular mobility, manifest in a rightward shift of high-temperature process connected with dipole-segmental relaxation, and a leftward shift of low-temperature dipole-group process. Molecular interpretation of these phenomena was based on ideas of adsorption limitation of the mobility of polymer chains close to the solid boundary and, as a result, worsening of the conditions of macromolecular packing. This causes an increase in mobility of the polymer chain kinetic units in the dipole-group relaxation process. This mechanism had been studied in detail and confirmed for many linear and crosslinked polymers. Thus, if aerosils in amounts of 20 wt% containing chemisorbed aliphatic alcohols on their surface are used as fillers for epoxy oligomer, the character of the effects depends on the length of the grafted chain. The shift of the high-temperature relaxation process toward higher temperatures is reduced in comparison with that observed for the filler without organic molecules on its surface, due to the plasticizing action of the grafted molecules in the oligomer boundary layer. At the same time, the low-temperature relaxation process is shifted less toward low temperatures. These observations are explained by the attainment of a more balanced polymer condition at the boundary layer.

The changes in polymer matrix relaxation behavior in the presence of surface-modified fillers are important when the filler is introduced into the polymer matrix to provide better adhesion to the solid surface. It was shown above that by introducing dispersed filler into a polymer composition (adhesive) it is possible to increase the polymer–substrate adhesive bond strength considerably with certain filler contents. From this it was discovered that the effect on adhesion strength increase is greater for a filler whose surface is chemically bound with organic molecules sufficiently long to display their own mobility. Thus, when aerosil modified by butyl alcohol or diethylene glycol is used as the filler for crosslinked polyurethane, the equilibrium force of breakaway of the filled polymer from metal substrate exhibits a maximum for certain concentrations, and the maximum is the same for both fillers (Figs. 6.1 and 6.2).

To establish the dependence of the filled polymer coating–metal substrate adhesive bond strength on the modified filler concentration, let us consider experimental data for adhesion strength in comparison with the results of molecular mobility investigations.

Figure 6.4 shows the dependence on ADEG concentration of the equilibrium force of breakdown from steel of linear polyurethane based on OTMG, 2,4-TDI, and 4,4′-diamino-3,3′-dichlorodiphenylmethane. At low filler content (up to 2%) the equilibrium breakdown force increases, reaches a maximum, and then begins to decrease with further increase of the filler content. Recall that in crosslinked polyurethane, as was shown above (Fig. 6.1), the strength of the adhesive bond with substrate falls more sharply than the linear decrease observed here.

However, crosslinked polyurethanes are not considered a good model for investigation of molecular mobility and its dependence on the amount of filler because at high temperatures there is no clear $\tan \delta$ maximum for dipole-segmental relaxation processes. For linear polyurethanes, the clearly expressed dipole-group relaxation process at 168–179 K and dielectric losses connected with macromolecule segments relaxation at 238–248 K are typical (Fig. 6.5). As seen from the

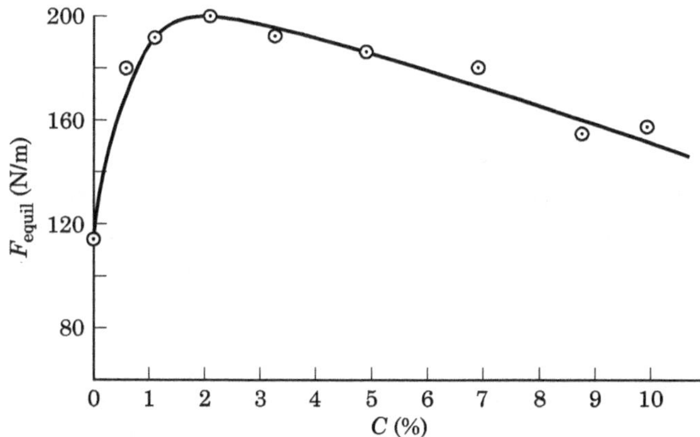

Figure 6.4 Dependence of equilibrium force of breakaway from steel (F_{equil}) of polyurethane films based on OTMG, 2,4-TDI, and 4,4′-diamino-3,3′-dichlorodiphenylmethane on ADEG concentration (C).

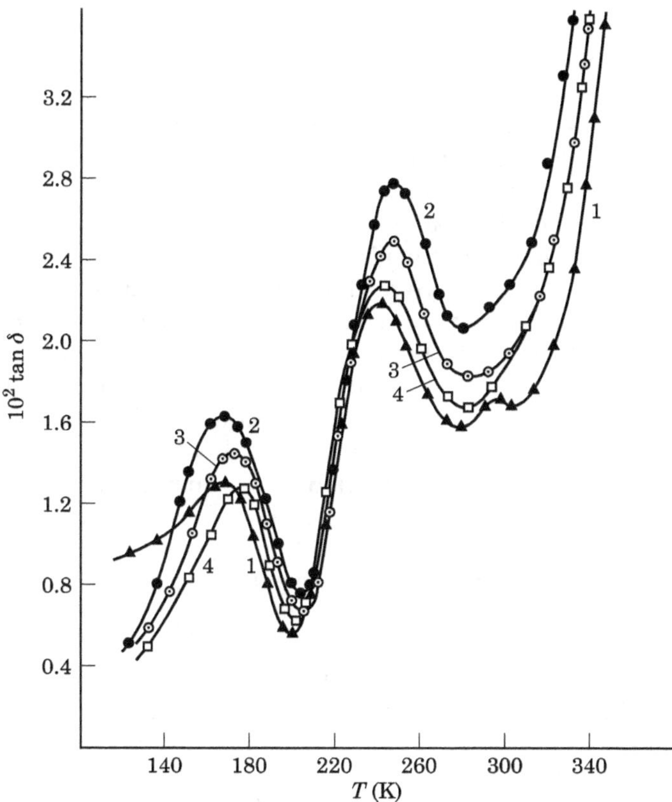

Figure 6.5 Dependence of $\tan \delta$ on the temperature (T) of polyurethanes based on OTMG, 2,4-TDI, and 4,4′-diamino-3,3′-dichlorodiphenylmethane at 1 kHz frequency: (1) unfilled polymer; (2) polymer with 0.5 wt% ADEG; (3) polymer with 1.0 wt% ADEG; (4) polymer with 3.2 wt% ADEG.

data presented, increasing ADEG content in the polymer results in a noticeable temperature shift of the tan δ maximum by nearly 10 K for both relaxation processes. Dipole-segmental relaxation processes shift toward higher temperature up to 1.0–1.2% ADEG content, after which glycol molecules evidently begin to exert a plasticizing influence, and the segmental relaxation process shifts toward lower temperatures, even below the temperature of the tan δ maximum for unfilled polyurethane. At the same time, the dipole-group relaxation process is characterized by nonlinear dependence of the tan δ maximum temperature shift toward high temperatures with ADEG concentration.

The data reveal previously unknown features of the relaxation behavior of filled polymer systems. In contrast to earlier observations, segmental mobility limitation was observed only at small amounts of filler. After reaching the maximum possible mobility limitation it begins to grow considerably (the tan δ maximum temperature decreases). Along with this, and in contrast to the effects observed earlier, there is a reduction of mobility of kinetic units in the low-temperature relaxation process, i.e., increase in the temperature of dipole-group losses. This unusual behavior can only be attributed to the fact that organic molecules grafted onto the filler surface act as the plasticizer. Typical effects of segmental mobility in the system are displayed at low content of filler particles, but on reaching a certain concentration the grafted molecules begin to play the role of "surface" plasticizer, reducing the temperature of the tan δ maximum. At some concentrations it becomes lower than that for unfilled polymer. Appealing to the ideas about the influence of the surface on relaxation processes, it is clear that plasticization and increase in mobility improve the conditions of macromolecular packing in the surface layer. As a result, temperature of the maximum for dipole-group losses increases. Both effects tend to saturation with increasing filler content.

Consequently, in respect of its influence on the relaxation processes in polymer matrix, filler modified with long organic chains should be considered as a component combining the functions of filler and plasticizer.

Referring now to the adhesion strength data, it can be seen that the maximum is located at a filler concentration exceeding that causing maximum segmental mobility restriction, and the most intensive increase in polyurethane–metal substrate bond strength occurs when the mobility of small kinetic units of the chain is not considerably reduced. It may be supposed that precisely this concentration range provides maximum interaction with substrate, i.e., adhesion, due to good contact of polymer chain groups as a result of their redistribution and changes in the conformation set of the polymer surface layers.

Evidently, a prerequisite of high adhesion is enough bonds with the surface under conditions of optimal molecular mobility. Together with this, the explanation offered does not exclude a contribution to adhesion from the cohesion strength of the filled polymer matrix [329]. In reality, the polymer strength does not grow with increase of ADEG content at low concentrations (up to 0.5%); it achieves its maximum at 1.0%, and is slightly reduced at higher degrees of filling. However, the most considerable growth of adhesion strength (from 35 to 50 MPa) at filler concentrations up to 0.5% indicates that the governing factor influencing polymer–substrate adhesive bond strength in this concentration range of dispersed phase particles is the change of polymer microheterogeneity due to formation of boundary layers, distinguished by the optimal correlation of kinetic unit mobility in both high-temperature and low-temperature relaxation processes.

It follows that for polymer interaction with the filler surface as a result of adsorption and orientation effects, caused by interphase boundaries, redistribution of the segments and small portions of macromolecular chains caused by conformation changes occurs in the polymer surface layers. At the same time, a boundary layer is formed, which might conditionally be termed “rigidly elastic.” Formation of this layer determines the maximum strength of the filled polymer–solid substrate adhesive bond.

There is thus a correlation between the adhesion strength of the filled polymer material and the molecular mobility of the polymer matrix. Taking into account that changes in the character of the relaxation processes reflect changes occurring in the polymer material structure with introduction of modified filler, we now consider the polymer structure, and in particular the structure of polymer layer formed in contact with metal substrate surface.

6.4 Influence of Aerosil Modification on the Aggregation of Particles in Oligomer Medium

The introduction of mineral or polymer filler into polymers is done to serve multiple purposes [339]. One of the essential effects of the filler particles is regulation of the structure formation process [340]. That is, the dispersed filler is used to create artificial structure formation centers. It has been demonstrated in practice that in most cases the optimal physical and mechanical properties of polymer materials are achieved when their structure comprises fine-scale and homogeneous elements. One of the factors for production of homogeneous structure in filled polymers is a uniform distribution of the filler in the polymer matrix. There is every reason to assume that changing the filler particles' surface energy through chemical modification of their surface by organic molecules comparable in size but differing in the nature of the end group will change the interaction between the particles and alter their degree of aggregation. As a result, the use of fillers varying in surface character will permit polymer materials to be obtained with different heterogeneities and, correspondingly, with different properties.

Filler particle distribution has been studied in the model systems of filled oligoester glycol, this being the initial component in the production of polyurethane elastomers. As the introduction of filler into oligoester glycol affects its rheological behavior most of all, we studied viscosity first. Filler dispersion in low-molecular weight glycol was used for comparison. Studies were done in three dispersed systems: ADEG in ethylene glycol (system No. 1), ADEG in OTMG (system No. 2), and ABA in OTMG (system No. 3).

The concentration dependence viscosity at small contents of solid phase are described by the Einstein equation:

$$\eta = \eta_0(1 + \beta\varphi_{\text{vol}}) \tag{6.1}$$

where η is the viscosity of the dispersed system; η_0 is the viscosity of the dispersed medium; φ_{vol} is the volume fraction of solid phase particles in the dispersed system; and β is a coefficient equal to 2.5 for spherical particles.

It was found experimentally that system 1 complies with the Einstein model up to volume content of solid phase particles not exceeding 0.07; system 2 does so up to $\varphi_{\text{vol}} = 0.034$; system 3 up to $\varphi_{\text{vol}} = 0.01$. The coefficient β, calculated from experimental results and representing the tangent of the slope of the dependence of relative viscosity η/η_0 on volume content of the filler particles, does not vary with temperature in the interval examined, but it is different for each of the three dispersions studied and takes values of 19.5, 29.6, and 78.8 for systems 1, 2, and 3, respectively.

Fine (highly dispersed) solid particles, including aerosil, are in an aggregated state in organic liquids [341–343]. The size of the aggregates and their packing density ratio depend on many factors: solid phase dispersity, interfacial tension, viscosity of the dispersion medium, and other parameters. The fact that β exceeds 2.5 is apparently explained not only by the presence of solvation spheres on the particle surfaces, as was mentioned in [344], but also by the aggregation of solid phase particles. The authors of [345] propose that $2.5/[\eta]$ correlation is the coefficient of packing density of monodispersed spheres in the aggregate and the characteristic viscosity $[\eta]$ is equivalent to the β coefficient in the Einstein model; consequently, $\rho = 2.5/\beta$. Calculation of ρ for the three dispersions examined gave values of 0.130, 0.084, and 0.032 for systems 1, 2, and 3, respectively. The radius r of the primary aerosil particles, equal to 7×10^{-9} m (calculated from the specific surface area), and R, the radius of the aggregate, are related by [345]

$$\rho = \left(\frac{r}{R}\right)^{3-1/z} \tag{6.2}$$

where $z = 0.429$, the statistical mean coefficient.

The respective aggregate dimensions calculated using this equation for systems 1, 2, and 3 are 1.5×10^{-7} m, 2.8×10^{-7} m, and 12.0×10^{-7} m; that is, ADEG is better distributed in low-molecular weight glycol than in an oligomeric glycol, and ABA is worse distributed in OTMG compared with ADEG.

To verify the results obtained, the mean aggregate dimensions were determined by turbidity spectral methods. The results of dispersed particle size measurements (Table 6.3) show that radii of the aggregates of the dispersions under study, obtained using rotational viscometry, are higher than those determined by turbidity spectral methods. This may be caused by the fact that in the conditions of shearing motion, aggregates move in a complex rotational and advancing manner, which leads to their expansion under the action of centrifugal forces, enhanced by the structural component of adsorption–solvate layer separation pressure.

Accordingly, it was established that aerosils modified by diethylene glycol and butyl alcohol are in an aggregated state in the oligomer glycol medium, and that the degree of ADEG distribution is higher than that of ABA.

Let us estimate the thickness of the adsorption–solvate layer, Δ, on which particles of radius r, composing an aggregate, are moved away from each other, causing increase in size compared with R values in

TABLE 6.3 Dimensions of Aerosil Particle Aggregates in Glycols of Various Molecular Masses

	Radius of aerosil particle aggregates $10R$ (m)	
Dispersed system	Opacity spectrum method	Viscometry method
ADEG-EG	0.9	1.5
ADEG-OTMG	1.2	2.8
ABA-OTMG	2.6	12.0

static conditions. It is easy to demonstrate that the density of packing of n particles in static conditions is given by the ratio

$$\rho_{\mathrm{st}} = \frac{nr^3}{R_{\mathrm{st}}{}^3} \tag{6.3}$$

Under shearing action, the particles repack in the aggregate with the formation of a looser structure with lower density of particles. We denote the increase in the aggregate radius as y. Then, Equation (6.3) becomes

$$\rho = \frac{nr^3}{(R+y)^3} \tag{6.4}$$

Taking account of the invariability of the particle number n, we assume the formal transition to random dense packing [334] of spheres of effective radius $r + \Delta$, where Δ is the effective thickness of the adsorption–solvate layer, giving

$$\frac{n(r+\Delta)^3}{(R+y)^3} = 0.64 \tag{6.5}$$

Combining the solutions of the last two equations gives the following expression for Δ:

$$\Delta = r\left(\sqrt[3]{\frac{0.64}{\rho}} - 1\right) \tag{6.6}$$

Calculation of adsorption–solvate layer thickness in conditions of simple shear for the dispersions under study shows that it is largest at 120×10^{-10} m on the surface of ABA particles; for ADEG particles in OTMG it is less, at 70×10^{-10} m; and for ADEG particles in EG it is 50×10^{-10} m.

Thus adsorption–solvate layers are formed on the aerosil surface on its introduction into the oligomer glycol, as a result of which the aggregates expand in shear conditions, and this expansion is greater for ABA and OTMG aggregates.

The results obtained show that a modified filler with a lyophobic surface in an oligomer medium is structured into aggregates that are larger than the aerosil modified by diethylene glycol. Due to this aggregation, for the same filler content in the oligomer the overall surface of oligomer interaction with lyophobic filler will be less than with use of a filler with higher surface energy. It may be supposed that the difference in the size of modified filler particle aggregates observed in oligomer glycol will be maintained after the formation of polyurethane elastomer, even taking into account some possible change in aggregate size. Therefore, the conclusion reached above will also be true in relation to filled polymer. These ideas about the aggregation ability of modified aerosil particles explain the occurrence of maximum strength values for adhesive bonds of polyurethanes containing ABA at higher filler content than for the polymer filled with ADEG (Figs. 6.1 and 6.2). This is because with higher aggregation of ABA particles for equivalent changing of the transition layer of the polymer phase boundary, the amount of filler should be larger.

Differences in the filler aggregation ability are certainly reflected in the heterogeneity of the polymer material, and thereby in its cohesion strength. Increase in cohesion strength in accordance with [329] promotes growth in adhesion strength. However, comparison with the results given in the previous chapter that we obtained in studies of adhesion and strength properties of polyurethane coatings does not permit us to refer the intense adhesion strength growth at low filler concentrations to cohesion strength increase. The important role of molecular mobility in the adhesive bond was also pointed out there. We may suppose that the increase in adhesion strength observed on introduction of modified filler is the result of variations in microheterogeneity of the polyurethane coating interphase layer on the metal substrate surface, leading to better adhesive–substrate interaction.

6.5 Structure of the Filled Polyurethane Interphase Layer at the Metal Substrate Boundary

The results of polyurethane elastomer structural studies are generalized comprehensively in several monographs [331, 346–348]. It was established that the physical-chemical and mechanical properties of polyurethanes are determined mainly by microsegregation of oligoester and urethane segments into two phases due to their thermodynamic incompatibility [349–352]. The degree of microphase separation of the flexible and rigid blocks in polyurethanes is influenced considerably by availability of hydrogen bonding and by the correlation of two types of it: rigid segment–rigid segment, and rigid segment–flexible segment [353, 354].

As shown above, introduction of filler into polyurethane results in the growth of adhesion strength, most intensively for small amounts of filler. This change in adhesion is certainly caused by reconstruction of polyurethane structure under the influence of the introduced filler surface (demonstrated by the data on molecular mobility variation), and in particular of the polymer adhesive layer structure. In this connection, it was of particular interest to study the influence of the surface of the solid particles of filler on the polyurethane structure and particularly on the microphase separation of flexible and rigid blocks. At the same time, it was important to clarify the influence of the polymer–substrate boundary on microphase separation in surface layers, which is essential from the point of view of adhesion.

The structure of the surface layers was investigated by IR spectroscopy of disturbed total reflection at the polyurethane film specimens based on OTMG, 2,4-TDI, and 4,4′-diamino-3,3′-dichlorodiphenylmethane, filled with ADEG. To describe the change in the absorption band with change of filler content, optical densities were determined in accordance with the baseline method. The valence and deformation fluctuation bands of CH_2 groups at 2860 and 1370 cm^{-1} and of benzene rings at 1600 cm^{-1} were used as an internal standard.

The presence of four peaks in the multiplet band of carbonyl group valence fluctuations (amide I band) demonstrates self-association of urethane and carbamide groups (bands at 1690 and 1660 cm^{-1}) leading to the formation of domain structure [355]. There are many non-self-associated groups recorded (bands at 1725 and 1710 cm^{-1}) and correlation of the corresponding band intensities indicates a degree of self-association of 50–60%. On the other hand, in the area of valence fluctuations of N—H bonds we observe a wide band at 3300 cm^{-1}, evidence that all NH groups participate in hydrogen bond formation. Free NH groups in similar systems appear near 3430–3450 cm^{-1} [332]. Along with this, C=O groups free of hydrogen bonding (1725 and 1710 cm^{-1}) show that a considerable portion of NH groups are bound with ester oxygen. The presence and concentration of the filler influence the correlation of the intensities of multiplet band components of carbonyl absorption. This is connected with redistribution of hydrogen bond types and change in the degree of segregation of blocks of similar chemical composition [356]. It is especially evident for specimens with small amounts of filler (0.5%).

Figures 6.6 and 6.7 present correlations of the optical densities of several absorption bands (located far or less far from each other) for specimens with various filler contents. Bands at 1220, 1540, 1600, and 3300 cm^{-1} are due to fluctuations in rigid blocks: amide III, amide II, benzene ring, and ν(NH), respectively [357]. Correlation of the intensities of bands due to flexible blocks does not change with filler concentration in specimens for filled polymer surface layers, formed both in contact with metal substrate (adhesive layers) and in conditions of no contact. The bands connected with fluctuations in groups participating in hydrogen bonding proved to be sensitive to the additives (Figs. 6.6 and 6.7).

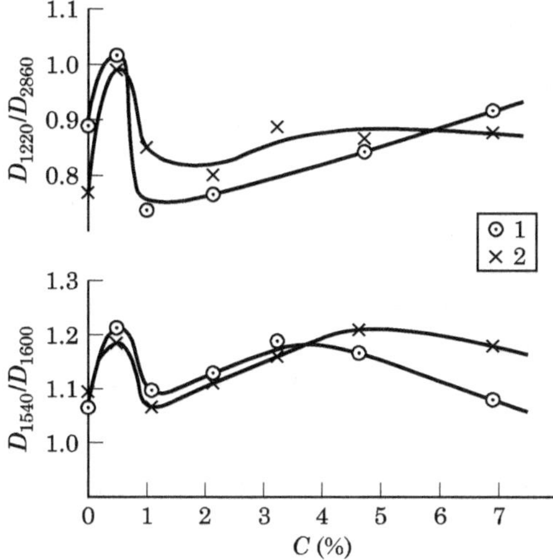

Figure 6.6 Dependence of relative intensities of absorption bands in IR spectra of MNPVO polyurethane films based on OTMG, 2,4-TDI, and 4,4′-diamino-3,3′-dichlorodiphenylmethane on the filler concentration (C): (1) polymer surface layer formed in the absence of contact with metal substrate; (2) surface layer formed in contact with metal substrate.

It is remarkable that very small amounts of filler change the 1220, 1540, and 3300 cm^{-1} band intensities. "Peaks" observed for small amounts of additive cannot be attributed to measurement errors, because they are not observed for bands connected with fluctuations in groups not participating in hydrogen-bond formation. On the other hand, while the decrease in intensity of the 3300 cm^{-1} band may be related to the reduction of hydrogen bond concentration, the increase in the intensity of amide II and amide III bands may be caused by the

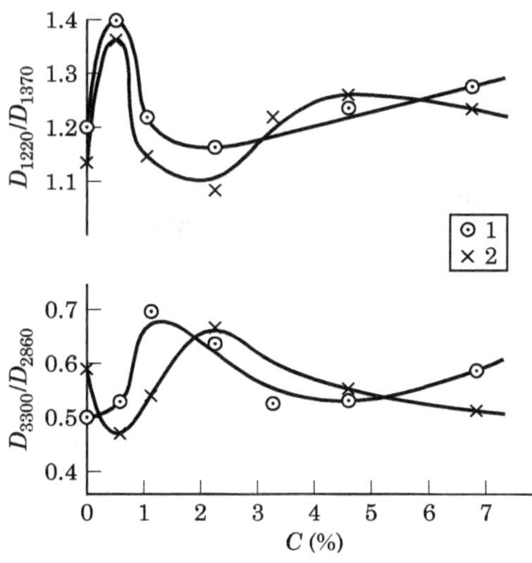

Figure 6.7 Dependence of relative intensities of absorption bands in IR spectra of MNPVO polyurethane films based on OTMG, 2,4-TDI, and 4,4′-diamino-3,3′-dichlorodiphenylmethane on filler concentration (C): (1) polymer surface layer formed in the absence of contact with metal substrate; (2) surface layer formed in contact with metal substrate

change in rigid block conformation, by higher association density, or by other factors (increased efficiency of dispersion forces may increase the index coefficient in the absorption band, necessarily leading to greater penetration depth in the MNPVO method and, correspondingly, to the growth of its intensity).

Without clarification of the real causes of these changes in the intensity of bands, connected with fluctuations in rigid segments, we can only assert that they are determined by the influence of optimal concentrations of chemically modified filler on the polymer structure as a whole, and in the case under discussion on the structure of permolecular formations of rigid blocks in particular. Alteration of the whole association structure is not necessary; the transition layer at the phase boundary is enough. Probably, the extremal character of the dependences of certain physical and mechanical properties of the filled polymers on the filler concentration (mainly for the extremum at small concentrations) is connected with this. In any case, the extremal values of adhesion strength that we observed may be explained from this starting point.

Where CH_2 group deformations appear at 1440–1490 cm^{-1} there is a tendency to redistribution of intensities between the bands at 1445 and 1480 cm^{-1}, indicating conformation changes in oligoglycol fragment chains toward *trans–trans* conformations [333], which is probably connected with regrouping of hydrogen bonds in rigid domains.

Let us consider what differences are observed in the case of polymer filling for the adhesion layers and those not subject to the influence of the high-energy surfaces. In general, the curves for polymer films that differ in the conditions of surface layer formation show similar trends (Figs. 6.6 and 6.7), but differences are also found. Thus, for the correlation D_{1370}/D_{2860} (the bands at 1370 and 2860 cm^{-1} correspond to flexible polymer components) the experimental points for both investigated surfaces with uniform scatter are located near the same straight line. For the correlation D_{1370}/D_{1600} (the bands for flexible and rigid blocks) we observe separation of the areas occupied by the experimental points for these layers, and the points for polymeric layers not formed in contact with metal surface are situated somewhat lower. It may be supposed that the concentration of benzene rings in this layer is slightly higher and, consequently, the content of flexible blocks in the polymer adhesive layer is increased.

It should also be noted that in the amide I region of the polymer adhesive layer spectrum the band intensity at 1725 cm^{-1} is noticeably increased, and that at 1690 cm^{-1} is decreased. At the same time, increase in the band intensity occurs at 1710 cm^{-1} and a decrease occurs at 1660 cm^{-1}. These changes indicate the interaction of both filled and unfilled polymer with substrate, resulting in decreased participation of carbonyl groups in hydrogen-bond formation in the polymer layer contacting the metal substrate surface—i.e., a decrease in self-association of urethane and carbamide groups in this layer. Correlation of the absorption band intensities at 1660, 1710 cm^{-1} and 1690, 1725 cm^{-1} is also altered with variation of the filler concentration in the polymer. The difference in optical densities of IR absorption bands of the surface adhesive and nonadhesive layers of polyurethane carbamide is evidently connected with the change of macromolecular conformation in the polymer layer bordering on the solid surface and, consequently, with changes in the conditions of hydrogen-bond formation.

The results presented show that polyurethane boundary layers change under the influence of aerosil surface modified by organic molecules and of the metal substrate surface, and this change consists in a reduced degree of self-association of urethane and carbamide groups as a result of cohesion redistribution. Owing to these changes, most pronounced at low filler concentrations, electron-donor or electron-acceptor groups of rigid and flexible segments, not mutually connected by hydrogen bonds, become capable of interacting with active centers of the filler surface (or its modifier molecules) and with the substrate surface, depending on the nature of the donor–acceptor centers on them.

Chapter

7

Criteria of Adhesive Joint Strength

7.1 Adhesive Joint Strength under Combined Action of Various Stresses

Let us consider the influence of various combinations of stress conditions, namely, normal fracture, shear, normal compression, and the combined action of direct and tangential stresses of specified ratio, on the strength of adhesive joints, as well as their dependences on materials, adhesion conditions, and polymerization kinetics. Adhesives that were analyzed in Chapter 3 were taken as the objects of investigation.

The specific character of adhesive contact consists in the time dependence of strength, starting from adhesive solidification to the beginning of the test. This dependence for various adhesive compositions may be manifested in different ways. However, the kinetics of strength variation with time is not always taken into account in research and engineering practice, which causes difficulties in comparison of test results and errors in calculation of the bearing capacity of adhesive joints. Thus, examination of the regularities of the variation of adhesion strength with time is a necessary condition for making a reasonable choice of adhesive composition and ensuring its working capacity.

The variation of specimen strength with the time of adhesion using the Sprut-5M composition is represented in Fig. 7.1a by the curves 1–3. As shown, strength of adhesion with Sprut-5M (28 and 24 MPa under normal fracture and shear) is considerably higher than that achieved by PN-1 and PU (18, 12, and 8.5 MPa, respectively). The strength of the specimens tested increases in the course of time, but

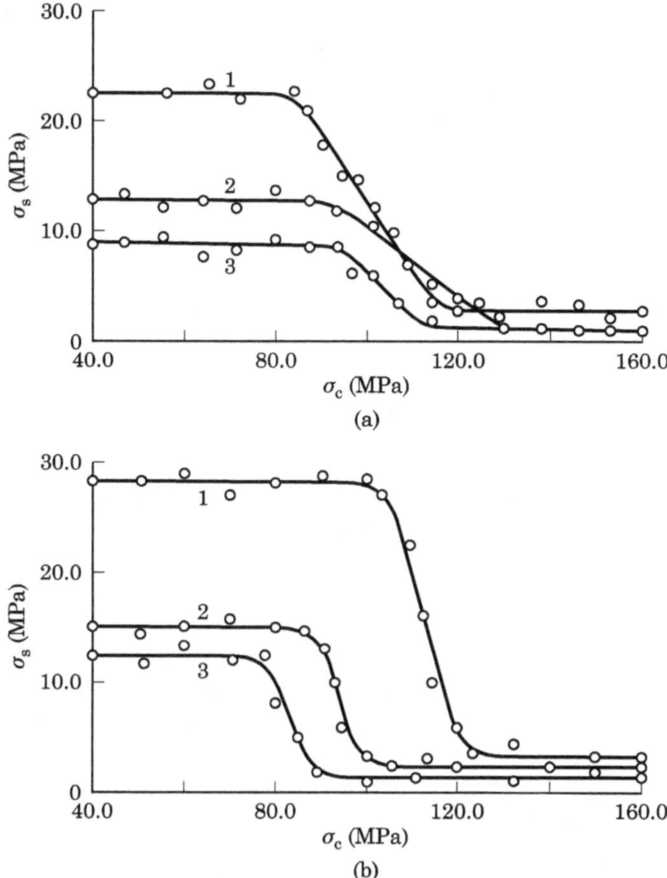

Figure 7.1 Strength of steel specimens under normal compression bonded with (a) Sprut-5M and (b) VAK: (1) in air; (2) under water; (3) in oil.

the intensity and duration of the increase differ. The strength of joints using PU is achieved slowly, during 24×10^2 h; with PN-1 it is far more intensive and ends in 6×10^2 h. The duration of adhesive strength increase in the Sprut-5M joint is comparable with the time of strengthening of PU. This may be explained by the fact that Sprut-5M is a mechanical mixture of two homopolymers [(PN-1 (polydiethylene glycol maleate phthalate solution in styrene) and PU (product of interaction of 2 moles of toluylene diisocyanate and 1 mole of polydiethylene glycol adipinate)] that have different polymerization rates. PN-1 polymerized first, with the formation of a strong network; PU cures much more slowly, with simultaneous occurrence of relaxation processes in PN-1.

The strength of VAK composition adhesion (33 and 32 MPa under normal fracture and shear) is considerably higher than the adhesion strength of AK mixture (solution of polybutyl methacrylate in methyl methacrylate) (25 and 22 MPa) (Fig. 7.1b). This is also explained by the effect of ATZh product action. The intensity and duration of the increase in strength with time differ slightly for the two compositions.

No great dependence of adhesion strength on the material of the specimens is observed. Specimens made of stainless steel are characterized by the highest value of adhesion strength, while those of aluminum alloy are the least strong. However, the difference in adhesion strengths does not exceed 20 MPa. The curves of strength variation with time for stainless steel and aluminum alloy specimens are virtually the same. These regularities are also observed for specimens cemented under water and in oil.

Visual inspection of the specimens' working surfaces after tests indicates that fracture takes place in the adhesion film (cohesion fracture pattern).

The kinetics of strength variation for adhesive joints, under water or oil, had not previously been investigated. When formed in liquids, PN-1, PU, and AK compositions do not display adhesion properties. The adhesion strength for underwater fixing with Sprut-5M is 14 and 12 MPa under uniform fracture and shear; for VAK adhesive the values are 13 and 12 MPa.

A gradual decrease in strength is noted for specimens cemented with Sprut-5M. The strength decreases under water over 24×10^2 h, by cohesion fracture pattern. The strength decrease practically stops then, but fracture occurs predominantly by mixed or adhesion pattern. This is probably connected with diffusion of water at the polymer–metal phase boundaries and with low water resistance, typical of polyester resin-based adhesives.

Over time, the underwater strength of specimens using VAK composition increases 1.4–1.5 times. The intensity of the strength increase under water is lower than that in air, and its duration is twice as long, with a cohesion fracture pattern. Strength increase with exposure under water ceases after 24×10^2 h.

The strength of steel specimens cemented in oil with Sprut-5M increases during 24×10^2 h from 8.2 MPa to 12.8 MPa, and the curve asymptotically approaches a constant value. Specimen fracture occurs by cohesion pattern. The mechanism of strengthening in oil is most likely similar to that in air [358, 359]; the high oil resistance of polyester resins is also an important consideration.

Reduction in the strength of specimens cemented with VAK composition in oil practically stops after 24×10^2 h. Gradual strength decrease is accompanied by a change in specimen fracture pattern: cohesion fracture turns into mixed and adhesion fracture. This is probably connected with the low oil resistance of acrylate compositions.

Thus, the experimental dependences show that the strength of these adhesive joints under normal fracture and shear depend substantially on the time from the moment of cementing to the beginning of the test. Variations in intensity and duration of the strength depend on the formulation of the adhesive composition and on polymerization conditions, which in principle allows for active control of these processes. The results enable the recommendation of Sprut-5M mainly for adhesive joints operating in oil media and the VAK composition for use in underwater conditions.

Given the observed dependence of specimen strength on time, it is necessary to standardize exposure in the test media from the time of adhesive film solidification to testing in order to be able to compare results. Statistical analysis of the results for strength tests under normal fracture and shear shows that the largest deviation from the average values is observed in the interval from 0.72×10^2 to 2×10^2 h of specimen exposure. During this period, the intensity of variation of adhesion strength in the air is so great that only a small discrepancy in the time to testing leads to significant errors. For specimens exposed for 2.4×10^2 h and more, the mean-square deviation does not exceed 2.4 MPa, making it possible to compare test results under various loads after exposure for 2.4×10^2 h. In view of this, investigations of adhesive joint strength under normal compression, as well as under the combined action of direct and tangential stresses, were performed after holding in media for 2.4×10^2 h.

The strength of glued specimens under normal compression was investigated by construction and analysis of curves of normal fracture stresses against values of specimen precompression stresses.

Results under normal compression for steel specimens glued with Sprut-5M and VAK compositions in air, under water, and in oil are given in Fig 7.1. As is obvious, the strength of adhesive joints under normal fracture is practically the same up to a certain critical value of precompression stress. Evidently, the applied normal compression stresses do not result in essential fractures in the adhesive joint; the fracture occurs in the elastic zone. Using Sprut-5M composition (Fig. 7.1a), these compression stresses remain within 0–43.0 MPa in air, within 0–50.0 MPa under water, and within 0–52.0 in oil; the corresponding values for VAK are 0–104.0 MPa, 0–90.0 MPa, and 0–84.0 MPa (Fig. 7.1b).

Further increase of normal compression stresses leads to a sharp reduction of the normal fracture stresses. Evidently, increase in the normal compression stresses causes an irreversible process of adhesive joint fracture. For specimens fixed with Sprut-5M these compression stresses are within the limits of 43.0–80.0 MPa in air; 50.0–90.0 MPa under water, and 52.0–75.0 MPa in oil; the corresponding values for VAK are 104.0–125.0 MPa, 90.0–104.0 MPa, and 84.0–91.0 MPa. Apparently, under normal compression stresses of 80.0, 90.0, and 75.0 MPa, specimens fixed with the Sprut-5M in air, under water and in oil fail completely; fracture of specimens fixed with VAK in corresponding conditions occurs at compression stresses of 125.0, 104.0, and 91.0 MPa.

Insignificant strength after the action of compression stresses, corresponding to practically complete failure of the adhesive joint, cannot be explained unambiguously. Under large enough compression stresses, the processes of mechanical penetration of adhesive into irregularities in the specimen, mutual diffusion, molecular and chemical interaction, and so on, may occur concurrently with adhesive film fracture, and probably ensure the adherence of secondary surfaces.

Upon analysis of the curves in Fig. 7.1, it is easily seen that they have various slopes at the sections of specimen fracture. The curve slope with respect to the abscissa axis may be taken as the measure of a glued specimen's brittleness under compression. The more brittle the joint, the steeper the angle. Evidently, the brittleness of specimens glued with Sprut-5M composition under normal compression is less than that of specimens glued with VAK, which may be explained by higher elasticity of the Sprut-5M as a result of the plasticizing effect of the PU compound.

The brittleness of specimens under normal compression depends also on the ambient medium and on exposure. Specimens fixed with Sprut-5M under water and in oil are less brittle than specimens fixed in air. Reduction of brittleness under water and in oil is somewhat characteristic of the specimens glued with VAK also. This may be explained by the plasticizing influence of the liquid medium on the adhesive joint borders, leading to partial elimination of the stress concentration factor [360] that occurs with adhesive shrinkage in the process of polymerization.

The absolute values of the strength of glued specimens under normal fracture and normal compression on the section, corresponding to the fracture initiation, differ substantially. The coefficient $x = \sigma_p/\sigma_c$, characterizing the different tensile and compression strengths of steel specimens fixed with Sprut-5M, is 0.53 for gluing in air, 0.22 under water, and 0.16 in oil; for specimens glued with VAK, the values are 0.27, 0.17, and 0.14, respectively. Evidently, the adhesive joints under consideration belong to systems of different resistance to the tension and compression stresses, regardless of the gluing medium.

Investigation of the adhesive specimens under the combined action of normal fracture and shear was made using fixed ratios of direct and tangential stresses (σ/τ) of 0.5, 1.4, 3.0, and 8.0 in the plane of the adhesive line. Values of strength under normal fracture ($\tau = 0$) and shear ($\sigma_p = 0$) were also used in constructing the limiting state diagrams. Results are presented in Fig. 7.2. Inclined lines from the origin correspond to the above-mentioned fixed ratios of direct and tangential stresses. Tests were performed after fixing and exposure of the specimens in air, under water, and in oil.

Figure 7.2 presents diagrams of the limiting state of specimens made of stainless steel and aluminum alloy fixed with Sprut-5M composition and VAK in air. Evidently, the values of adhesion strength for specimens made from different materials differ only slightly, and their limiting state diagrams are practically equidistant.

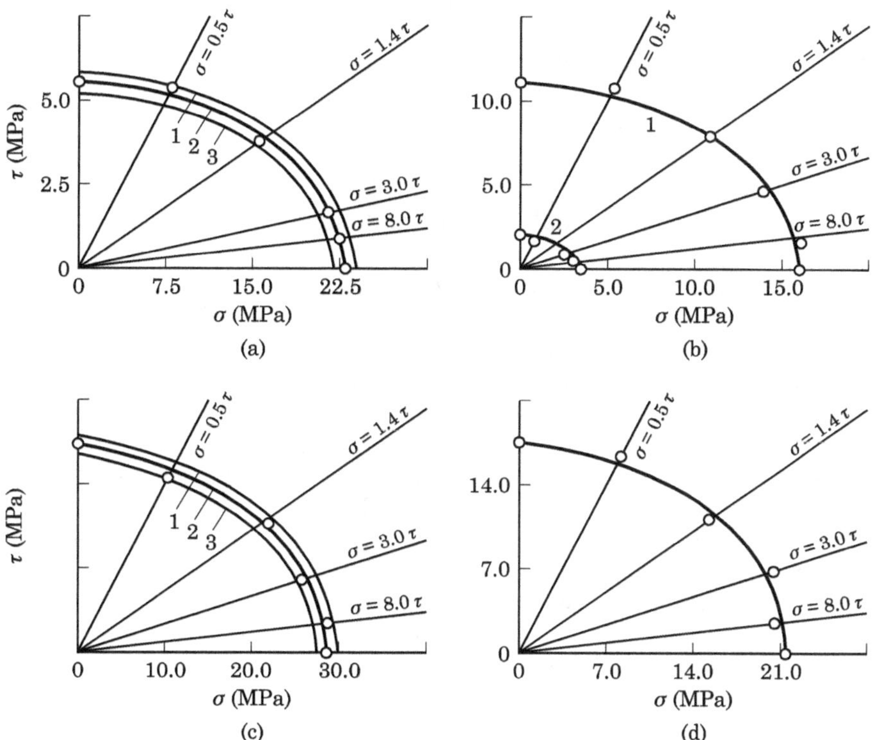

Figure 7.2 Diagrams of the limiting stressed state of adhesive specimens under the combined action of normal fracture and shear (bonded in air): (a, c) bonding of specimens made of stainless steel (1), steel (2), and aluminum alloy (3) using Sprut-5M and VAK; (b) bonding of steel specimens with PN-1 (1) and PU (2); (d) the same as (b) for AK adhesive.

Diagrams of the limiting stressed state for steel specimens fixed with PN-1 and PU compounds in air are presented in Fig. 7.2b, and for those glued with VAK in Fig. 7.2d. Evidently, the strength of these specimens under the combined action of the direct and tangential stresses is less than that of specimens fixed with Sprut-5M and VAK compositions. This is in conformity with the results of separate tests for normal fracture and shear.

Changes in the strength of specimens fixed under water and in oil, obtained with various combinations of the normal fracture and shear stresses, are similar to the analogous changes for specimens fixed in air. In the areas limited by the reference axes and curves of the limiting state, short-term static strength is guaranteed with any combination of the direct and tangential stresses. Estimation of the bearing capacity of the adhesive assembly from the results of separate tests of specimens for normal fracture ($\tau = 0$) or shear ($\sigma_p = 0$), may lead to considerable errors. For example, as shown in Fig. 7.2a, strength of St3 steel specimens fixed with Sprut-5M in air is 23.0 MPa under normal fracture and 16.5 MPa under shear. At present, only these values are taken into account for evaluating the potential compositions for gluing, ignoring the fact that combined action of the stresses will inevitably result in joint fracture.

Special attention was paid to the kinetics of short-term strength variation for adhesive specimens under the combined action of breaking stresses of normal fracture and shear after exposure to the test media. Tests were performed under direct and tangential stress ratios in the plane of the adhesive line of 0.5, 1.4, 3.0, and 8.0. Surfaces of limiting stressed state were constructed in σ_p–τ–time coordinates (time in hours). Curves of short-term strength change kinetics of adhesive specimens from separate tests of breaking stresses of normal fracture ($\tau = 0$) and shear ($\sigma = 0$) were used in constructing the surfaces.

The kinetics of short-term strength change with the time under the combined action of normal fracture and shear is presented in Fig. 7.3. The oblique planes going through the time axis correspond to the above-mentioned ratios of direct and tangential stresses.

Surfaces of the limiting stressed states for steel specimens fixed in air using Sprut-5M, PN-1, as well as VAK and AK, are presented in Fig. 7.3. As is obvious, specimen strength increases with time. The kinetics of short-term adhesion strength change with time under the combined action and under the separate actions of normal fracture and shear breaking stresses shows similar patterns. Thus, the increase in the strength with Sprut-5M and PU stops in 24×10^2 h, with PN-1 in 6×10^2 h, and with VAK and AK in 8–12×10^2 h. In all the cases failure of specimens occurs by cohesion pattern.

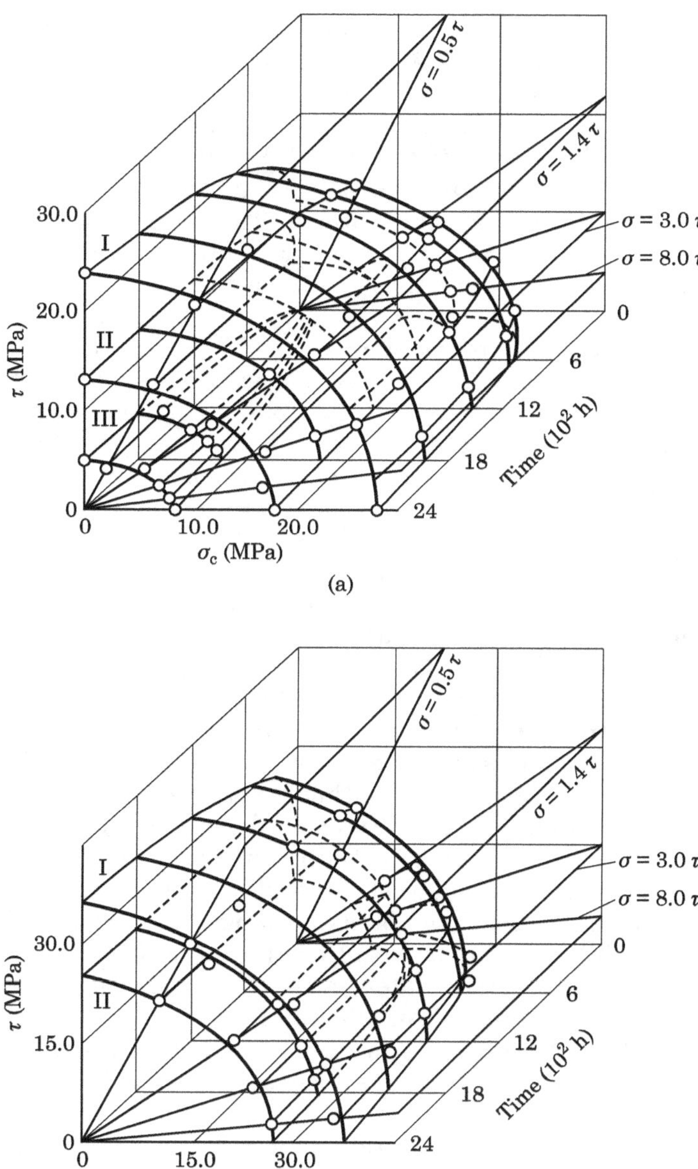

Figure 7.3 Kinetics of strength change for steel specimens fixed in air and tested under the combined action of normal fracture and shear. (a) Sprut-5M (surface I), PN-1 (surface II), and PU (surface III); (b) VAK (surface I), AK (surface II).

The strength of steel specimens fixed under water using Sprut-5M under the combined action of breaking stresses decreases over 24×10^2 h, after which it stops decreasing. Toward the end of the exposure time, transition of cohesion fracture to adhesion and mixed pattern is observed. Specimens fixed with VAK also exhibit the strength increase, which again stops in 24×10^2 h, but in this case the joint fracture occurs by cohesion pattern. The dependences of short-term strength with underwater fixation are similar for specimen breakdown under the combined and the separate actions of direct and tangential stresses.

For specimens fixed with Sprut-5M in oil and held in oil until strength testing under normal fracture and shear, gradual strength increase throughout the time of holding in oil is observed; the strength of specimens fixed with VAK decreases. For the separate tests also, this process is practically finished in 24×10^2 h. Specimens fixed with Sprut-5M maintain the cohesion fracture pattern throughout holding in oil, whereas specimens fixed with VAK display a transition from cohesion fracture to adhesion and mixed pattern.

Specially developed methods were used for investigation of the strength of adhesive specimens under normal compression and shear. Specimens fixed with Sprut-5M and VAK in air, under water, and in oil remained in these media until tests began. The specimens were exposed to compression in the stresses range $\sigma_c = 0$–28.0 MPa and failed under the action of shear stresses τ.

Figure 7.4a shows diagrams of the limiting stressed state of specimens fixed with Sprut-5M. Values of tangential stresses at $\sigma_c = 0$ correspond to the data given previously. With increase of the normal compression stresses, shear breaking stresses also increase, regardless of the gluing medium. Thus, the breaking tangential stresses of specimens glued in air (curve 1) increase from 16.5 MPa at $\sigma_c = 0$ to 46.0 MPa at $\sigma_c = 28.0$ MPa; under water (curve 2) the corresponding values are from 9.4 MPa to 37.0 MPa; and in oil (curve 3) they are from 7.3 MPa to 34.0 MPa. Increase in shear breaking stresses is proportional to the normal compression stress in the interval considered.

Diagrams of the limiting stressed state of specimens glued with VAK composition are presented in Fig. 7.4b. In this case there is also a practically proportional increase in the shear breaking stresses with increasing compression stresses, regardless of the gluing medium.

It should be noted that proportional increase in the shear breaking stresses with compression stresses is probably the characteristic feature of adhesive joints. Evidently, adhesive joints resist the combined action of normal compression and shear much better than the actions

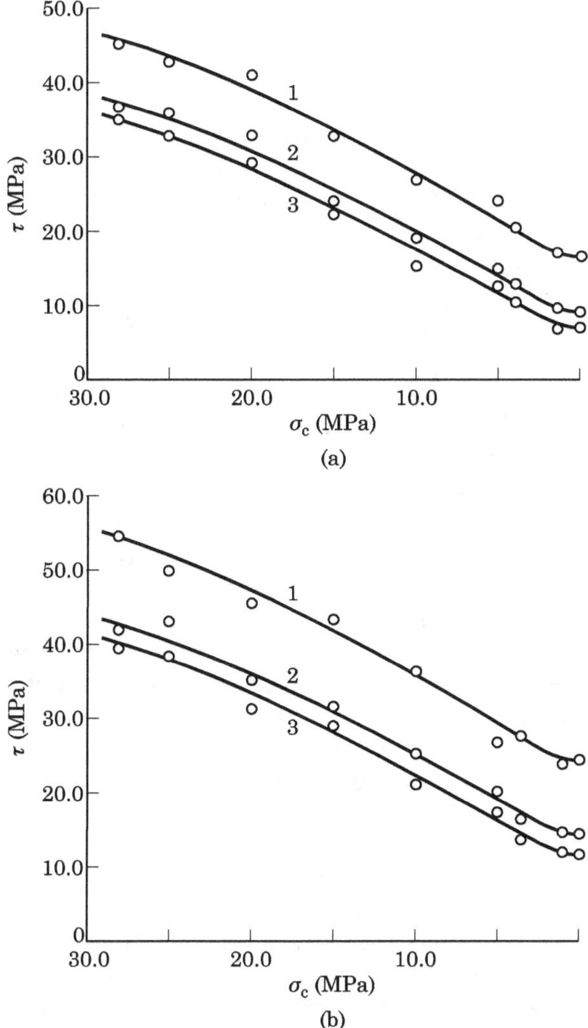

Figure 7.4 Diagrams of the limiting stressed state of adhesive specimens made of steel under the combined action of normal compression and shear. (a) Sprut-5M and (b) VAK; (1) in air; (2) under water; (3) in oil.

of normal fracture and shear. Further investigation and elucidation of the regularities of the resistance of adhesive joints to the combined action of normal compression and shear will facilitate their use for bonding of structures in conditions of all-round compression in liquid media, as well as of structures made of materials that cannot be joined using traditional methods (welding, bonding, riveting, bolting, etc.) not only in air but directly in liquid media as well.

7.2 Analysis of Strength Criteria as Applied to Adhesive Joints

Much research effort is aimed at the accumulation of experimental data on the strength of adhesive joints, and attempts are made to explain the phenomena accompanying their fracture process. However, as shown by the experience of adhesive joint usage, apparently highly encouraging results from wide-ranging laboratory strength tests under normal fracture or shear are often far from guaranteeing the efficiency of adhesive assembly. This causes distrust of the adhesive method of joining; sometimes it is used only when other methods are impossible.

These contradictions can to a considerable degree be resolved by a successfully chosen or properly developed theory of limiting stressed states, with the help of which it should be possible to find an explanation of the discrepancies between design, laboratory, and operating values of breaking stresses in adhesive joints.

Investigation of strength under the action of normal or tangential stresses only is common to all the existing theories of adhesion and the majority of experimental developments. At the same time, the increase of the areas and volumes of adhesive joints is interrelated with the need to develop engineering methods of calculation of their strength taking account of various kinds of stressed state.

The work [361] presents investigation of applicability of various strength criteria for anisotropic materials to a widely used type of adhesive joint, namely, glass-reinforced plastic, by comparison of experimental and design contours of breakability under biaxial compression. Design contours of breakability were constructed by criteria due to Mises [367], Fisher [369], Hu and Marin [370], Norris [371], and Warren [372], developed for materials possessing the same resistance to compression and tension, as well as by the criteria of Marin [373] (generalization of anisotropic material strength condition on the basis of the Mises–Genky plasticity theory), Zakharov [374], Malmeister et al. [375], and Goldenblatt and Kopnov [376]. Ashkenazi [377] developed criteria for materials unequally resistant to compression and tension. It has been shown that for the glass-reinforced plastic CAST-V, with the ratio of compression strength limits along the main axes of anisotropy equal to 1.0:1.11:3.67, none of the stated criteria permits construction of a breakability contour that deviates positively from the experimental points by less than 10%. Also, unacceptability of the I and III strength theories, producing deviations up to 19%, was established. For the glass-reinforced plastic PPN with a ratio of ultimate compression strength values along the main axes of anisotropy of 1.0:1.35:1.51, the I and III strength theories give up to

26% deviation from the experimental points. Application of the criteria developed by Prager, Fisher, Hu–Marin, and Norris leads to considerable overestimation of design strength, and use of the Mises and Marin criteria result in strength margins of up to 11% and 18%. For glass-reinforced PPN, the strength criteria of Warren, Zakharov, Goldenblatt–Kopnov, and Ashkenazi are the preferred ones, as the design contours deviate from the experimental points by 4–6% at most in the direction of increased safety margin.

Adhesive joint strength criteria developed recently do not have sufficiently rigorous substantiation and are based on the fact that the limiting surface in six-dimensional space of stresses is described by an equation of the second or higher order. By tailoring the equation, any degree of approach to the real strength surface is possible, but this necessitates determining a large number of constants experimentally. Thus, the criterion of [374] is represented by six input equations; the criterion of [376], taking account of dependence of strength on the sign of tangential stresses, by seven equations; the criterion of [378], including the value of elastic potential for anisotropic material, by eight equations; and so on. These criteria are given in general form, and their applicability is shown only for certain kinds of adhesive joints: paper–bakelite, textolite, and glass fiber-reinforced plastic joints.

Equations for the layered medium for investigation of the interlaminar fracture and shear stresses in glass-reinforced plastic with fabric reinforcement are given in [379]. The medium considered is a set of orthotropic glass-reinforced plastic plates; each of them includes one layer of reinforcing fabric and is joined without intermediate layers, and mechanical characteristics of separate plates and the whole set are assumed equal. Experimental validation of the equations is not produced.

On the basis of the equations [379], it is possible to determine [380] critical stresses of the reinforcing layer of a plate made of orthotropic glass-reinforced plastic under uniaxial and biaxial compression. Compression failure is considered as the result of violation of the operational compatibility of reinforcing material layers under the action of interlaminar fracture and shear stresses, occurring due to loss of separate layer stability. Since compression strength is equal to or less than the critical stress calculated by the formulas obtained, only the maximum possible strength value for the glass-reinforced plastic is determined. Experimental validation of the analytical dependences was not done.

The work reported in [381] deals with interlaminar fracture and shear stresses under compression in the plane of the layers, which occurs due to natural bending of reinforcing fibers as a result of their interweaving. Similarly to [379], the equations of a layered medium are derived, obtained with the use of an energetic smoothing principle and a hypothesis of straight lines for the layers. The form of the curving of the fibers is assumed sinusoidal. It was shown that the binding agent/glass-reinforced plastic ultimate strength relation changes insignificantly. On this basis, it was concluded that fracture occurs due to interlaminar fracture. The investigation is reduced to the development of analytical dependences.

In [382] the results are reported of an investigation of the strength of glass-reinforced plastic under the combined action of radial normal and interlaminar tangential stresses. The investigations were performed on ring specimens with one cross-cut under conditions of non-axisymmetrical loading. Maximum radial and tangential stresses were determined by the formulas

$$\sigma_{r,\max} = \frac{3}{2}\frac{M}{bhR} \qquad \tau_{r\varphi,\max} = \frac{3}{2}\frac{Q}{bh}$$

where M = the bending moment in the section considered; Q = the transverse force; b and h are the thickness and width of the ring; and R = the sum of the internal and external radii of the ring.

A plot of the limiting state for specimens of glass-cloth T-I impregnated with binding agent IF-ED-6 approximated a straight line, and the strength condition was

$$\frac{\tau_{r\varphi}}{\tau_{r\varphi,\mathrm{b}}} + \frac{\sigma_r}{\sigma_{r,\mathrm{b}}} \le 1$$

where $\tau_{r,\varphi,\mathrm{b}}$ and $\sigma_{r,\mathrm{b}}$ are ultimate strengths under simple shear and normal fracture. It was shown that the Mohr criterion is acceptable for the estimation of strength of the given glass-reinforced plastic. The method offered does not permit investigation of the material strength under the combined action of the stresses of shear and radial compression, as the reason for fracture lies in the circumferential direct stresses.

The criterion of short-term adhesion strength for the case of interphase failure under combined action of direct and tangential stresses is suggested in [382]. This criterion presumes the existence of a definite dependence between octahedral stresses, which is assumed linear in the first approximation. The hypothesis of the octahedral direct and tangential stress dependence was proposed by Nadai for continuous isotropic materials and was further developed by Balandin [384] into the strength criterion, presenting a linear function from the component of the spherical stress tensor.

Criteria for the contact layer strength have the form

$$\tau \le \frac{\sqrt{3}}{3}\sqrt{\left[\left(\sqrt{3}\frac{T_{\rm adh}}{R_{\rm adh}}-1\right)^2-1\right]\sigma^2-2\sqrt{3}T_{\rm adh}\left(\sqrt{3}\frac{T_{\rm adh}}{R_{\rm adh}}-1\right)\sigma+3T_{\rm adh}^2}$$

where T_{adh} and R_{adh} are the adhesion strength under shear and normal fracture.

Under the action of tangential stresses τ and normal tensile stresses σ:

$$\sigma \le \frac{\sqrt{3}T_{\rm adh}\left[\sqrt{3}\frac{T_{\rm adh}}{R_{\rm adh}}-1\right]}{\left[\sqrt{3}\frac{T_{\rm adh}}{R_{\rm adh}}-1\right]^2-1}\left[1-\sqrt{\frac{\left[\left(\sqrt{3}\frac{T_{\rm adh}}{R_{\rm adh}}-1\right)^2-1\right](T_{\rm adh}^2-\tau^2)}{T_{\rm adh}^2\left[\sqrt{3}\frac{T_{\rm adh}}{R_{\rm adh}}-1\right]^2}}\right]$$

Under the action of tangential stresses τ and normal compressive stresses σ:

$$\tau \le \frac{\sqrt{3}}{3}\sqrt{\left[\left(\sqrt{3}\frac{T_{\rm adh}}{R_{\rm adh}}-1\right)^2-1\right]\sigma^2+2\sqrt{3}T_{\rm adh}\left(\sqrt{3}\frac{T_{\rm adh}}{R_{\rm adh}}-1\right)\sigma+3T_{\rm adh}^2}$$

The given dependences allow calculation of the diagrams of limiting state from three parameters determined experimentally (T_{adh}, R_{adh}, and σ or τ) and are useful only for the case of adhesion fracture, which is not typical for high-quality adhesive joints. The criteria have been validated experimentally for polyester resin/glass film adhesion systems under the action of normal fracture and shear (quadrant I of the diagram of limiting states). The applicability of the criteria for the case of normal compression and shear is not shown.

Considerable attention is given to the criteria for estimating the strength of wood as a laminated material under the action of load at various angles to the fiber direction. In this case, direct and tangential stresses occur simultaneously on areas parallel to the fibers. Ashkenazi [385] developed a criterion for the strength of anisotropic wooden and synthetic materials in the following form:

$$\sigma_{\rm b} = \frac{\sigma_0}{\cos^4\alpha + \sin^2 2\alpha + C\sin^4\alpha} \qquad C = \frac{\sigma_0}{\sigma_{90}}$$

where σ_0, σ_{90} are the ultimate strengths along and across the fibers.

Diagrams of the limiting state of pinewood under the action of normal fracture and shear are presented in [364]. These are adequately approximated by the expression

$$0.67\sigma_y{}^2 - 0.94\sigma_y\tau_{xy} + 0.32\tau_{xy}{}^2 + 40\tau_{xy} + 28\sigma_y \leq R_{a=90^\circ}$$

where σ_y, σ_{xy} are the actual stresses on the area parallel to the fiber and $R_{a=90^\circ}$ is the ultimate strength under extension across fibers. It was shown that design diagrams of the limiting state constructed by the criterion of Ashkenazi [385] for wooden and synthetic materials disagree with the experimental ones considerably. A generalized strength criterion is not produced.

Diagrams of the limiting state of veneer sheet and delta-wood, depending on the slope angle of the wood fibers, are given in [386]. Correspondence of the experimental diagrams to design diagrams constructed by criteria in [395] was demonstrated. Limiting state curves for glass-reinforced plastic on epoxy–phenol binder were not checked for correspondence with the strength criteria.

Strength criteria for anisotropic materials do not directly take into account the dependence of the adhesion–cohesion interaction. At the same time, increase in cohesion strength does not always result in adhesion strengthening, and vice versa. An attempt to optimize the ratio of adhesion and cohesion strengths for composite materials reinforced with fibers was made in [378]. For complete realization of cohesion strength σ_τ of the reinforcement fibers of diameter d and length l, the minimum adhesive resistance under shear τ_{adh} should fulfill the correlation

$$\tau_{\text{adh}} \geq \sigma_\tau \frac{d}{2l}$$

A criterion of adhesion and cohesion strength [360] based on mechanical concepts of the optimal nature of the structure with full strength elements was developed for the coating–substrate system. Moskvitin had already stated the principle of full strength adhesion, cohesion, and self-adhesion, but it was not presented analytically. A physical sense of the criterion of full strength adhesion and cohesion is found in the fact that the coating should not separate from the substrate being deformed before complete exhaustion of its capacity in respect of cohesion strength. The condition of full strength adhesion and cohesion presents a linear dependence between deformation of the substrate ε_{cr} and adhesive resistance τ_{adh} taking into account the thicknesses of the coating and substrate, as well as their elastic properties, which favors the criterion described in [360] over that of [378].

The possibility of generalizing this criterion for other types of adhesive joints is not shown.

There are thus only few investigations of the applicability of existing strength criteria, and the number of new ones developed for adhesive joints is limited. Investigation of the strength of conical adhesive joints with epoxy compound UP-5-169c in the system of metal–glass-reinforced plastic resulted in the following dependences for calculation of the normal compression and shear stresses [362]:

$$\tau = \frac{G \cos\alpha\, Q}{(G\cos^2\alpha + E\sin^2\alpha)F}$$
$$\sigma = \frac{E \sin\alpha\, Q}{(G\cos^2\alpha + E\sin^2\alpha)F} \tag{7.1}$$

where E and G are moduli of normal elasticity and shear of the compound ($E = 1.9 \times 10^2$ MPa; $G = 0.2 \times 10^2$ MPa); Q is the load on the metal cone; F is the aea of the conical surface; and α is the slope angle of the conical joint.

For tubular specimens loaded with axial compression force and torque, the stresses are defined by

$$\sigma = \frac{Q_{comp}}{\pi(r_{out}^2 - r_{in}^2)}$$
$$\tau = \frac{M}{2\pi r_{out}^2(r_{out} - r_{in})} \tag{7.2}$$

where Q_{comp} is the axial compression load; M is the limit torque; and r_{out} and r_{in} are the outer and inner radii of the specimen.

It has been shown that the strength of an adhesive joint is best characterized by the Mohr criterion, and experimental data are approximated using the following expression:

$$\tau \le \sqrt{\frac{\sigma_p^2 - k\sigma_p\sigma + \sigma_p\sigma - k\sigma^2}{(k+1)^2}}$$

where $k = \sigma/\tau$.

It should be noted that the diagram of the limiting state constructed using formulas (7.1), including the elastic constants of the compound, agrees with that constructed using formulas (7.2). Evidently, this agreement will not be observed with substitution of values of the elastic modulus of the compound in the joint that differ quantitatively and qualitatively from those assumed.

An attempt was made in [360] to generalize the criterion presented there for adhesive joints on the assumption that the thickness of the second substrate $h \to 0$. However, the applicability of the analytical dependences obtained is restricted by the necessity of defining the modulus of elasticity of the adhesive layer, which, as was shown in Chapter 1, may differ quantitatively and qualitatively from that of the material in the free state.

The paper [363] gives general formulas for the calculation of stresses in the adhesive layer of a bushing-pivot joint loaded with tensile forces, transverse forces, and bending moment. For the case of combined tension and shear, estimation of the strength of the joint using VK-1 adhesive was been made on the basis of the unified theory of Davidenko–Friedman [388], permitting definition of the fracture type depending on the type of stressed state. Plots of the mechanical state indicate that the III strength theory is applicable to the adhesive cylindrical joints. Hence, the condition of adhesive layer strength is represented in the following form:

$$\tau_0 \leq \frac{\tau_{\text{cp}}}{n\sqrt{\left(\frac{k_\sigma}{2}\right)^2 + k_\tau^{\,2}}}$$

where n is the safety factor; k_σ and k_τ are coefficients of concentration of direct and tangential stresses, depending on the geometric dimensions of the joint, the Poisson ratio, and the modulus of elasticity of the adhesive layer.

It should be noted that the mechanical state diagram by Davidenko and Friedman is constructed using the values E_k and μ_k for the free state of the adhesive layer. Severe deviations of these parameters for the layer in the joint may influence the conclusions about applicability of the III strength theory considerably.

This brief review of attempts at estimation of criteria of adhesive joint strength illustrates the variety of approaches to this problem, which lead to a similar diversity of results, sometimes conflicting with each other. This is a reflection of the complex nature of the adhesive contact process. Development of the criteria of adhesive joint strength is restricted to only a small number of investigations, which need more rigorous experimental validation and cannot yet pretend to completeness. In addition, to date there are no sufficiently detailed investigations in respect of the applicability of existing strength theories to estimate the bearing capacity of adhesive joints.

We will consider further the principal features of the criteria of adhesive joint strength.

7.3 Applicability of the Limiting Stressed States Theories for Materials Unequally Resistant to Tension and Compression

As is well known, classic strength theories are based on the hypotheses of continuity and uniformity of substance distribution in the body being deformed. It is supposed that any arbitrarily small solid particles possess the same properties. However, this does not correspond to reality. The heterogeneity of the structure consists in local disturbances of chemical composition, the presence of various impurities, polycrystalline material structure, microcracks, and other defects causing considerable concentration of stresses. This applies equally to adhesive joints of various materials.

Restricted applicability of the classic strength theories (I–IV theories) is to a great extent the result of simplified interpretation of the structure and properties of real materials. Various measures are taken to eliminate this contradiction. For example, Coulomb long ago considered maximum tangential stress as a linear function of the average direct stress, to extend the maximum tangential stresses theory to materials possessing anisotropy of properties expressed as different resistance to tension and compression. In addition, there were theories developed by Mohr, reduced to the empirical determination of the curve $\tau = f(\sigma)$, turning round the family of main limit circles; the energetic strength theories of Beltramy–Heig and Guber–Genky; the latest energetic theories; and more.

As shown earlier, adhesive joints are considered to be systems unequally resistant to the stresses of normal fracture and compression. Thus, it is expedient to limit our analysis to strength theories and the Mohr theory as the pure experimental one.

The majority of the latest energetic theories are based on the hypothesis of functional dependence of octahedral tangential and direct stresses proposed by Nadai [384]. According to Nadai, the condition for occurrence of the limiting state takes the form

$$\tau_{\text{oct}} = f(\sigma_{\text{oct}})$$

According to Schleicher [384], the limiting state occurs at a specific value of total specific potential energy. The strength condition for the flat stressed state similar to the stressed state of adhesive joints being investigated ($\sigma_2 = 0$) is of the form

$$\sigma_1{}^2 + \sigma_3{}^2 - 2\mu\sigma_3\sigma_1 + (\sigma_c - \sigma_p)(\sigma_1 + \sigma_3) \leq \sigma_p\sigma_c \qquad (7.3)$$

This criterion is applicable to materials with any correlation of $X = \sigma_p/\sigma_c$ within the limits $1 > X > 0$, as well as the other latest energetic theories. For $\sigma_p = \sigma_c$, the criterion of Schleicher (together with those stated below) corresponds to the theory of Beltramy–Heig.

Schleicher's hypothesis is poorly validated by experimental data and is therefore not widely used. It is impossible to investigate the criterion's applicability for adhesive joints, since at present there are no methods available for determination of the Poisson ratio for the adhesive layer in the joint.

Buzhinsky's hypothesis [384] is more universal. According to this, the failure condition occurs when the energy of deformation and a part of the energy of volume change reach definite critical values, being a linear function of a spherical tensor. The design equation for the volumetric stressed state is

$$\sigma_1^{\,2} + \sigma_2^{\,2} + \sigma_3^{\,2} - 2\left(\frac{\sigma_c\sigma_p}{2\tau_k^{\,2}} - 1\right)(\sigma_1\sigma_2 + \sigma_2\sigma_3 + \sigma_3\sigma_1) + (\sigma_c - \sigma_p)(\sigma_1 + \sigma_2 + \sigma_3) = \sigma_c\sigma_p \tag{7.4}$$

and for $\sigma_2 = 0$,

$$\sigma_1^{\,2} + \sigma_3^{\,2} - 2\left(\frac{\sigma_c\sigma_p}{2\,\tau_k^{\,2}} - 1\right)\sigma_3\sigma_1 + (\sigma_c - \sigma_p)(\sigma_1 + \sigma_3) = \sigma_c\sigma_p \tag{7.5}$$

Yagn's strength theory [390] agrees with the hypothesis put forward by Buzhinsky, although it is based upon entirely different concepts. Yagn proposed finding the condition of occurrence of the limiting state in the form of a second-order equation, symmetrical in all three main stresses. The following equation meets this requirement for $\sigma_2 = 0$:

$$\sigma_1^{\,2} + \sigma_3^{\,2} + (\sigma_3 - \sigma_1)^2 + \frac{6\tau_k^{\,2} - 2\sigma_p\sigma_c}{\sigma_p\sigma_c}(\sigma_1 + \sigma_3)^2 + \frac{6\tau_k^{\,2}(\sigma_c - \sigma_p)}{\sigma_p\sigma_c}(\sigma_1 + \sigma_3) = 6\tau_k^{\,2} \tag{7.6}$$

As was shown in [384], Yagn's theory is based on a more detailed physical conception of the mechanism of material fracture than is the Buzhinsky hypothesis.

Balandin's strength theory [384] is based on assumptions about the functional dependence of the specific potential energy of deformation on the spherical stress tensor. In first approximation, this function is assumed linear and for $\sigma_2 = 0$ it results in the following strength condition:

$$\sigma_1{}^2 + \sigma_3{}^2 - \sigma_3\sigma_1 + (\sigma_c - \sigma_p)(\sigma_1 + \sigma_3) = \sigma_p\sigma_c \tag{7.7}$$

As shown in [384], Balandin's hypothesis is a particular case of the Buzhinsky hypothesis, and it is interpreted by a paraboloid of rotation in the stress space that, in accordance with Equation (7.4), is observed for the condition

$$\tau > \sqrt{\frac{\sigma_p\sigma_c}{3}}$$

Results of adhesive joint strength tests under normal fracture and compression, given earlier, prove observance of this condition.

Several strength theories [384, etc.] are based on the linear dependence between tangential and direct stresses in the octahedral plane. According to Botkin [389], one constant—the traction coefficient—is required for estimation of strength of plastic materials whose bonding strength considerably exceeds internal friction. For the estimation of loose materials' strength, only the coefficient of internal friction is required. Since most real materials occupy an intermediate position between plastic and loose ones, their strength should be estimated using two coefficients. Fracture occurs precisely at the moment when the shear stresses are sufficient to overcome the forces of friction and cohesion between these particles. The strength condition $\tau_{oct} < m(n + \sigma_{oct})$, with the coefficient of internal friction m and traction coefficient n expressed through the values of ultimate strength under tension and compression, and transformation to main stresses ($\sigma_2 = 0$), takes the form

$$\sqrt{\sigma_1{}^2 + \sigma_3{}^2 + (\sigma_3 - \sigma_1)^2} \le \frac{2\sqrt{2}\sigma_c\sigma_p}{\sigma_c + \sigma_p} - \frac{\sqrt{2}(\sigma_c - \sigma_p)}{\sigma_c + \sigma_p}(\sigma_1 + \sigma_3) \tag{7.8}$$

Investigation of the applicability of Botkin's theories to adhesive joints is interesting also because one of the constants determining material strength is the traction coefficient.

Miroliubov [391] considers that the strength of a material is defined by its shearing resistance in the octahedral plane. The criterion is a quadratic function of the components of the spherical stress tensor. This function for $\sigma_2 = 0$ gives

$$\sqrt{\tfrac{1}{2}[(\sigma_{L1} - \sigma_{L2})^2 + (\sigma_{L2} - \sigma_{L3})^2 + (\sigma_{L3} - \sigma_{L1})^2]} + \frac{1-X}{1+X}(\sigma_{L1} + \sigma_{L2} + \sigma_{L3}) = \sqrt{\frac{2}{1-X}}\sigma_p \tag{7.9}$$

where σ_{L1}, σ_{L2}, σ_{L3} are the ultimate stresses of the similar stressed state for $\sigma_2 = 0$:

$$\sqrt{\tfrac{1}{2}[\sigma_{L1}{}^2 + \sigma_{L3}{}^2 + (\sigma_{L3} - \sigma_{L1})^2]} + \frac{1-X}{1+X}(\sigma_{L1} + \sigma_{L3}) = \sqrt{\frac{2}{1+X}}\sigma_p$$

As mentioned earlier, the criterion of adhesion strength under the action of direct and tangential stresses due to Skudra and Kirulis is also based on the functional dependence between octahedral stresses at the moment of fracture.

The strength criterion proposed by Pisarenko and Lebedev [383] suggests that the occurrence of limiting state is controlled by the ability of the material to resist both tangential and direct stresses. Actually, the limiting state is defined by the appearance of cracks (under shearing resistance) and by crack propagation (under resistance to normal fracture). Thus, the strength criterion is in the form of tangential stress functions independent of the stressed condition, and [383] represents this in the following expression:

$$\tau_{oct} + m_1\sigma_1 \leq m_2 \tag{7.10}$$

The constants m_1 and m_2 expressed through the ultimate stresses σ_p and σ_c, under uniaxial tension and compression, allow the criterion to be written in the form [365]

$$\frac{3}{\sqrt{2}}X\tau_{oct} + (1-X)\sigma_1 \leq \sigma_p \tag{7.11}$$

This brief analysis of the latest energetic strength theories suggests the possibility of investigating their applicability to adhesive joints from the following considerations:

(a) The theories are applicable for materials unequally resistant to tension and compression ($1 > X > 0$). The objects considered—adhesive joints of metals—meet this condition.

(b) The equations formulating the theories (apart from Schleicher's) do not include material constants (for example, modulus of elasticity, Poisson ratio, etc.), and methods for their determination for adhesive layers in joints are under development.

(c) The analytical expressions of the theories permit calculation of the diagrams of limiting states from the results of tests under normal fracture, shear, and normal compression of adhesive joints and their comparison with experimental diagrams obtained under combined action of direct and tangential stresses.

Thus, investigation of the applicability of strength theories of continuous uniform materials to adhesive joints can be restricted to the theory of Mohr [384] as a purely experimental one, the energetic theories of Yagn [390], Balandin [384], Botkin [389], Miroliubov [391], and Pisarenko–Lebedev [383], as well as the adhesion strength criterion of Skudra–Kirulis [382]. The accuracy of this correspondence may be estimated by comparison of design and experimental diagrams of the limiting stressed state.

7.4 Analysis of Design and Experimental Diagrams of the Limiting Stressed State

Calculation of the limiting stressed state diagrams of adhesive joints has been performed using strength criteria [361, 384, 389–391] for materials unequally resistant to tension and compression. Initial formulas of these criteria, obtained by substitution of values of τ_{oct} and $\sigma_{1,3}$, and algebraic transformations, are presented in Table 7.1, in the form of corresponding equivalents for calculation on adhesive joints.

For the construction of design diagrams of the limiting state of adhesive joints it is enough to have the results of strength tests under normal fracture, compression, and shear. These values of strength for adhesive compositions Sprut-5M and VAK are given in the following figures and in Table 7.2. Experimental diagrams of the limiting state of adhesive joints are constructed by the results of tests under combined action of the direct and tangential stresses, and are given in Figs. 7.2 to 7.4.

Design and experimental diagrams of the limiting stressed state of the specimens fixed in air, under water, and in oil using Sprut-5M are shown in Fig. 7.5; those for VAK are shown in Fig. 7.6. Diagrams were computed using the following strength theories: 1, Balandin; 2, Miroliubov; 3, Botkin; 4, Yagn; 5, Pisarenko–Lebedev; 6, Skudra–Kirulis; 7, Mohr; diagram 8 is the experimental one.

Analyzing Figs. 7.5 and 7.6, it should be noted that under combined action of the stresses of normal fracture and shear (quadrant I of the diagram), the design and experimental curves of the limiting state are similar, regardless of the medium in which specimens are fixed. Under combined action of normal compression and shear (quadrant II) this similarity is not observed.

The largest relative errors of the design diagrams of quadrant I were determined by statistical processing (Table 7.3). Evidently, considerable deviations (up to 74.9%) are typical for diagrams constructed by the Balandin theory, and it probably should be considered un-

TABLE 7.1 Formulas for Calculation of Diagrams of Limiting State of Adhesive Joints

Strength theory	Initial formula for materials unequally resistant to tension and compression ($\sigma_2 = 0$)	Equivalences (after substitution of $\tau_{oct}, \sigma_{1,3}$ and algebraic transformations)
Yagn	$\sigma_1{}^2 + \sigma_3{}^2 + (\sigma_3 - \sigma_1)^2 + \frac{\sigma \tau_k{}^2 - 2\sigma_p \sigma_c}{\sigma_p \sigma_c}(\sigma_3 - \sigma_1)^2$ $+ \frac{\sigma \tau_k{}^2 - 2(\sigma_p \sigma_c)}{\sigma_p \sigma_c}(\sigma_1 + \sigma_3) = \sigma \tau_k{}^2$ where $\sigma_k = \sigma$ at initiation of fracture	$\sigma_{eq} = -\frac{\sigma_c - \sigma_{p1}}{2}\sqrt{\left(\frac{\sigma_c - \sigma_p}{2}\right)^2 + \frac{\sigma_c \sigma_p}{3\tau_b{}^2}(\sigma^2 + 3\tau^2)}$ $+\left(1 - \frac{\sigma_p \sigma_p}{3\tau_b{}^2}\right)\sigma^2 + (\sigma_c - \sigma_p)\sigma$
Balandin	$\sigma_1{}^2 + \sigma_3{}^2 - \sigma_3\sigma_1 + (\sigma_c - \sigma_p)(\sigma_1 + \sigma_3) = \sigma_p \sigma_c$	$\sigma_{eq} = \frac{1 - X}{2}\sigma + \frac{1}{2}\sqrt{(1 - X)^2\sigma^2 + 4X(\sigma^2 + 3r^2)}$ where $X = \sigma_p / \sigma_c$
Miroliubov	$\sqrt{\frac{1}{2}[\sigma_{L1}{}^2 + \sigma_{L3}{}^2 + (\sigma_{L3} - \sigma_1)^2]} + \frac{1 - X}{1 + X}(\sigma_{L1} + \sigma_{L3}) = \sqrt{\frac{4}{1 + X}\sigma_p{}^2}$ where σ_{L1}, σ_{L3} are the principal stresses similar to specified ones	$\sigma_{eq} = \frac{1 - X}{2}\sigma + \frac{1 - X}{2}\sqrt{\sigma^2 + 3\tau^2}$

Botkin	$\sqrt{{\sigma_1}^2+{\sigma_3}^2}+(\sigma_3-\sigma_1)^2 \le \frac{2\sqrt{2\sigma_c\sigma_p}}{\sigma_c+\sigma_p}-\frac{\sqrt{2}(\sigma_c-\sigma_p)}{\sigma_c+\sigma_p}(\sigma_1+\sigma_3)$	$\sigma_{eq}=\frac{1+\sqrt{\sigma^2+3\tau^2}+(1-X)\sigma}{2}$
Pisarenko–Lebedev	$\frac{3}{\sqrt{2}}x\tau_{dev}+(1-x)\sigma_1 \le \sigma_p$	$2\sigma_c(\sigma_c-\sigma\alpha)+\sigma^2(\alpha^2-2)+2\tau^2(\alpha^2-3)$ $=\sqrt{\sigma^2+4\tau^2}\,(\sigma\alpha^2-2\sigma_c\alpha)$ where $\alpha=(\sigma_c/\sigma_p)-1$
Skudra–Kirulis	(a) Tension and shear: $\tau \le \frac{\sqrt{3}}{3}\sqrt{\left[\left(\sqrt{3}\frac{T_{coh}}{R_{coh}}-1\right)^2-1\right]\sigma^2-2\sqrt{3}T_{coh}\left(\sqrt{3}\frac{T_{coh}}{R_{coh}}-1\right)\sigma+3T_{coh}^2}$	—
	(b) Compression and shear: $\tau \le \frac{\sqrt{3}}{3}\sqrt{\left[\left(\sqrt{3}\frac{T_{coh}}{R_{coh}}-1\right)^2-1\right]\sigma^2+2\sqrt{3}T_{coh}\left(\sqrt{3}\frac{T_{coh}}{R_{coh}}-1\right)\sigma+3T_{coh}^2}$	—

TABLE 7.2 Initial Data for Calculation of Diagrams of Limiting Stressed State of Adhesive Joints

Gluing medium	Strength (MPa)					
	Sprut-5M adhesive			VAK adhesive		
	σ_p	σ_c	τ	σ_p	σ_c	τ
Air	23.0	43.0	16.5	28.5	104.0	23.8
Water	10.8	50.0	9.4	15.0	90.0	12.8
Oil	9.5	52.0	7.3	12.0	84.0	10.9

acceptable for adhesive joints of metals with polyester and acrylate compositions.

The largest relative errors of diagrams constructed from the theories of Miroliubov, Botkin, Yagn, Pisarenko–Lebedev, and Skudra–Kirulis, do not exceed 9.6%. The diagrams constructed from the criteria of Miroliubov and Botkin are practically the same. Deviation of the diagrams based on the theories of Miroliubov, Botkin, Yagn, and Skudra–Kirulis from the experimental values occurs toward decrease of the safety margin of the adhesive joint, and that of the diagrams

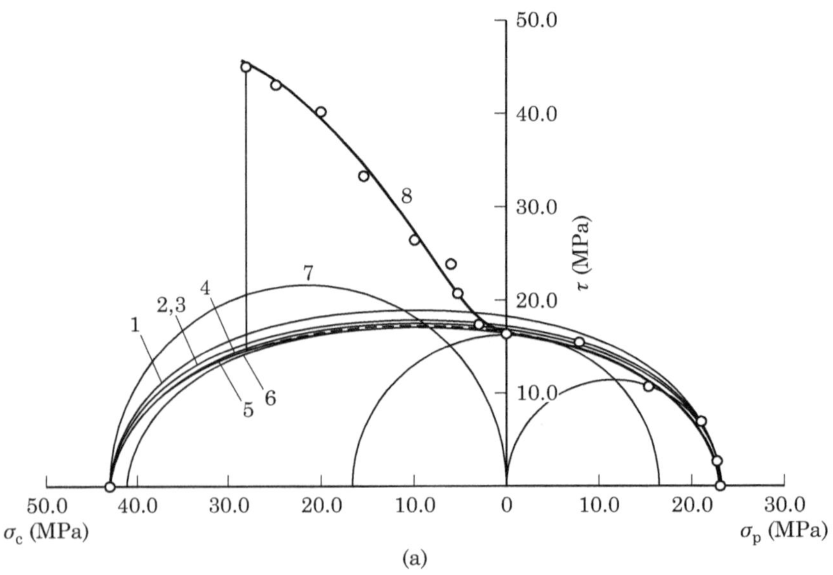

Figure 7.5 Diagrams of limiting stressed state of specimens made of steel St3, fixed with Sprut-5M (a) in air, (b) under water, (c) in oil. (1–7) Design diagrams on the criteria of Balandin (1), Miroliubov (2), Botkin (3), Yagn (4), Pisarenko–Lebedev (5), Skudra–Kirulis (6), and Mohr (7). Curve (8) is the experimental diagram.

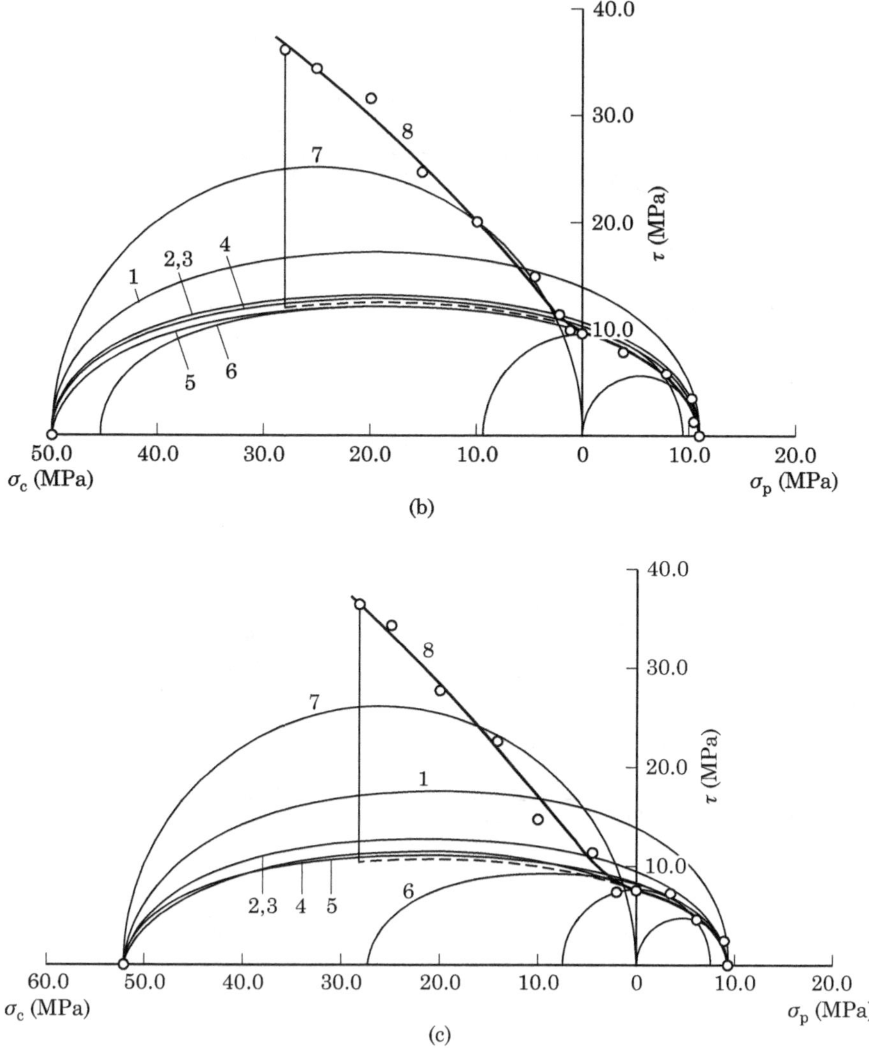

Figure 7.5 *(Continued)*

constructed on the Pisarenko–Lebedev theory in the direction of increase. These strength theories are acceptable for the adhesive joints investigated.

Let us consider quadrant II of the limiting state diagram. With increase of the normal compressive stresses, the experimental diagrams deviate sharply from the design ones toward increase of the shear breaking stresses. Long ago, Amonton and then Coulomb suggested that the critical value of shearing stress τ in the plane

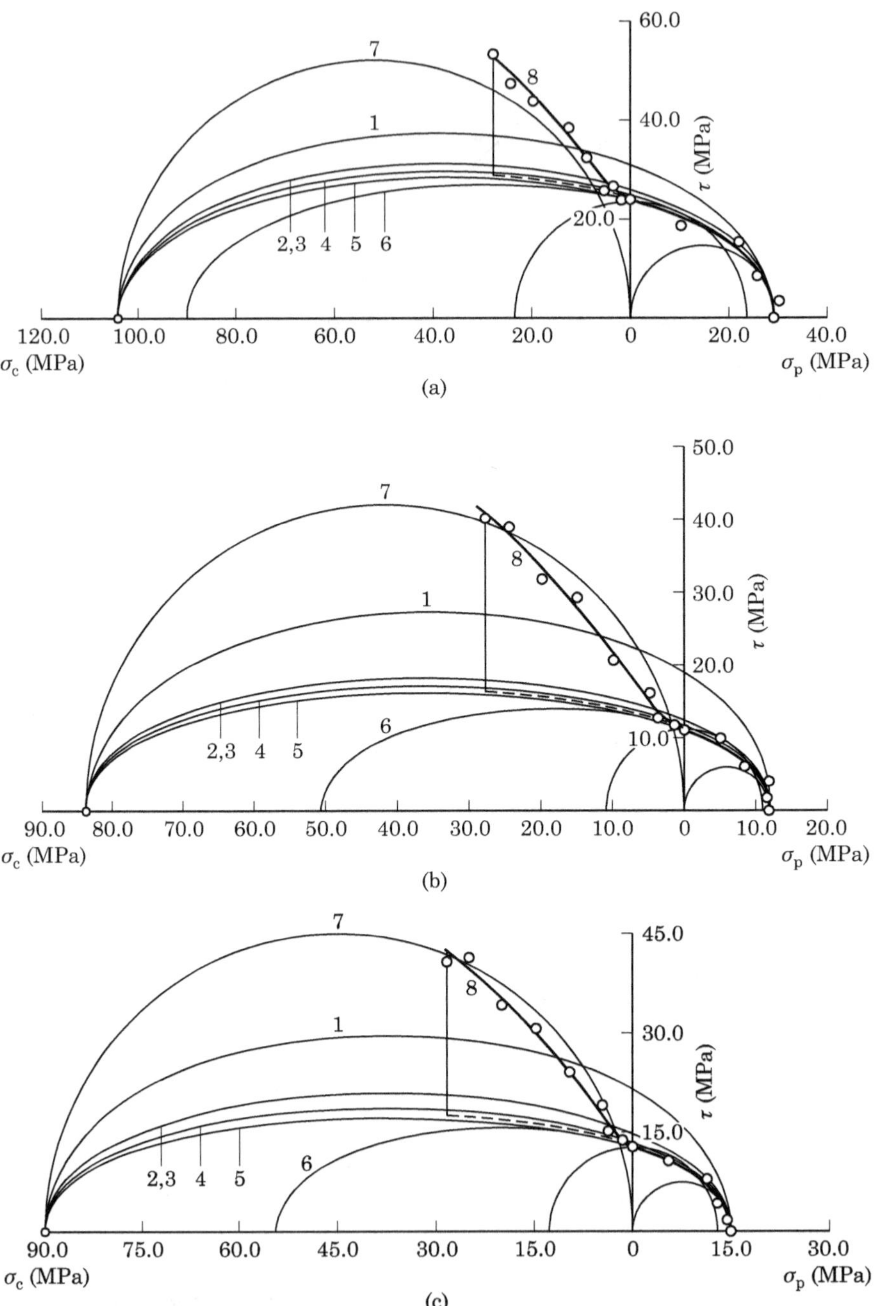

Figure 7.6 Diagrams of limiting stressed state of specimens made of steel St3, fixed with VAK (a) in air, (b) under water, (c) in oil. (1–7) Design diagrams on the criteria of Balandin (1), Miroliubov (2), Botkin (3), Yagn (4), Pisarenko–Lebedev (5), Skudra–Kirulis (6), and Mohr (7). Curve (8) is the experimental diagram.

TABLE 7.3 Results of Calculation of the Largest Relative Errors for Limiting State Diagrams of Adhesive Joints

Composition and gluing medium	Largest relative error ε_{max} (%)				
	Balandin	Miroliubov–Botkin	Yagn	Pisarenko–Lebedev	Skudra–Kirulis
Sprut-5M adhesive					
Air	9.4	6.1	3.0	−2.1	4.9
Water	38.2	8.5	4.2	−3.8	6.4
Oil	74.9	9.6	9.5	−3.7	7.6
VAK adhesive					
Air	25.0	7.9	5.0	−4.2	4.6
Water	61.5	8.7	3.8	−5.3	4.8
Oil	59.1	9.1	5.4	−5.2	5.9

increases in proportion to the normal stress of compression σ_N applied to this plane:

$$\tau = \tau_c + \mu\sigma_N$$

where τ_c is the value characterizing the cohesion strength of the material and μ is the friction coefficient.

Word [387] considers that the Coulomb criterion is of great interest for description of the behavior of polymeric materials. For uniaxial loading of monoliths under the action of compressive stress σ_1, if the fracture takes place in a plane the normal to which forms the angle Θ with the direction of action σ_1, the critical value of tangential stress is

$$\tau_c = \sigma_1(\cos\Theta\sin\Theta - \mu\cos^2\Theta)$$

However, as seen in [392] and [393], the Amonton law and the Coulomb criterion hold when the forces of intermolecular interaction between the shifted bodies are not taken into account. This interaction was taken into consideration by Derjagin [392, 393] in a binomial friction law. Analyzing the applicability of a binomial friction law, Derjagin notes that it should be used when equilibrium conditions are considered between two adjacent bodies with large area of actual contact, which are subjected not only to the forces perpendicular to the contact surface but also to forces parallel to it. An adhesive joint is just the contact of at least two surfaces by means of adhesive that fills the microrelief and ensures a large area of the true contact.

The binomial law for two bodies with large areas of true contact has the form

$$P = \mu(N + F_0 P_0) \tag{7.12}$$

where P is the friction force corresponding to the beginning of movement; N is the loading; F_0 is the area of true contact, as against the geometric one, which considerably exceeds the true area; μ is the true friction coefficient, as against the design one, equal to the ratio P/N; and P_0 is the specific cohesion at the section F_0.

Dividing Equation (7.12) by F_0, we get

$$\tau = \mu(\sigma + P_0) \tag{7.13}$$

where τ is the tangential stress at the start of slippage and σ is the normal stress, which according to Derjagin [392] may have negative sign under tearing forces.

According to Derjagin, the area of true contact should increase with increase of the compression stress, and checking the binomial friction law in normal conditions (i.e., direct contact of two bodies) is difficult. This difficulty does not exist for adhesive joints, since the adhesive layer impedes direct contact of the substrates. For example, Derjagin and Lazarev, in [393], deal with friction of paraffin with glass. Equality of the area of true contact and the geometric value was provided by building up paraffin on the glass surface, which also caused independence of the contact area from the value of normal loading. It was established that the friction force is in direct proportion to normal load, and the curve slope angle is equal to the true friction coefficient μ. In this case the straight line $P = f(N)$ does not pass through the origin of coordinates, making an intercept on the ordinate equal to μP_0, similar to the diagrams presented in Figs. 7.5 and 7.6.

An investigation of the boundary friction of paraffin on glass in the presence of a stearic acid soap intermediate layer is also very important [393]. This showed that with increase in the number of monolayers applied, the true friction coefficient does not depend on where in the lubricating layer the slip plane is.

From the above, an adhesive joint may be considered as a pair of solids joined by an adhesive layer that acts as a so-called solid lubricant, initiating adhesion interaction with the pair. For relative movement of the solids, both this interaction and the friction force have to be overcome. Under compression of the adhesive joint, the friction force increases as well as the shearing stress; under tension, the converse process takes place. Thus, by the action of the friction force we can explain the sharp deviation of the experimental diagrams of limiting state (Figs. 7.5 and 7.6) from the design diagrams. However, this explanation needs to be validated experimentally. The task consists in determining the strength components (adhesion and friction components) of the adhesive joint subjected to simultaneous action of both normal compression and shear.

It should be noted that the binomial friction law of Derjagin states that the adhesion bonds of two bodies and the friction forces determine a tangential load. The aggregate of these values is represented by the concept of static friction. Experimental validation [392, 393] of the binomial friction law allows the determination of this sum, but not of each item separately. As has been noted [392], this method serves only as an indirect estimation of adhesive joint strength.

In [366] an attempt was made to define the adhesion component under the combined action of normal compression and shear on the contacting metal surfaces. The cut-off strength [394] of the adhesive line under normal load is assumed as the initial dependence:

$$\tau_{\mathrm{n}} = \tau_0 + \beta N$$

where N is the actual pressure on the contact; τ_0 is the adhesive line cut-off strength in the absence of normal stresses; β is a coefficient characterizing the influence of normal stresses on tangential strength. This method is also indirect and permits determination only of the torque value required for overcoming static friction. The adhesion component is found by calculation.

Components of the adhesive specimen strength—adhesion component (τ_{adh}) and friction component (τ_{fr})—were determined in accordance with the method we have developed, using diagrams of shearing deformation (Fig. 7.7). If the glued specimen is subjected only to shear (Fig. 7.7a), the most typical deformation diagram looks like curve 1. If the specimen is compressed perpendicularly to the plane of adhesive layer, and the diagram of shearing deformation is recorded in compressed state, it will be in the form of curve 2, which may be explained by the following considerations. On a certain level of shear stress (point A) being achieved, the combined action of the friction and adhesion forces is overcome. With a generalization of the binomial friction law of Derjagin, this may be presented as overcoming of the static friction forces. The diagram of deformation displays a sharp stepwise reduction of the shearing stress until the point B is reached. The AB ordinate represents the value of the adhesion (τ_{adh}) component of adhesive joint strength. The ordinates of the sloped portion of the section BC are determined by the forces of boundary friction, which are caused by the compression stress value and decrease with the reduction of contact area.

To validate the method described, it is necessary to match the fracture surfaces of the adhesive specimen after recording the deformation diagram, to exert compression until the previous level of stress is

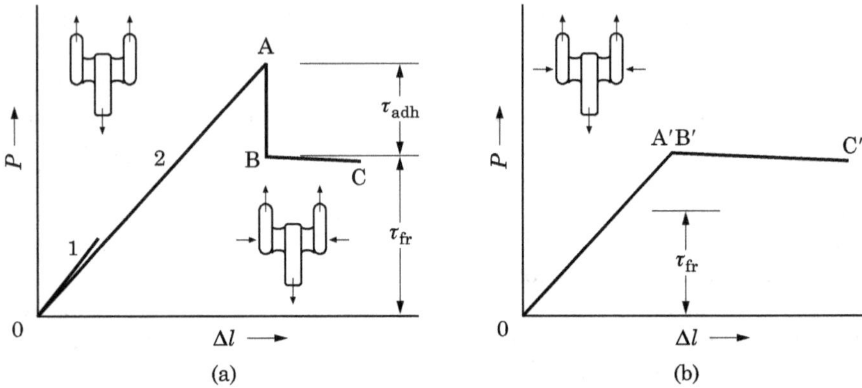

Figure 7.7 Methods of determining components of the strength of adhesive joint adhesion (τ_{adh}) and friction (τ_{fr}) under the combined action of direct compression and shear. (a) Shear deformation of (1) initial specimen; (2) precompressed specimen. (b) Shear deformation of the specimen joined by fractured surfaces and repeatedly compressed to the initial level.

reached, and then to record the shearing deformation diagram once more (Fig. 7.7b). Then the point A should practically coincide with the point B and occupy a position A′B′ due to insignificant influence of adhesion forces. The ordinates of the points B and A′B′ should be the same.

The presence of the friction forces component and its considerable influence on the strength of the adhesive joint are confirmed by the following experiments. Steel specimens were fixed with Sprut-5M and VAK in air, under water, and in oil. Using the Instron test unit, the shearing diagrams of specimens were recorded, as well as diagrams of specimens precompressed to 28.0 MPa. In this case the values of the adhesion component of the strength of compressed adhesive specimens defined by the steps in the deformation diagrams coincide with the strength values under shear of adhesive specimens given in Figs. 7.1 and 7.2 and in Table 7.2. Recording of the deformation diagrams of specimens compressed to stresses of 5.0 and 15.0 MPa also continues this coincidence. The ordinates of the deformation diagrams determine the dependence of the strength components of specimens on friction forces after failure of adhesion bonds. For specimens fixed with Sprut-5M the resistance from friction forces amounts to 28.5 MPa in air, to 27.5 MPa under water, and to 26.5 MPa in oil; for specimens glued with VAK the corresponding values are 29.0, 29.5, and 30.0 MPa, which agree with the data given in Figs. 7.1 and 7.2. Recording of the deformation diagrams for specimens compressed to stresses of 5.0 and 15.0 MPa has validated the above coincidence, as well as the facts that the sharp increase in the specimen strength under the combined action of normal compression and shear occurs due to increase in the friction forces, and that resistance from the adhesion forces is practically the same.

The recording of the adhesive joint deformation diagram is attended by certain difficulties and errors, since substrates are deformed together with the adhesive layer. In this connection, we developed a simpler method for determination of the strength components under the combined action of normal compression and shear. This is based on the fact that specimens compressed to various levels of stress fracture with tangential stresses, and the dependence of the breaking stress on compression value is constructed. The fractured specimen surfaces are then matched, compressed to the initial levels, and repeatedly fractured with tangential stresses. Due to the insignificant influence of adhesion forces, the ordinate of any point on the curve of shear breaking stress versus compression value in repeated tests represents the friction force component of the adhesive joint strength (τ_{fr}). The adhesion component of the adhesive joint strength for any specified value σ_c is found as the difference of the ordinates of point A in Fig. 7.7. The tangent of the angle α for the adhesive joint is a true coefficient of friction. The usual tearing machines may be used for this technique.

Figure 7.8 presents curves 1–3 of the strength dependence for steel specimens glued with Sprut-5M and VAK in air, under water, and in oil, under the combined action of normal compression and shear (see Figs. 7.5 and 7.6), as well as curves 1′–3′ obtained from the results of

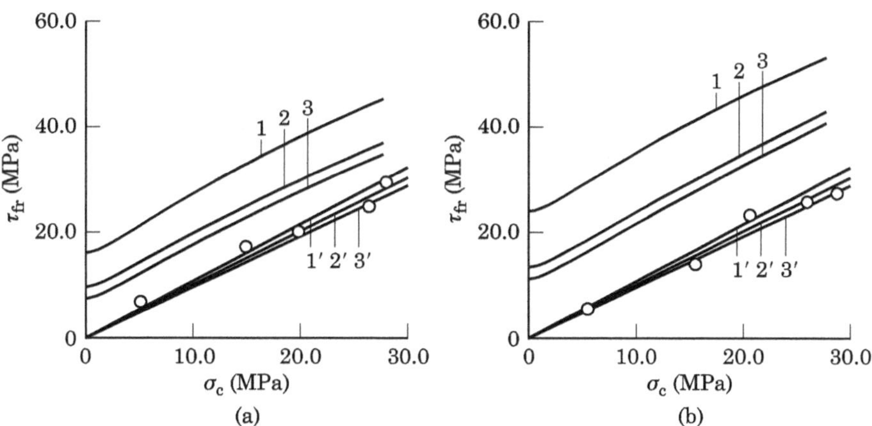

Figure 7.8 Determination of the friction component in shearing stength tests of fractured, matched working surfaces (1′–3′) and repeatedly compressed specimens (1–3) fixed in air (1), under water (2), and in oil (3) using (a) Sprut-5M and (b) VAK.

tests on specimens with matched working surfaces compressed to various levels of stress and fractured by shear. As is obvious, the friction force component of strength is essentially independent of the gluing medium, and the curves 1–3 and 1′–3′ are similar. From this one can conclude that the value of τ_{adh} depends insignificantly on compression stress, and only τ_{fr} increases proportionally with σ_c. For any value of σ_c in the interval 0–28.0 MPa, simple subtraction of the ordinates of curves 1–3 and 1′–3′ in accordance with the method described gives values of shear breaking stresses similar to those given earlier. The values of the true friction coefficients are characterized by the angle α.

Subtracting the friction component (τ_{fr}) from corresponding ordinates of experimental diagrams of limiting state obtained under the combined action of normal compression and shear (Figs. 7.5 and 7.6, quadrant II), we can construct the diagram sections (dashed line); the maximum relative deviation from the diagrams calculated from strength theories [382, 384] does not exceed 9%.

On the basis of the above, we conclude that a sharp increase in tangential breaking stress of the compressed adhesive specimen is the result of the friction force component.

Chapter

8

Control of Polymer Properties for Impregnation of Porous Materials

8.1 Introduction

In their end uses, structural units and buildings made of concrete, brick, or wood are subject to aging, during which their strength and/or hardness are reduced and their decorative properties deteriorate. Remediation is usually effected by means of replacement and reinforcing, and the cost of such work often exceeds the cost of new construction. Recently great interest has been shown in the problem of renovating structures by impregnation, both with solutions of inorganic salts and with organic monomers and oligomers [403, 404].

Impregnation with monomers and oligomers that achieve a polymeric state within the structure is preferable, as it permits the treatment of practically all types of porous materials, such as concrete, soil, wood, brick and other ceramic materials, asphalt, and so on, and the impregnated material exceeds the original in strength. The treated materials' capacity for deformation without breaking also increases, and density, tightness, chemical stability, and immunity to insects is improved.

The main obstacle to wide employment of organic compositions for impregnation purposes lies in selective adsorption of certain components by the pores of the surface of the impregnated materials. If this happens, the components are separated, leading to disturbance of the system's stoichiometry and to lack of curing. For example, when concrete is impregnated with epoxy resin/amine hardener mixture, the upper part of the concrete becomes impregnated with epoxy resin only, and the deeper parts with hardener.

To avoid selective separation of the components, one-component systems are usually used for impregnation; their cure is effected by interaction of monomers and oligomers with substances contained in the material being impregnated, mainly water. Unsaturated compounds or compounds with isocyanate groups are most often used. Polymerization of cyanoacrylates in the bulk of an impregnated material occurs by the catalytic action of water, which is always present on the surface of pores. Polyisocyanates react chemically with water, producing polyureas; acrylate polymerization, because of the necessity for high-purity materials, is achieved with ionizing radiation.

The use of one-component organic systems for impregnating structures is attended by considerable difficulties: for example, the cost of cyanoacrylate monomers is high; they can be used only for the impregnation of absolutely dry structures; they are quickly polymerized and hence cannot penetrate deeply into materials. The use of radiation for acrylate polymerization is hazardous, expensive, and technically complicated. When polyisocyanates cure by interaction with water, carbon dioxide is liberated and comes out to the surface where it forms continuous pores:

$$\mathrm{NCO{-}R{-}NCO + HOH \rightarrow {-}NH{-}CO{-}NH{-} + CO_2}$$

In addition, gas-generated pressure impedes deep penetration of the structure.

The problem faced was the development of organic materials and technologies for impregnation of porous structures, including those saturated with water, that would possess high permeability, cure with specified speed inside the structure, and provide considerable increase of strength, air-tightness, and chemical stability. We approached this probem with the help of the following techniques:

1. Changing the surface energy of impregnated materials to provide selective wetting of pore surfaces by polymer compositions in aqueous medium.
2. Use of one-component compositions, including functional groups and crown-ethers, for impregnation. These crown-ethers form complexes with metal atoms located on the surface of the materials being impregnated. The complexes possess catalytic activity and can initiate and accelerate polymerization of the compositions.

8.2 Physical-Chemical Aspects of the Impregnation of Porous Materials

Two main factors exert direct influence on the impregnation of porous materials—thermodynamic and kinetic. Whereas thermodynamic factors define the conditions of adsorption of compositions onto the surface of material pores, kinetic factors influence the speed of impregnation, which is determined primarily by the size of the pores in the material, the viscosity of the composition, and the wetting of the pore surface with the composition.

If the pores in the material are too small, the impregnation process is governed mainly by composition–material interaction; large-sized pores are mechanically filled with the composition.

Adsorption interaction largely controls such processes as formation of adhesive bonds between the material surface and the composition, selective adsorption of the composition components by the surface of material pores, and displacement of water by the composition from the pores of the impregnated material. Let us consider these processes in detail, as they are essential for determination of the properties of impregnated materials.

8.2.1 Adhesion of the composition to impregnated materials

The level of adhesion of the composition to the material largely determines the strengthening of the material after impregnation. Adhesion work is thermodynamically defined by the well-known Dupret equation [405] and depends on surface tension and interphase tension of both the material and the composition. As the material being impregnated possesses high surface tension and the organic composition is distinguished by a low one, the composition will wet materials well, ensuring high adhesion. Two exceptions are when the porous material is previously impregnated with surfactant solution to reduce selective adsorption of composition components, which decreases surface tension, and when the impregnation of wet material is undertaken. Both cases will be described below.

8.2.2 Selective adsorption of components of the composition

Substances such as concrete are characterized by high sorption activity, leading to separation of composition components during the impregnation process and, consequently, to disturbance of the system's stoichiometry.

We have investigated selective adsorption processes in concrete using compositions consisting of epoxy resin and amine hardener; polyurethane compositions of polyisocyanate and polyols based on polyethers and polyesters; and polyester compositions consisting of unsaturated polyester resin and a curing initiator. The polyester resin was a styrene solution of polydiethylene glycol maleate phthalate. Common chromatographic methods were used for the investigation. It was shown that in the process of penetration through a layer of concrete, the epoxy composition is divided into epoxy resin (upper layer) and hardener (lower layer); the polyurethane composition divides into polyisocyanate and polyol, and the polyester composition into polyester–initiator–styrene. The sorption capacity of concrete varies greatly depending on the type of concrete and its method of preparation. In impregnation of porous concrete or structures with badly damaged concrete showing numerous pores, adsorption separation of components is observed to a lesser degree.

To investigate the possibility of eliminating selective sorption of components in impregnation, we studied the properties of polymer boundary layers on the surfaces of glass and granite as model materials. The glass-transition temperature T_g characterizing the polymer condition was the parameter investigated. The glass-transition temperature in boundary layers was determined using reversed-phase gas chromatography by the dependence of the logarithm of retention volume on inverse temperature. As is obvious from Fig. 8.1, the polymer glass-transition temperature is reduced with the decrease of its thickness at the glass and granite boundary, indicating incomplete polymer curing due to disturbance of the equimolar stoichiometry. A

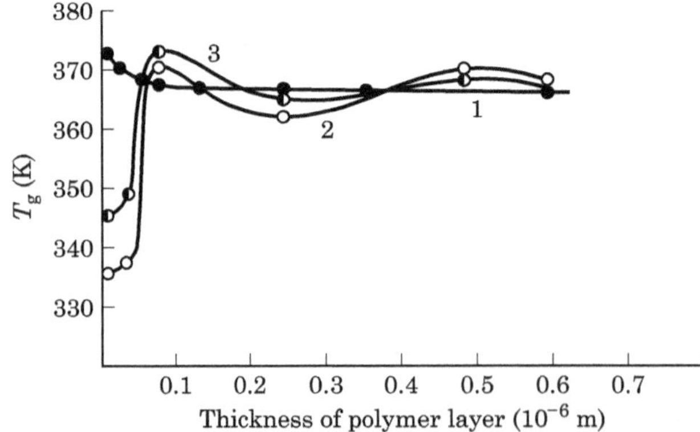

Figure 8.1 Dependence of glass-transition temperature (T_g) on the thickness of the epoxy polymer layer applied to a granite surface waterproofed with surfactant (1), glass (2), and granite (3)

composition boundary layer of up to 0.03×10^{-6} m thickness does not go over to polymer and its properties are similar to those of the initial epoxy resin. The data were confirmed with infrared spectra of polymer boundary layers [406].

There is no sorption separation of components when the material's surface tension is reduced. As is obvious from Fig. 8.1, treatment of the granite surface with surfactant excludes selective adsorption. We also observed these patterns in earlier investigations of iron and basalt [407, 408], and surfactants were used extensively in the modification of adhesives and coatings [409, 410]. Accordingly, preliminary impregnation of the material to be impregnated with the surfactant solution should eliminate the separation of components and ensure constancy of the properties of the impregnated material at all depths of impregnation.

Our experiments have shown that substances containing ester groups are the most effective for preliminary impregnation of concrete. As was shown by Adamson [413], ester groups undergo hydrolysis on alkaline surfaces and the hydrolysis products are chemically bonded with the surface. This reduces the free energy of the surface and at the same time prevents surfactant desorption during material impregnation.

A chromatographic column of 1 cm diameter filled with concrete prepared from a moist sand/cement mixture in 3:1 ratio was used. The concrete was characterized by 10% porosity and its compression strength was 18 MPa. The concrete was impregnated with a composition consisting of epoxy resin and amine hardener. After resin curing, the concrete-filled column was cut into specimens 1 cm long. In the other experiment, the concrete was impregnated beforehand with polyester-based surfactant solution. Test results for these specimens are presented in Table 8.1. It is obvious that preliminary processing of

TABLE 8.1 Influence of Impregnation on the Strength of Concrete Specimens

Specimen no.	Specimen length from the column top (cm)	Concrete impregnation with the surfactant	Compressive strength of concrete (MPa)
1	1	No	86
2	2	No	32
3	3	No	23
4	4	No	20
5	6	No	20
6	10	No	21
7	1	Yes	113
8	2	Yes	107
9	6	Yes	95
10	10	Yes	91

concrete with surfactant solution leads to considerable increase in its strength after impregnation.

8.2.3 Impregnation of wet materials

There are at least two problems in impregnating wet materials. First, water is a mechanical obstacle for permeation of the organic composition into the volume of the material. Second, water impedes the wetting of the surface of the pores of the material by the composition. If a water film exists between the cured composition and the solid, the strengthening effect of the impregnation will be minimal.

Theoretical analysis shows that to ensure the wetting of pore surfaces by the composition, it is necessary to achieve a composition–solid interphase tension less than the water–solid interphase tension. Preliminary impregnation of the material with surfactant solution reduces tension at the composition–solid boundary, improving wetting of the solid by the organic composition. In gluing of materials underwater, surfactant may be introduced directly into the composition. Introduction of only 1% of alkylphenol oxyethylized ester or fluorinated alcohol ($CF_3(CF_2)CH_2OH$) into a composition based on methyl methacrylate leads to reduction of the limiting angle of granite surface wetting from 64° to 30–34°.

Ensuring selective wetting of the pore surfaces of the material by the composition in aqueous medium is not alone sufficient for displacement of water by the composition. Application of mechanical work to the phase boundary is also required. The specific mechanical work necessary for complete water displacement by the composition on a solid surface is given by [410–412]

$$A = \gamma_{c,w}(1 + \cos\theta) + \gamma_{c,w}\frac{S_1 - S_2 - S_3}{S_2} \tag{8.1}$$

where $\gamma_{c,w}$ is the composition–water interphase tension; θ is the limiting wetting angle; and S_1, S_2, and S_3 are the solid–composition, solid–water, and composition–water boundary surfaces.

Equation (8.1) shows that spontaneous (without work applied to the boundary) solid surface wetting by the composition in aqueous medium is possible for equality of surface tension values in adhesive and liquid, i.e., for $\gamma_{c,w} = 0$. The maximum attainable $\gamma_{c,w}$ is advisable, however, as it is in direct proportion to thermodynamic work of adhesion. Therefore, it is impossible to ensure high-quality impregnation of porous materials with organic compositions without application of work to the composition–solid boundary, even if the thermodynamic conditions for selective wetting of material pores by the composition are provided. Mechanical work can be applied by feeding the composition into the material volume under pressure, or effecting the impregnation process using ultrasound.

Water can exert a direct influence on the process of composition curing. For example, impregnating wet concrete with a composition based on methyl methacrylate, the speed of curing is 1.2–1.5 times higher than in the impregnation of dry concrete, and the molecular mass almost doubles—from 3.68×10^6 to 7.13×10^6. This may be explained by the fact that a small amount of water in methyl methacrylate enhances the possibility of folding of the growing chains, decreasing the rate of their breaking. Water can inhibit the curing of unsaturated polyester resins and sharply increases the curing speed of cyanacrylate; water reacts chemically with certain groups that may be included in compositions, for example, isocyanate groups.

8.3 One-Component Organic Compositions for Impregnation of Porous Materials

Curing of one-component compositions in the material volume is achieved by physical or chemical means. Physical factors include primarily heating or the use of penetrative radiation. Curing can occur through the action of substances located on the surface of the pores of impregnated materials. In the case of concrete impregnation these substances comprise water, alkalis, and cations of mainly alkaline-earth and alkali metals. Water hardens compositions containing isocyanate groups with the liberation of carbon dioxide, leading to generation of a polymer foam. These compositions are sometimes used as so-called fast blocking compositions. Alkali under certain conditions may cause polymerization of composites containing epoxy groups. The possibility of curing compositions by the action of metal compounds is most interesting. Metals can cause polymerization of compounds containing unsaturated bonds, including isocyanate and alkylene oxide groups. However, salts contained in concrete are characterized by low degrees of dissociation and are not very active as catalysts. We investigated the possibility of creating catalytic complexes based on these metals and crown-ethers introduced into the oligomer chain. As shown in Section 3.4, these complexes are highly effective interphase catalysts and catalyze polymerization reactions. Specific catalytic action was provided by the particular crown-ether used to form complexes with the required metal atom.

Consider isocyanate trimerization as an example. This reaction in the presence of alkali and alkaline-earth metals was described in detail earlier. With crown-ether in contact with the concrete surface, it forms a crown-complex with the calcium cation of calcium oxide hydrate. A loose ion pair is formed and the released anion catalyzes the process of trimerization of isocyanate groups:

$$3\,O{=}C{=}N{-}R \rightarrow (O{=}C{=}N{-}R)_3$$

These isocyanurate groups are the points of oligomer binding and, due to isocyanate group trimerization, the diisocyanate-based oligomer turns into a strong, chemically stable network polymer. The formation of isocyanurate groups under the influence of crown-complexes can proceed so actively that water–isocyanate group interaction is not practically observed, and similarly the liberation of carbon dioxide in impregnating concrete structures.

The purposes of impregnating structures may vary from increasing bearing capacity to strengthening surface layers and enhancement of their chemical stability. Because of this, it is necessary to be able to control the properties of organic compositions for impregnation. For example, for deep impregnation of dense concrete structures, the composition should have low viscosity, good wetting ability, and very low rate of polymerization. When it is required to protect ferroconcrete-reinforcing elements from corrosion, the composition should possess inhibiting properties. At the same time "fast blocking" compositions should have curing times of tens of seconds or minutes. The creation of one-component compositions with controlled properties became possible with introduction of crown-ethers into them, and resulted in complexes with various catalytic activities.

The main obstacle to wide usage of crown-ethers in composition formulations is their comparatively high cost. One way of reducing costs is replacement of traditional crown-ethers with open-chain "pseudo-crown-ethers." Furthermore, the use of crown-ethers allows the utilization for impregnation of very cheap monomers that may even be wastes from chemical production.

8.4 Influence of Impregnation on Material Properties

Wooden structures are impregnated to protect them from various environmental factors such as water, aggressive chemicals, ultraviolet radiation, and so on. Impregnation can reduce wood's combustibility, increase its strength, protect from decay and the depradations of insects, and impart desirable decorative properties. After impregnation, softwood, such as lime-tree or aspen, becomes similar to tropical hardwood and can be used in novel fields, for example, as machine elements operating in aggressive conditions. Impregnated wood chipboards and fiberboards can be used in humid conditions or as structural materials.

Impregnation of wood may be effected by applying the polymer composition onto the wood surface with a brush, by spraying, or by dipping of a wooden item into the composition. The degree of impregnation depends on the type of wood, its quality, the time of contact of the wood and composition, and the composition curing speed.

The changes in properties of concrete upon impregnation are considered below. Dry, wet, or water-saturated concrete of 400 grade was used and impregnation was performed without preliminary treatment of the concrete with surfactant. An isocyanate-containing composition (variant 1), a composition with glycidyl groups (variant 2), and an acrylic composition (variant 3) were used. Impregnation was effected by immersing concrete specimens into the composition for 6 h. All the tests on concrete specimens were made in conformity with standard methods seven days after impregnation.

1. *Water absorption*. Before impregnation specimens were dried to constant mass. Tests showed that water absorption of impregnated specimens was zero, and check samples were characterized by 4.5–4.7% water absorption.
2. *Radiation stability*. Radiation stability was checked by exposing impregnated specimens to γ-radiation from a ^{60}Co source. Tests showed that variants 1 and 2 are the most radiation-stable compositions. At doses of less than 120 Mrad, no change was observed in the properties of the impregnated specimens.
3. *Chemical stability*. Chemical stability was determined by immersing the impregnated specimens in various chemicals. At 6 and 12 months after exposure in different media, specimens were removed, dried, and subjected to compression tests. The tests indicated that impregnated specimens are stable to the action of solutions of salts, acids, and alkalis and unstable to the action of concentrated nitric, sulfuric, and hydrochloric acids. The most chemically stable is the acrylic composition.
4. *Permeability to chloride ions*. Of various chemicals, concretes come into contact with chloride-containing salts the most often, especially in places where streets are cleaned of snow with the help of salts. This leads to intensive deterioration of the concrete elements of roads and underground service lines.

   Impregnated concrete specimens, simultaneously with check specimens, were immersed in 10% NaCl solution for 5 days. Dry, wet, and water-saturated specimens were impregnated to a depth of 3 mm. After testing, the specimens were split, and $AgNO_3$ solution

TABLE 8.2 Adhesion of Protective Coatings to Concrete

Specimen no.	Composition	Concrete condition	Adhesion (MPa) Number of freeze/thaw cycles 0	10	15	20
1	1	Dry	3.3	3.0	2.7	2.4
2	2	Dry	3.1	2.9	2.8	2.7
3	3	Dry	2.9	2.6	2.4	2.4
4	1	Wet	3.1	2.8	2.6	2.5
5	1	Water-saturated	3.1	2.7	2.6	2.5

was applied onto the chip surface. The presence of chlorides was determined visually by the appearance of a white sediment. No chlorides were found in concrete impregnated by any of the three compositions whereas in check samples chlorides were found at a depth of 2–3 mm.

5. *Freeze resistance*. Freeze resistance was determined by a rapid method involving freezing of concrete specimens in 5% NaCl solution, at −50°C and thawing in the same solution. Impregnated specimens displayed no changes after 200 cycles of freezing and thawing, featuring F 200 grade freeze resistance. Check samples displayed strength losses, and flaking and chipping were found at specimen surface ribs.

TABLE 8.3 Influence of Impregnation on Concrete Properties

Property	Initial concrete	Concrete after impregnation
Ultimate strength (MPa)		
Compressive strength	40	160
Tensile strength	2.5	16
Bending strength	5	22
Elastic modulus under compression (MPa)	3×10^4	4×10^4
Ultimate strain under compression	0.001	0.002
Fitting bond resistance (MPa)	1	13
Dynamic tensile strength (τ) (10^3 MPa · s)	6	27
Shrinkage deformation	50×10^{-5}	4×10^{-5}
Creep deformation	50×10^{-5}	7×10^{-5}
Electrical resistance (Ω)	10^5	10^{14}
Water absorption (wt%)	4	0.2
Freeze resistance (cycles)	200	8000
Stability to acids	Low	High

6. *External crack resistance under alternating loads.* Impregnation of specimens to a depth of only 2 mm results in 3-fold to 8-fold increase of external crack resistance. The best results were obtained for specimens impregnated with isocyanate-containing compositions.
7. *Adhesion of the protective coating to concrete.* To explore the possibility of using polymeric compositions as protective coatings, the adhesion of all the compositions to dry, wet, and water-saturated concrete was studied, as well as adhesion changes after specimens were frozen and thawed. B30 class heavyweight concrete was used in the studies.

Table 8.2 shows that adhesion of coatings to concrete exhibits low dependence on concrete moisture content and the number of freezing and thawing cycles, thus significantly increasing the requirements shown by specifications.

Table 8.3 contains data on the influence of impregnation on concrete properties. Composition No. 3 was used for impregnation.

Chapter

9

Practical Applications of Polymer Adhesion Studies

The practical use of polymeric composite materials (PCMs) in specific branches of industry requires resolution of the problem of control of their properties. Means of achieving this control may be found when the regularities outlined under the descriptions of studies of phenomena connected with polymer adhesion strength are analyzed. It should be noted, however, that despite the fact that the principal property of PCMs is polymer adhesion stength, their use requires the control of a great number of other polymer properties and the development of special technical and operational methods, comprehensive testing, and so on. Thus, the Adhesive Research and Development Center includes basic departments, as listed below, dealing with the introduction of high technology into the field of PCMs.

1. Fundamental Research Department, whose employees contributed a great deal to these studies.
2. Applied Research Department, established to focus on the control of PCM properties, taking into account the results obtained in the Fundamental Research Department, and to study the use of PCMs for specific industrial and civil engineering branches, as well as to develop PCMs.
3. Pilot Plant Department develops "know-how" for manufacturing PCMs and produces pilot plants to undertake in-situ testing.
4. Technological Department develops methods of operation in PCM usage.
5. Certification Department is engaged in full-scale tests of PCMs and items made of them.

Here we describe the possibilities of utilizing the regularities already described for the development of PCMs featuring novel properties that help provide solutions currently unattainable with PCMs now available on the market.

9.1 Adhesives for In-situ Maintenance and Repair Work

Adhesives of this type were developed for repair and maintenance in transport facilities, plants, pipelines, ships, etc. in the open air. In-situ conditions require development of adhesives usable at all ambient air temperatures, in wet weather, in deep waters, and in oil product media and other liquids. Surfaces to be glued together may be covered with corrosion layers, and glue line thickness cannot be limited. There is no need for any additional alteration of the joints.

9.1.1 Ship repairs

9.1.1.1 Hull repairs. The main costs incurred when ships are repaired in shipyards are due to replacements of corroded hull parts, i.e., deck sections, partitions, pipes, holds, drinking water tanks, drainage installations, etc. Large costs are caused by certain features of currently used repair procedures, involving electric welding and entailing huge amounts of auxiliary operations. Thus, to replace a corroded upper deck, the complete cleaning of rooms underneath is required, i.e., dismantling and removal of heat insulation coverings, furniture, and instruments. Afterwards, using electric welding, the deck is removed and replaced with a new one. These procedures for ships of 20–60 thousand tonnes of displacement take 4–6 months.

We have proposed specially developed polymer compositions for use in ship repair. Repairs are performed by impregnation of corroded structures with a composite involving reinforcing element layers and a face layer. Glass fabric is mainly used as the reinforcing element. When basalt flakes are used as reinforcing elements, they are applied onto metalwork surfaces together with atomized polymer composition. The surface of these flakes is specially modified to promote formation of strong flake-to-PCM ahesion bonds, permitting significant increase of the compositie material strength. The properties of these flakes will be discussed in a later section.

High cohesion and adhesion bond strength to corroded metals, high water resistance, low levels of interior stress, and glue line resistance to vibration stresses are basic requirements for polymer composites utilized for these operations. To achieve these goals, a polymer mixture featuring interpenetrating polymer networks has been used. The composition includes two types of surfactant—reactive (RS) and chemically indifferent (IS)—the latter playing the role of corrosion inhibitor. In addition, the composition also contains corrosion protectors.

This method is most widely used to repair fishing vessels. This type of vessel often operates thousands of kilometers from home ports. When back home, crews are comparatively idle and may be engaged in repair and maintenance works. Six to seven days is the average time required to repair a fishing boat's upper deck while the vessel is afloat.

We offer an attractive procedure for corrosion-resistant protection of tanks, cisterns and holds. A polymer composition layer is poured to cover a tank bottom, and a water solution of density greater than the composition's density is delivered below it. The composition floats up slowly, coating interior tank surfaces thoroughly. This operation may be repeated several times to increase the coating layer thickness.

9.1.1.2 Gluing of pilothouse to the deck. Ships' decks are manufactured of steel, and light-metal alloys are the preferred materials for construction of the pilothouse. The pilothouse is fastened to the deck by electric welding with the help of a bimetal plate, one side made of steel and the other of light alloy. This plate is essentially a galvanic cell. When it is exposed to seawater, intensive metal corrosion occurs, sharply reducing durability of the fastening to the deck.

To eliminate this phenomenon, we propose gluing the pilothouse using Sprut-12 adhesive. Tests performed immediately after gluing have shown that the durability of the pilothouse fastening is not inferior to that made by electric welding; and after one year the durability of the glued pilothouse fastening was already twice that of the welded fastening.

9.1.1.3 Wreckage work. Hull repair in the case of a breach is performed in dry dock, but ships often cannot reach dry dock themselves. Use of adhesives in this case obviates the problem of transporting ships to dry dock and can eliminate dock service entirely for minor damage. Divers using Sprut-4, TSAU, or LING adhesive types and special tools make glued repairs to breaches. These tools provide for fixation of plasters on the breach with adhesive curing and safe plaster fastening. The simplest tool comprises a plastic bag inflated from the scuba, and more complex devices are equipped with special suction cups. This combination of tools and polymeric composition properties promises the possibility of overcoming such frequently encountered problems as breaches of considerable size, badly deformed edges of hull plating, and flow of water into the ship.

The most interesting example of such work took place in the Caspian Sea. A floating drilling rig produced at the shipbuilding yard in Sevastopol, for oil recovery in the Caspian Sea, navigated the Black Sea, the Sea of Azov, and the river Don, but on going over to the river Volga it was found that the lock was 30 cm narrower than the width of the rig. It was resolved to cut the rig into two parts and to convey them separately to the river Volga, where they were to be glued; the rig was then supposed to be transferred to the town of Astrakhan on the Caspian Sea, where the parts were to be welded together in dry dock.

To achieve this, a metal box was glued underwater onto the rig bottom. Water was removed from the box, allowing the rig body to be cut into two parts by electric welders. After passage of the rig halves through the lock, the body was glued together again, allowing the rig to arrive at the dock unaided.

In the 1980s a Bulgarian merchant ship sank in the Black Sea near the town of Odessa. It was proposed to use adhesive to haul it up. For this purpose the breach, holes, and all the sites of leakage in the hull were thoroughly sealed with VAK adhesive, whereupon compressed air was pumped into the hull. Once all the water had been expelled from the ship, it floated to the surface.

9.1.2 Damage control in the oil and gas industry

9.1.2.1 Oil tank reconstruction. Oil recovered in the field is characterized by a high content of water in which salts, acids, and hydrogen sulfide are dissolved: such oil results in metal corrosion. The service life of corrosion-resistant coatings in use does not exceed 2–4 years when the oil contains large amounts of hydrogen sulfide.

At the same time, replacement or repair of oil tanks at the site of oil recovery is usually excessively complicated, since tanks are often located in regions difficult to access. In addition, preparation of the tank for application of corrosion-resistant coatings involves labor-intensive operations such as thorough cleaning of oil products from metal surfaces, degreasing and sandblasting, as well as electric welding and steaming of the tank when necessary.

The properties of Sprut-12 and Sprut-14 polymeric compositions developed for tank repair allow these labor-intensive operations to be avoided. The compositions can be applied onto the metal surface without thorough cleaning off of oil products, water, and metal corrosion products. The speed of curing of the compositions can be controlled over a wide range, even at temperatures below 0°C.

If intensive corrosion damage of the tank body is observed, the body is strengthened by means of reinforcing elements, namely, glass fabric or basalt flakes applied onto the inner surface of the tank together with the polymeric composition. These operations are most often performed on the lower part of the tank, where aqueous solutions of aggressive substances accumulate, and on the roof, which experiences high concentrations of hydrogen sulfide vapors.

The most secure corrosion-resistant protection of the metal is observed with use of basalt flakes. These flakes of 2–3 mm length and 1–2 mm diameter are laid parallel to the metal surface with the application of the coating and form a durable corrosion-preventive barrier of "cold ceramics" type. There are 200–300 flake layers in a typical coating layer.

9.1.2.2 Repair of pipelines. Problems emerging during pipeline operation include failure of rust-preventing coatings on the internal and external parts of a pipe, decrease in pipe strength due to corrosion failure, appearance of flute cracks in the pipe body, and diffusion of gas through leaks in welds.

Corrosion inhibition of internal surface of pipelines. Corrosion of internal surfaces is observed, primarily in pipelines used in crude oil production, where the oil contains a great amount of water. An interesting feature, so-called "stream" corrosion, characterizes the corrosive damage to pipes when water and solid suspended matter make grooves in the bottom of a pipe.

For corrosion inhibition and strengthening of such pipes we have proposed the use of polymer compositions that provide high adhesion strength in application to metals in water–oil mixtures. A plunger stretched lengthwise and made of rubberized tissue is placed into the pipe; the length of the plunger is generally 2–5 times the pipe's diameter. Water is pumped in between the plunger walls until they close, providing secure tightness of the pipe section. If pressure is applied, the plunger starts to roll along the pipe like a volumetric "caterpillar." Compared with pistons (pigs), usually used for work inside pipes, plungers ensure considerably better air-tightness of the pipe cavity even if there are foreign objects in it; they are not worn in the course of moving along the pipe, as there is no friction between plunger and pipe. The plunger may be filled with water or oil after being placed in the pipe.

After introduction of the plunger into the pipe, a second one is placed immediately after it, and the space between them is filled with polymer composition. The required amount of composition is calculated from the length of pipe to be processed and the thickness of the coating layer to be applied. After the second plunger, a metallic brush (hog) is introduced, supplied with an electric voltage. The main function of the hog is to provide curing of the coating and high coating–metal adhesion strength.

On application of pressure in the pipe using water, oil, or air, the plungers start to move along the pipe, leaving on the pipe walls a layer of composition. Electrical pulses cause curing of the coating even at low temperatures. The polymer composition contains basalt or ceramic flakes which, by the action of the plunger, are deposited parallel to the pipe wall surface, ensuring high strength and chemical stability of the coating.

Corrosion inhibition of the pipeline's outer surface. We have developed a composition based on nonsaturated polyester resin (Sprut-9) for inhibition of corrosion in pipelines and, in particular, in gas-lines. This composition contains surfactants allowing application of the composition onto metal surfaces without the need for thorough cleaning from corrosion or onto wet surfaces; as well as corrosion inhibitors, a controller allows variation of curing from 5 minutes up to 2 days. A high speed of composition curing is required for repair work or insulation of pipe joints during laying pipelines, and low speed is needed to allow the mechanization of procedures during insulation. Pipeline insulation is done by winding onto the pipeline fiberglass wetted with polymer composition, or by spraying of composition including basalt flakes.

Properties of some polymer compositions are given in Tables 9.1 and 9.2.

The most interesting example of the use of Sprut-9 took place in Kazakhstan on the Mangyshlak oil field. Very viscous oil is produced on this field, and it has to be preheated for pipelining. Part of a pipeline passed along the bottom of a saline lake, and the water produced rapid breakdown of the thermal insulation and antirust coating of the pipe, sharply decreasing the pipe's transfer capacity. A group of divers

TABLE 9.1 Studies of the Stability of 2.3 mm Thickness Coatings to Cathodic Lifting

Exposure duration (days)	E_2 (V)	E_1 (V)	I_1 (μA)	Lifting area (cm^2)
1	−1.45	−1.0	1.5	0
7	−1.45	−1.0	2.0	0
14	−1.45	−1.1	1.5	0
30	−1.40	−1.0	3.0	0

Note: No corrosion was found under the coating.

TABLE 9.2 Physical-Mechanical and Dielectric Properties of Coatings Before and After Tests for Stability to Cathodic Lifting

Absence of breakdown under test voltage (kV)	Transient electrical resistance ($\Omega \cdot m^2$)	Impact strength (kgf·cm)	Adhesion, number
Before			
8.5 ± 0.2	1.2×10^8	125	1
After			
8.5 ± 0.2	3.7 ± 10^4	125	1

from the Adhesive Research and Development Center applied an anti-rust coating to the submerged line and with Sprut-9 adhesive glued hermetic heat-insulating elements to the pipe surfaces, providing a thermal shield for the pipeline. These operations allowed the volume of oil transferred to be increased threefold and gave secure protection of the pipe from corrosion.

Now let us consider some properties of basalt flakes widely used by us in reinforcement of polymer compositions.

Effect of basalt surface on the properties of boundary layers of network polymers. Epoxy-resins ED-20 and MP-400 were used, with hardeners polyethylene polyamine (PEPA), dicyanethyl ethylenetriamine (DDETA), and monocyanethyl diethylenetriamine (MDETA). Samples were made from polyester resin IIH609-21M, which is a 60% solution of oligodiethylene glycol maleate naphthalate (DEGMF) in triethylene glycol dimethacrylate (TGM-3). The resin was hardened by an oxidation–reduction system in the form of methylethylacetone peroxide–cobalt naphthenate. Octyltrimethylammonium bromide (OAB) and cetyltrimethylammonium bromide (CAB), the product of interaction of allyl alcohol and 2,4-toluylene isocyanate (ATI), were used as surfactants.

The hardening of the boundary layer of epoxy and polyester compositions was studied by IR spectroscopy (of repeatedly disturbed total internal reflection—RDTIR) on a UR-20 spectrometer. A KRS-5 prism ($N = 18$, $C = 45^\circ$) and a germanium prism ($N = 14$, $C = 45^\circ$) were used for simulation of the substrate surface. The thickness of a polymer layer was 0.5–1 nm for the element made of germanium and 2–3 nm for KRS-5. Inhibition of polyester resin polymerization by atmospheric oxygen was eliminated by applying a polyethylene film to the surface of a system to be hardened. The thickness of the applied epoxy and polyester composition was up to 20–30 nm. To evaluate the glass-transition temperature and the polymer–solvent interaction parameter, we used reversed-phase gas chromatography. Basalt-1 and basalt-2 (see below) served as fillers and as a model substrate. The glass-transition temperature T_g was determined from the experimental curve of the logarithm of retention volume against inverse temperature ($\log V_g$–$1/T$). Mercury was used as the high-energy surface. The basalt disks and flakes were obtained from melt by the method of inflation; the two types of basalt flakes were basalt-1 from ordinary basalt and basalt-2 from activated basalt produced by adding surface-active elements to the melt. The surface tension was 430 mN/m and 405 mN/m, respectively. The surface of the basalt flakes was treated with bianchor urethane-containing surfactants.

To determine the adhesion strength, basalt disks were cemented to each other and then the package obtained was cemented using a stronger cement between steel "mushrooms"; this specimen was ruptured at a normal tension.

The properties of both types of basalt flakes were investigated using electron paramagnetic resonance (EPR) and electron-energy spectroscopy. EPR spectra were measured at 300 K on a YES-ME-3X spectrometer, the first derivative of absorption lines being recorded. The elementary composition of basalt flakes and of the variation of the content of elements through the thickness of a specimen during activation was studied using a Came-Bacs scanning electron microscope equipped with a Link-860 energy dispersion system intended for X-ray microanalysis.

Investigations included qualitative analysis of the distribution of elements from a predetermined area of a specimen; qualitative analysis of the distribution of the elements through the cross-section of a specimen; and qualitative analysis of several local areas of a specimen with respect to the scanning line.

The results of the electron microscopy examinations confirm that the elemental distribution in ordinary basalt flake is characterized by uniform distribution of elements in the field and through the cross-section of a flake. The uniformity of distribution of Si, Fe, and Ca is confirmed by the characteristic Si $K\alpha$, Fe $K\alpha$, and Ca $K\alpha$ radiation as well as by qualitative analysis by scanning a probe across the plate and by quantitative analysis in local areas of a specimen at various distances from the flake edge.

The data show that, unlike the initial specimen, the flake structure is heterogeneous. Near the surface of the flake we observe thin (15 nm) layers of a phase that differs in composition from the central part. Increase in the content of Si and decrease of such elements as Fe and Ca is typical for these sections. The content of Si is greatest near the surface of the flake. In this case we observe slow disappearance of silicon from one surface to the center and rapid increase in the narrow (1.5 nm) zone at the other surface of the specimen. The increased content of Fe and Ca is observed on one of the flake surfaces.

TABLE 9.3 Quantitative Analysis of Initial and Activated Specimens

		Elemental content (mass %)		
Specimen	Analysis	Si	Fe	Ratio of elements C_{Si}/C_{Fe}
Initial	Mean	17	6.6	2.6
Activated	1	15	6.5	1.8
	2	16.7	8.6	1.9
	3	21	8.7	1.5

Data from the quantitative analysis in local areas of the initial and activated specimens are reported in Table 9.3. It is obvious that the content of Si at the side surfaces of the activated specimen is greatest and reaches 21%, which is higher than the content of Si in the center of the flake and 2% lower than in the initial specimen. The content of Fe through the cross-section of the activated specimen is higher than in the initial specimen. The relationship of Si content to Fe content is also not retained through the specimen thickness.

The data are indicative of the change of stoichiometric composition of the specimens through the cross-section during their activation. The redistribution of elements in the boundary layer of basalt-2 flake results in considerable change of its surface tension and other properties. The basalt flake matrix includes about 52% of silicon oxide (quartz and cristobalite), 16% Al_2O_3, 15% Fe_2O_3, 3% MgO, and 8.37% CaO. In accordance with [4] and [5], coordinatively unsaturated metal atoms can be active centers of polymerization on metal oxides. During polymerization the coordinatively unsaturated metal atom concurrently retains the growing chain of polymer and the monomer molecule. Activation of the basalt flake results in redistribution of the elements in the volume and on the surface. On the flake surface we observe increased content of Si and Fe, which can serve as active centers of polymerization.

EPR spectroscopy of the specimens is indicative of the activity of the basalt flakes. For ordinary basalt flakes we observed a slightly anisotropic EPR spectrum that consists of two intense absorption lines in the region $H_1 = 1.4K_e$ and $H_2 = 3K_e$ and a series of weaker lines. The character of the spectrum shows the presence of short-range order in the specimen structure. Analysis of the specimen composition and EPR characteristics (factor value, linewidth) enables us to assume that the EPR spectrum is determined by Fe^{3+} ions in two different positions with concentrations $N_1 = 3 \times 10^{18}$ spin and $N_2 = 1 \times 10^{19}$ spin. These can be the positions with different local symmetry—rhombic and axial [419]. Absorption in small fields ($H = 1.4K_e$) can be altered by crystal fields of axial symmetry [420]. After activation, the intensity of EPR lines in the region $H_2 = 3K_e$ increases sharply. This is caused by increase of the proportion of rhombohedral phase in the flake structure. In this case the total concentration of the paramagnetic centers reaches a value of 6×10^{19} spin. The presence of unpaired electrons with concentration of 6×10^{19} spin in the activated basalt flakes suggests that the surface of the activated basalt flakes possesses initiating centers responsible for the polymerization in the absorption layer [417].

It was earlier shown that a layer of epoxy polymer on a metal surface does not change the polymer condition [422, 423]. Treatment of the basalt surface with surfactant affects the glass-transition temperature of the polymer. As seen from Fig. 9.1, for a low-energy surface (basalt, treated with surfactant) the polymer glass-transition temperature does not depend on variation of the thickness of the polymer layer.

The increase in the glass-transition temperature T_g of a film 0.01×10^{-6} m thick is associated with limitation of the mobility of the polymer chains near the solid surface. The character of T_g change for the surfaces of basalt-1 and basalt-2 is indicative of a complex structure of the boundary layer. For films $(0.01–0.03) \times 10^{-6}$ m thick on the log V_g–$1/T$ curve, no fracture is observed in the investigated temperature range. We assume that a boundary layer of epoxy composition 0.03×10^{-6} m thick does not change the polymer condition. The high-energy surface can selectively sorb the epoxy resin, as a result of which the adhesive layer is improved by hardener, and the stoichiometry of the composition becomes disturbed. The polymer layer enriched in

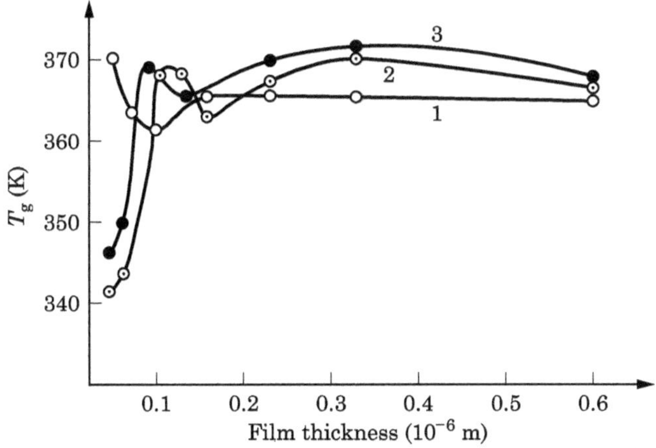

Figure 9.1 Dependence of glass transition temperature T_g on film thickness for ED-20–PEPA on basalt-2 with a layer of surfactant (1), basalt-2 (2), basalt-1 (3).

epoxy resin has a considerable expansion. In addition, the incomplete hardening of the boundary layer of the adhesive can be caused by decrease of the mobility of polymer chains due to the interaction energy limitation of the conformation set and blocking of the active groups of components of the composition by the solid surface.

Heating specimens at 423 K for 5 h results in the appearance of a break on the log V_g–$1/T$ curves, corresponding to the glass-transition temperature of the polymer in films $(0.01–0.03) \times 10^{-6}$ m thick. In this case the surface nature of basalt-1 affects the glass-transition temperature more than that of basalt-2. Heating at the same temperature for 10 h results in increase of T_g. Further heating for 15 h causes no increase of T_g. The increase of the glass-transition temperature of the polymer in the 0.5×10^{-6} m thick film can be explained by the fact that this layer contains an excessive quantity of hardener, which reacts with resin during long-term thermal treatment.

Starting from a thickness of 0.3×10^{-6} m, the properties of the boundary layer of the polymer approach the volumetric values. The data enable determination of specific changes of the retention volume V_g with change in the film thickness. For the polymer located on the surfactant-treated basalt surface we observe a negligible change of V_g with variation of the film thickness. When the untreated basalt flakes with a film thickness of 0.03×10^{-6} m are used, we observe a sharp increase of V_g that can be explained by the decrease of structure density. Figure 9.2 presents the dependence of the thermodynamic interaction parameter of polymer solvent $\chi_{1,2}$ on the film thickness. As is

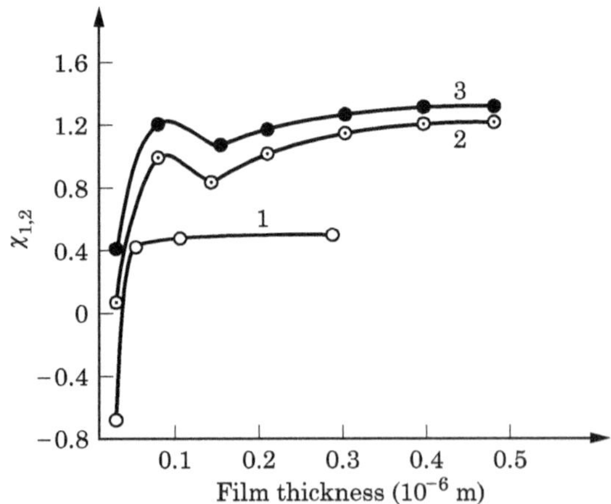

Figure 9.2 Dependence of the polymer solvent interaction parameter $\chi_{1,2}$ on film thickness for ED-20 (1) and ED-20–PEPA (2, 3). Supports: (1) basalt-2 with a layer of surfactant; (2) basalt-2; (3) basalt-1.

seen, the values of the parameter for an epoxy film 0.03×10^{-6} m thick with and without hardener are approximately the same, indicating incompleteness of the gel-formation process. Heating the 0.03×10^{-6} m thick film on the basalt surface results in increase of $\chi_{1,2}$, which is associated with additional hardening reaction of the polymer during heating. With increase of the film thickness, the effect of the polymer support on the value of $\chi_{1,2}$ becomes much less, though it is still observable up to 0.3×10^{-6} m thickness.

Increasing the polarity of the hardeners (i.e., their surface activity at the polymer–high-energy surface boundary upon replacement of MDETA by DDETA) results in the attainment of approximately equal surface activities of resin and amine, such that the interfacial tension of the mercury–ED-20 boundary is 342 mN/m, of that with MDETA is 372 mN/m, and of that with DDETA is 354 mN/m. The increase of the surface activity of the hardener results in increased degree of conversion of the epoxy groups in the boundary layer (Fig. 9.3). It is known that considerable increase of the polymer adhesion strength can be attained by the addition of surfactants. With this aim, the halogen salts of quaternary ammonium bases have been added to the epoxy compounds. The adsorption of surfactant molecules on the interface partly eliminates the preferential sorption of epoxy resin and formation of an insufficiently hardened boundary layer. While increas-

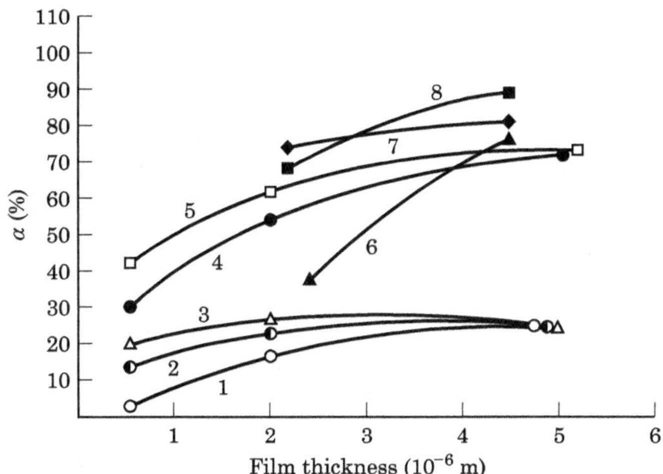

Figure 9.3 Dependence of the degree of conversion of epoxide (1–5) and polyester (6–8) compositions on the film thickness one day after mixing (1) ED-20 + PEPA, 293 K; (2) ED-20 + PEPA with 2% CAB; (3) ED-20 + PEPA with 3% CAB; (4) ED-20 + MDETA, 293 K; (5) ED-20 + MDETA, 423 K; (6) PN-609-29M + 2% PMEK + 1% NK, 293 K; (7) composition 6 with addition of 10% ATJ, 293 K; (8) composition 7 with addition of 10% ATJ, 357 K

TABLE 9.4 Strength of Adhesion Joint of Basalt Disks

Resin	Hardener	Additive	Hardening temperature	Adhesive strength (MPa)	
				Basalt-1	Basalt-2
ED-20	PEPA	–	293	11.2	16.8
ED-20	PEPA	CAB	293	22.8	24.7
ED-20	PEPA	OAB	293	23.3	24.9
ED-20	MDETA	–	423	59.6	67.4
ED-20	DDETA	–	423	69.1	69.6
II H-609-21M	PMEK + IK	–	293	28.0	30.6
II H-609-21M	PMEK + IK	ATJ	293	40.0	44.6
II H-609-21M	PMEK + IK	ATJ	353	38.6	38.7

ing the degree of hardening of the epoxy compound in a thin layer (Fig. 9.3), the surfactants exert practically no influence on the degree of hardening of the polymer in the bulk, and the adhesive strength of cemented basalt disks varies in parallel with the degree of conversion of the epoxy groups in the thin layer (Table 9.4). The adhesive strength increases especially sharply upon treatment of the basalt surface with surfactant. Treatment of the surface of the disks made of basalt-2 with bianchor urethane containing surfactant has enabled an adhesive strength of 56 MPa to be obtained in application of the polyester base.

As is seen in Fig. 9.3, the degree of conversion of the polyester composition at the boundary layer is also considerably lower than in the volume. To clarify the causes of this, the composition of the ingredients of polyester resin with TGM-3 and DEGMF was investigated by RDTIR. It was established that DEGMF is practically not polymerized either in the volume or on the boundary with the KRS-5 elements. Upon polymerization of TGM-3, the degree of transformation of double links in a 2–3 nm thick layer and the rate of hardening in the thin layer differ from the values in the volume.

While investigating the dependence of the rate of polymerization of the composition in relation to the copolymerizing components, we established that insufficient hardening of the boundary layer of polyester resin is caused by the concentration of STGM-3.

The increase of the content of DEGMF in the boundary layer is confirmed by the 4.5-fold rise in the intensity of the benzene ring absorption band in the 1580 cm^{-1} region for hardened resin IIH-609-21M in the layer 2–3 nm thick compared with the value in the volume. We concluded that the insufficient hardening of polymer near the highly effective solid surface is associated with preferential sorption of DEGMF at the phase boundary. Heating the polymeric film increases its degree of hardening. However, even then, the degree of transformation of double bonds near the solid surface is lower than in the volume of the composition. Heating for 6 h at 353 K increased the degree of transformation of polyester resin in the 2–3 nm thick layer to 65% and in the volume to 87%.

Adding ATJ, which displays surface activity at the polymer–high-energy surface boundary, to the polyester resin increases the degree of polymer hardening in the boundary layer to 15–20% (Fig. 9.3). ATJ is presumed to have dual action at the phase boundary. First, the formation of an adsorption layer of ATJ must change the conditions of adsorption interaction of the composition components on the interface, which has been observed upon adding surfactant to the epoxy resin. Second, the allyl groups of ATJ can be copolymerized with the double links of DEGMF, increasing the degree of hardening of the system in the boundary layer. Thus, the degree of system hardening at a ratio of 40:60 has been as high as 97%.

The increase of the degree of transformation of the reaction system in the boundary layer increases the strength of cementing of the basalt. The strength of cementing of basalt-2 is in all cases higher than that of basalt-1. The adhesion strength of polyester polymer upon warming to 352 K also increased by more than 30%, and the addition of surfactant (ATJ) resulted in still greater strength increase.

Pipeline strengthening. In the case of partial damage to a pipeline as a result of metal corrosion, it is preferable from the economic point of view to strengthen the pipeline rather than replace it with a new one. The method of pipeline strengthening by winding onto it glass fabric with polymer composition is of low efficiency, as the modulus of elasticity of glass-reinforced plastic is approximately 10 times lower than that of steel. A more effective method consists in winding the surface of the pipe with steel wire. For this method, a 1 mm thick layer of Sprut-9 composition was first applied onto the surface of a corroded pipe of 600 mm diameter. To increase the viscosity of the composition, 10% hydrophobic aerosil was introduced into it. Then 5 mm diameter steel wire was wound onto the pipe with a tension of 200 kgf. Antirust coating was applied over the wire. Hydrostatic tests show that whereas the source pipe ruptured at a stress of 9 MPa, the repaired tube withstood a stress of 16 MPa.

Repair of underwater pipelines. Failures of underwater pipelines are observed most often because of damage by ships or anchors. In rivers with strong currents, scours may appear beneath pipelines: the pipelines sag, and as a result of the vibration and stresses generated, cracks may appear, most often observed in the weld area. The pipe is usually collapsed in the area of flute cracks; it is often ellipsoidal, which impedes the process of sealing it.

Two other specific problems complicate the process of sealing pipelines with adhesives. One is that gas in gas mains is transported under 70–75 atmospheres pressure, and noticeable increase of gas main diameter is observed under this pressure. Thus, on the adhesive–metal interphase boundary heavy internal stresses occur, which can lead to strength reduction or to failure of the adhesive joint. Joint failure will also be promoted by vibration, which occurs with passage of fluid or gas through a deformed section of the pipe.

To solve this problem, we have suggested the use of two adhesives when sealing a flute crack. High-viscosity adhesive, which will form a highly elastic polymer after curing, is applied to the flute crack. High-strength reinforced adhesive is laid over the elastic adhesive; along the edges of a crack this adhesive forms pockets or gulleys filled with highly elastic adhesive. The system is fastened from above with a secure metallic or glass fabric clamp. The highly elastic adhesive seals the edge of the adhesive joint and transmits pressure to the pockets or gulleys. In this case, the high-strength adhesive is pressed to the body of the pipe, ensuring increase of adhesion strength. Thus, the larger the pressure in the pipe, the more securely it is sealed.

The only alternative to the adhesive method for underwater pipelines repair is replacement. The cost of construction of underwater mains crossing large rivers amounts to 10–50 million dollars. The cost of repair by the adhesive method is 10–50 thousand dollars. These data were obtained for repair of pipeline junctions through the Volga, Yenisei, and Moscow rivers, and for operations in the Okhotsk Sea, in swamps near the river Goryn (Belarus), in Bulgaria (Stevis Corporation), and from other projects.

9.1.3 Reconstruction of structural units and buildings

Reconstruction of structural units and buildings is a comprehensive problem, as they usually consist of different materials, such as concrete, metal, wood, gypsum, ceramics, and so on. These materials are subject to various environmental factors that result in their breakdown. These factors and, therefore, the nature of failure differ depending on their location. In this connection, let us consider the possibilities of using polymer composites for restoration of ground-based, underground, and underwater structural units and buildings.

9.1.3.1 Reconstruction of ground-based buildings and structures. The exploitation of industrial and engineering structures results in corrosion of metal and concrete and incurs considerable expenses for maintenance procedures. Among the most rapidly damaged buildings one can identify industrial buildings, effluent treatment plants, drinking water treatment plants, bridges and overpasses, crossings over railroads, road coatings, and the foundations of buildings.

In the overwhelming majority of cases the basic causes of damage are corrosive processes developed as a result of unfavorable environmental impacts. So, most industrial buildings deteriorate due to aggressive gases and vapors in the air. The breakdown of floors is caused by spillage of aggressive fluids; the concrete structures of water treatment plants are destroyed by chlorine and coagulants. The majority of overpasses and bridges, as well as road coatings, break down due to the application of anti-icing reagents and because of atmospheric nitric oxides, and sulfur dioxide emitted by automobile engines and industrial plants. Frost greatly reduces the durability of concrete. The failure of constructions through the action of mold is often encountered, which unfavorably affects people's health especially that of children. Considerable trouble is connected with the appearance of salt on the surfaces of concrete and brick buildings. Fluctuating loads strongly affect the service life of concrete bridges and overpasses. Carbon dioxide in the atmosphere affects reinfored concrete, resulting in carbonization—the transformation of CaO into $CaCO_3$. Thus, concrete loses its anticorrosive properties and corrosion of metal armatures in the bulk of the concrete begins. Such corrosion causes increase of armature volume, the appearance of stresses in the concrete, and ultimately its breakdown. Thus, the properties of polymer materials used for recovery of industrial and engineering structures should take into account features of the corrosive breakdown process.

Silor is most widely used for impregnation of damaged concrete and repair of corrosion. Regulation of the impregnation depth is achieved by variation of the speed of curing of the composition. It was found that to increase concrete's hardness, its chemical stability, and its freeze resistance, and to decrease water absorption, it is sufficient to impregnate only 0.5 mm of the concrete. At a concrete porosity of 4%, the composition penetrates to this depth in 4 h. It is interesting to compare the economic efficiency of concrete protection by impregnation and by application of polymer coating to the surface of the concrete. At 0.5 mm coating thickness, 500 g of polymer is required for protection of 1 m^2 of concrete. With impregnation of the concrete to the same depth, only 40 g of polymer is required.

We consider some examples of the use of Silor composition for the solution of particular problems of reconstruction of industrial and engineering structures.

Reconstruction of bridges. Let us consider the particular case of reconstruction of the Harbor Bridge (length 400 m) over the river Dnieper in Kiev, Ukraine. The bridge was closed to traffic in 1994 when it was in danger of falling down. Inspection showed that concrete piers and bridge superstructures were cracked; part of the concrete had fallen out, and in the openings the corroded metal armature was visible. The design cost of recovery was 1.2 million dollars. The repair cost using polymer compositions was calculated as only 0.3 million dollars.

Portions of concrete that were falling out were removed first. Then the rusty armature and concrete were impregnated with Silor composition, air-placed concrete was formed onto defective sites, and after curing the surface of the reconstruction was impregnated by Silor composition with a pigment additive. Tests following these procedures showed that impregnation of defective concrete increased its hardness to a level exceeding that of the source concrete. The adhesion strength of the concrete with air-placed concrete bonding was higher than the cohesive strength of the concrete.

Repair of systems for potable water treatment. Water treatment plants are built of reinforced concrete and processes employed in water purification cause breakdown of concrete. These processes include preparation of coagulant and its mixing with water, chlorination of water, and preventive bactericidal treatment of reservoirs. Leakage of water, especially at sites of wall and floor joints, is frequently observed.

For repair, the concrete of reservoirs is impregnated with Silor composition to seal the concrete, strengthen it, and increase its chemical stability. Polyester cloth is glued at wall/floor joints using the one-component polyurethane adhesive UTK-M.

Repair of drain siphons—pipes of 3 m diameter and 21 km length—has also been undertaken. The pipe was intended for water supply at 0.6 MPa pressure.

At the start of the drain siphon operation it was found that certain parts of the pipe were of poor quality and that its walls leaked. Leaking water washed away the foundations of the pipe, which had resulted in its deformation and the appearance of slots at sites of pipe joints. To eliminate defects the concrete of the pipe was sealed with Silor, and at pipe joints corrugated polyester tissue impregnated with UTK-M composition was pasted. This ensured air-tightness of the drain siphon despite variation of width of gaps between pipes.

Rehabilitation of industrial structures. A major factor causing damage to industrial constructions is the influence on them of aggressive gases and vapors that are released into the air during technological processes. Thus in metallurgical, chemical and galvanization shops, sulfur dioxide gas and hydrochloric acid vapors, salts formed in the production of mineral fertilizers, and so on, lead to both metal and concrete failure. Floors are often damaged due to spills of acids, alkalis, or compositions used for washing machinery. The most dangerous effect is corrosion of pillars, girders, and floor slabs and numerous examples of building failure are encountered worldwide.

The repair of pillars and floor slabs of buildings does not differ essentially from repair of bridge structures. Silor composition is used for protection of concrete against chemical corrosion, applied to the surface of concrete structures. For the protection of floors against chemical corrosion, the joints between acid-resisting tiles are impregnated with Silor. Such procedures are often performed at plants manufacturing beer or wine.

When buildings are covered with ceramic tiles bonded using customary cement formulations instead of a pozzolanic one, falling of tiles is inevitable due to water freezing in the gaps between tiles. Impregnation of gaps with Silor eliminates this.

Repair of power stations. Various tanks for chemicals, cooling towers, and chimney flues are frequently in need of repair at power stations. To repair concrete cooling towers, the pores and fractures in the cooling tower body are sealed with Silor, whereby the concrete is strengthened and its freeze resistance is increased. For repair of cooling towers constructed of miscellaneous materials—wood, metal, asbestos-cement—a set of polymer materials including Silor, UTK-M, and LING will be used.

Using Silor composition it is very easy to repair concrete and brick chimney flues. The outside parts of chimney flues are impregnated until complete filling of pores and defects in the pipe body is achieved. In the upper part of a damaged pipe, in addition to impregnation, defects are filled with sand and cement impregnation, laid immediately onto a layer of Silor. After setting, the repair is impregnated to full saturation.

Repair of aerodromes and roads. The primary problem in the use of aerodromes is the condition of the runways. A pebble that breaks away from the concrete of the runway can enter the turbine of an aircraft and damage or destroy it; air service providers incur heavy expenses for the use of special vacuum cleaners for removal of such debris.

Impregnation of runways with Silor eliminates such breakaway of stone fragments, and increases the strength and freeze resistance of the concrete. In addition, it is possible to increase considerably the coefficient of friction between aircraft wheels and runway. For this purpose, the runway is covered with sand following impregnation; the sand adheres securely to the runway and enhances the coefficient of friction.

Taxi-ing tracks where concrete is intensively damaged by antifreeze are also impregnated with Silor, and the holes and potholes are filled with sand mixed with the same composition. Potholes on asphalt–concrete roads are filled similarly.

9.1.3.2 Reconstruction of underground structures. Development of underground space is now of growing importance all over the world in the design and layout of large cities. Underground objects, buildings, and engineering systems include:

- Items concerned with urban “off-street” high-speed passenger rail transport, especially subway tunnels and stations
- Underground structures connected with urban “street” traffic—traffic and pedestrian tunnels
- Sewage collectors
- Structures connected with storage and maintenance of urban transport, including underground automobile garages
- Multifunctional and all-level units for various purposes—warehouses, shopping centers, stations, etc.

The underground structures are built in soil masses having various physical, and mechanical, and hydrogeological performances and properties that can negatively influence construction. Soil mineralization and underground waters present a special hazard to underground structures, especially by corrosive activity in relation to concrete and metallic constructions; and cases of biological corrosion are increasingly often encountered. For sewage collection tunnels these circumstances are aggravated by the fact that large masses of wastewaters are transferred through them that differ in chemical composition, with predominance of hydrogen sulfide, carbon dioxide, and sulfur dioxide, causing so-called “gas” corrosion. Hydrogen sulfide is the most dangerous of these; it reacts with alkaline components of cement with the formation of hydrated gypsum and unconnected masses of silicic acids, and hydroxides of aluminum and iron.

Corrosion is sharply accelerated by the vital activity of aerobic carbothionic bacteria, transforming hydrogen sulfide into sulfuric acid, and destroying concrete and reinforcement elements. As an example, during tunnel operation a case of local corrosion failure piercing tunnel walls was found in the Blackwall tunnel of the London underground, and was fixed. Underground waters containing oil wastes and heavy metals salts destroyed a Tashkent subway section, resulting in serious contamination of the environment by toxic substances. Similar events took place when a subway in Los Angeles, California, was constructed. There are many examples of sewage collectors having been fixed.

One of the basic factors causing concrete collapse in transportation tunnels is the increased content of carbon dioxide. Carbon dioxide concentrations continuously exceed tolerances and cannot be reduced by ventilation because of significant aerodynamic resistance. The protective layer of concrete undergoes carbonization and no longer contains calcium oxide hydrate, resulting in accelerated corrosion of reinforcements. Water oozing into underground constructions from the surface contains dissolved salts used for cleaning highways in winter, or as lawn fertilizers. As a result, chloride concentrations in concrete constructions fluctuate within 0.11–2.07% of the cement mass, often exceeding the limit of 0.4%, which is sufficient to initiate steel corrosion.

The features of corrosive destruction of concrete in underground conditions require the use of polymer materials exhibiting specific properties. Thus, when underground spaces are sealed, it is necessary to apply polymer compositions onto wet concrete walls or leaking concrete surfaces. We have developed two sealing variants for such spaces. With the first, surfactant solution containing crown-ethers and ester groups is applied onto a wall surface. The surfactant penetrates into the concrete balk, hydrophobicizing the concrete pores and giving the opportunity to use Silor for the final hydrophobicization. In the second variant, a modified Silor composition is used that is able to coat a wet concrete surface. The first variant is operationally more complicated, though it permits deeper impregnation of concrete, which is often important since the impregnated concrete must have sufficient strength to withstand the hydrostatic pressure of water on the other side of the wall. Concrete cohesion strength, especially in sewage systems, may be very poor and sealing of this concrete without strengthening may lead to collapse of the wall.

Such work frequently has to be performed when sealing submerged pumping stations. If a pumping station pumps fecal material, the concrete loses density and strength because of gas corrosion. In emergency switch-off of power, the pumps overflow with fecal matter, resulting in serious damage. Impregnation of the concrete wall should therefore be done to the full depth. The strength of concrete thus impregnated exceeds that of standard concrete.

In the case of large water flows, so-called fast-blocking foaming polyurethane compositions are used worldwide. Shot holes in the wall or tunnel lining are bored to allow composition to be pumped in. We have developed a cheaper variant involving drilling the wall, to allow the water to leak. The wall is sealed and the hole may be glued afterwards.

The North Muy tunnel on the Baikal–Amur railroad illustrates a most interesting case of sealing near Lake Baikal. At a depth of 1 km a water flow at up to 40 atmospheres of pressure intruded into the tunnel. Hundreds of pumps expelled water and there was no possibility of shutting off the flow. To eliminate the damage, a polymer composition was promptly manufactured featuring high underwater adhesion capacity to granite. Crown-complexes were developed as curing catalysts, ensuring a 5-second time of composition curing. Access to the site from the ground surface was impossible because there were no roads and there was a large lake at the top of the mountain. A shot hole was drilled to the flow-bed; composition was pumped in through the shot hole and component mixing was done directly before pumping into the shot hole. Composition entering the flow adhered to the walls and cured immediately. The flow was blocked within several minutes.

In early June, 2000, a large sewage collector collapsed near the town of Cherkassy (Ukraine). It was decided to repair it without interrupting the water supply to the town. In 2 hours a group of divers using LING adhesive directly in the contents of the sewers had sealed the collector.

Silor was widely used for the first time in Holland in 2000 for the repair of sewer systems, not only strengthening and sealing concrete but also protecting it from biological destruction.

Repair of transportation tunnels, underground passages, underground structures, and storm drains is traditionally undertaken with the help of Silor composition: this is applied onto the surface of concrete structures until its penetration into the concrete ceases. If the concrete surface is wet, a two-stage method of concrete hydrophobicization and sealing is commonly used. In cases of partial breakage of structures, concrete or organic concrete is applied to defective sites through a Silor layer after impregnation.

9.1.3.3 Reconstruction of submerged structures. Water, particularly salt water, leads to breakdown of submerged concrete structures, piers, piles, foundations, bridge footings, dikes, and dams. The most rapid breakdown of concrete is observed in variable dampening zones and at sites of contact between old and newly laid concrete under monolithic concreting.

Airtightness and surface strength of concrete structures are optimally reconstructed with the polymer composition LING. Glass web is impregnated with this composition, laid onto polyethylene film, and transferred to a diver who attaches this web to a specific section of concrete surface with the help of a roller. The operation is then repeated on the adjacent section. For repair of piles, the pile is spiral-wrapped with the web. The adhesive cures in 40 min at water temperatures of 0–30°C. The film is removed after curing of the adhesive. For cracks in concrete, Silor composition is injected into the body of the concrete; if the crack reaches the surface, for example, in the case of old–new concrete contact, it is glued with LING composition before injection.

The largest volume of work on sealing of concrete and brick surfaces under water is in relation to protective casings located along main lines. Here cathodic protection from corrosion is located. When water gets into the casing, often through its walls, cathodic protection ceases and there is danger of pipeline corrosion. If the casing is filled with water, a diver goes down and puts glass web onto the walls; the water is pumped out after the adhesive is cured.

Underwater functional joints are filled with UTK-M composition and if the joint width varies considerably, polyester cloth impregnated with UTK-M is glued on.

9.2 Manufacture of Pressware from Cellulose-Containing Materials

Wood chip produced by grinding wood, with binding agents based on urea–formaldehyde resin, is used for the manufacturing of pressware termed "wood-base plastic laminate" (WPL). The amount of binding agent used is 12–15%. The main problems connected with WPL manufacture consist in the necessity to use wood of a comparatively good grade for making the chips and the requirement for drying to 2–2.5% moisture; thus the cost of WPL is relatively high. Furthermore, the strength and water-resistance of WPL are low and after the pressing process, toxic formalin impurities remain in the product.

To overcome these problems, we have developed a new binding agent that allows gluing of wood particles containing up to 50% moisture. Usually wood with equilibrium moisture content (12–20%) is used for pressing, and in this case there is no possibility of pressware casting due to their dampening or drying. The binder composition includes a substance that sharply reduces the stiffness of cellulose-containing materials for the time of pressing and adds plasticity. It allows binder content to be reduced 6–7 times (to 2%) and the use for pressing of wastes from woodworking and agricultural production, such as sawdust, bark, and chaff, wastes from cotton and linen making, and flour industries, leaves, etc. For example, using sawdust, products obtained are characterized by the following indices:

Density (g/cm^3)	0.85–0.95
Transverse stress under breaking (MPa)	21.7
Bending modulus (MPa)	2080
Failure stress under compression (MPa)	143
Impact elasticity	7.4
Water absorption over 24 h (%)	3.5

By pressing, irregularly shaped items can be produced in the form of boxes, tubes, or finished house structures. As well as flat presses, worm presses may also be used. If wet veneer, for example, of oak, is pressed in a tool whose die is made in relief, the pressware will be very similar to carved items manufactured by a cabinet-maker.

9.3 Adhesive for Fixing Organic Soft Tissues: KL-3

The cyanacrylate adhesives Eastman (USA), “Aron-Alfa” (Japan), Kanokonlit (Bulgaria), and MK (CIS) are widely used in medicine, but they have disadvantages: decrease of adhesive properties with excess moisture, insufficient film elasticity, comparatively high histotoxicity, and non-universal applicability as medical adhesives.

The polyurethane adhesive KL-3 developed by us does not have any topical or systemic toxic effects. Its polymeric film is porous and of high elasticity, and the highly developed surface of the polymer facilitates rapid healing. KL-3 combines the properties of a medical adhesive and of a filling material. Its use is especially indicated when it is necessary to seal tissue defects of considerable size or for closing pathological cavities.

9.3.1 Biodegradation of KL-3 polyurethane adhesive

The biodegradation of KL-3 as a hardened composition was studied [419–423]. The degradation of KL-3 in animal tissues is facilitated by two mutually complementary mechanisms: nonenzymatic hydrolysis, and cellular processes.

The acellular (hydrolytic) mode of biodegradation is identical in model media (*in vitro*) and in living organisms (*in vivo*). It consists in breaking ester urethane, and then ether bonds [423, 424]. Biodegradation of polyurethane is also accompanied by breaking of the secondary biurethic and allophane bonds, a decrease in the effective density of the polymer suture, and a change in its supramolecular structure [422–424].

Study of the cellular type of biodegradation was possible as a result of investigations on the specific staining of polyurethane fragments in archived preserved preparations with the use of Sudan III-IV stain [425, 426]. Binding of polymer fragments to this stain produces a red-orange color, which makes it possible to identify not only free polyurethane particles but also intracellular particles.

The resorption of the KL-3 polymeric film begins soon after implantation and is connected with the productive response *in situ* [419, 420, 426]. Cellular resportion depends on two groups of cells: macrophages, which absorb the finest particles of polymer, and the so-called giant cells that absorb foreign bodies. Cellular biodegradation plays the more important role when the surface of an implant is more developed [420, 427]. Its importance increases with advancing destruction of the polymeric implants.

The products of biodegradation of KL-3 polymeric films are excreted by the organism via the urinary system and the gastrointestinal tract and do not accumulate in tissues and organs [428].

Response of tissues to KL-3 adhesive. The use of medical adhesives on organs and tissues represents a reliable methodology that allows amelioration of the traumatic consequences of surgical operations.

The response of tissues to KL-3 was studied on the surface of liver, kidney, small and large intestines, and gluteal muscle of rabbits and white rats [429]. Asepsis was observed in the tissues adjacent to the KL-3 polymeric film, with the typical sequence of leukocyte, macrophage, and fibroblast phases. Each of these phases has its own features, which depend upon the polymer structure. The leukocyte and macrophage phases are characterized by considerable duration.

None of the organs examined showed any histological changes due to the polymeric film. Such a favorable tissue response [419, 429] to the adhesive was achieved by using ingredients of high purity. It was facilitated in particular by the development of a method of purifying the polymerization accelerator that was used in the adhesive composition [430].

9.3.2 Use of KL-3 in experimental and clinical surgery

Joining striated muscle using KL-3 adhesive. The gluing together of cross-sectioned rabbit's skeletal muscles using KL-3 adhesive was described in [431–437]. The study of energy and protein turnover in muscles rejoined with KL-3 or MK-2 or sutured with silk has showed them to occur in ways typical of the process of reparative regeneration. A porous polymeric film of KL-3 adhesive does not prevent connective tissue growing through. Moreover, it facilitates the regeneration of such highly specialized structures as newly formed striated muscles and peripheral nerve endings [437].

The antibody composition of skeletal muscles joined with KL-3 does not differ from that of conventionally sutured muscles [438], which indicates that the preparation has no allergenic effects.

The KL-3 adhesive allows reliable gluing together of the musculo-aponeurotic leaves of anterior abdominal wall in rabbits under conditions of high mechanical loading [438], which allows use of this preparation in the clinical plasty of massive hernias.

Closing fistulas and filling cavities. Closing fistulas of various etiologies is one of the most complicated problems in modern surgery. The polyurethane composition KL-3, which combines the properties of a medical adhesive and a filling mass, shows high efficiency in managing intestinal and bronchial fistulas.

In experiments on rabbits and dogs, a method was worked out for closing colonic fistulas with the KL-3 adhesive [439] as follows. A polyvinyl chloride probe with 1–1.5 ml of the adhesive at its end is drawn into the fistula canal. Then the probe is gradually drawn out and simultaneously the adhesive fills the fistula canal, being pressurized by means of a syringe. Adhesive polymerization occurs immediately in the fistula canal.

Pneumopression tests showed that there is no loss of integrity of the intestine in the region of a colonic fistula closed with the adhesive [441]. Tests using the adhesive labeled with ^{14}C showed that the biodegradation products do not cumulate in organs and tissues [442].

Fragmentation of the adhesive mass and the through-growth of connective tissue occurred in 14 days [443]; 2–3 months later, connective tissue scarring was formed at the site of inflammatory infiltrate in the gut wall. Replacement of the adhesive filling for the connective tissue was necessary only after one year. This method of closing intestinal fistulas has been successfully used in clinical practice [444, 445].

KL-3 adhesive was successfully used in managing bronchial fistulas [446, 447]. The adhesive composition was supplied to the damaged area via the drawtube of a bronchoscope with the use of specially designed apparatus. This method allowed closure of bronchial fistulas without thoracotomy.

Filling of the pathological pleural cavity for chronic emphysema of the chest was conducted successfully with twenty 50-year-old patients [448]. Positive results were noted in 80% of the cases. The accessibility, simplicity, and safety of the method recommend it as promising in patients for whom surgical treatment is impossible or is connected with high risk.

Use of KL-3 adhesive in abdominal surgery. KL-3 adhesive was used clinically for resection in ulcerative disease of the stomach and duodenum, in appendectomies, for sealing sutures of the bile duct in choledochostomies, and to close the gallbladder bed following cholecystectomies [449]. The use of KL-3 adhesive is proven in traumatic injuries of the liver [450].

Techniques have been developed for termino-terminal anastomoses with KL-3 adhesive covering, an incomplete eversion of the mucous membrane and the use of a loop-shaped intestinal suture [451, 453]. Such anastomoses are simple to make and lack the disadvantage inherent in an operation carried out by the routine method: the development of an adhesive process in the region of anastomosis with subsequent deformation of the anastomosis because of eversion of the mucous membrane. It was noted that the KL-3 adhesive, used for the "peritonization" of the zone of classic everted anastomosis, prevents the development of adhesive processes and the deformation in the region of the anastomosis reliably restricts the everted mucous membrane from the organs of the abdominal cavity, consolidates the zone of anastomosis, and induces minimal inflammatory response in the adjacent tissues.

The method has been widely used in clinical practice. No postoperative complications associated with these techniques were noted in either the short or the long term.

Management of ulcers of the stomach and duodenum using KL-3 adhesive has been practiced on a large scale [452, 454, 455]. The adhesive is supplied to the site of the ulcerative defect via a gastroscope; following its polymerization an elastic porous "dressing" is obtained under which the regeneration of tissues occurs. This porous elastic dressing is permeable to secretions of mucous membranes. Any influence of the active gastric juice upon the ulcer is eliminated owing to the permanent flux from the mucous membrane to the stomach lumen.

The method has been used in more than 200 patients; the average term of scarring of the duodenum amounted to 15.3 ± 0.5 days and that of the stomach to 20.8 ± 0.8 days. This method of closing ulcerative defects can be widely used in the endoscopy rooms of hospitals and polyclinics.

The KL-3 adhesive is successfully used to stop bleeding from the bed of the gallbladder [456]. Following resection of the gallbladder, thermocoagulation of the wounded surface of the liver is carried out and then KL-3 is additionally introduced into the liver parenchyma in the vicinity of both the bleeding vessels and the open bile ducts until they are completely closed. This method was used in 48 patients with resections of the gallbladder; no complications were noted.

In pancreatic surgery, KL-3 adhesive has been used to fill the duct system in cases of cancer of the head [457], allotransplantation of the gland [458], and acute [459, 460] and chronic pancreatitis [461]. Occlusion of the pancreas leads to rapid atrophy of the exocrine tissue of the pancreas while saving its endocrine part. Ablation of the most affected part of the gland and processing of the stump by occlusion of the pancreatic duct allows a patient to be spared pain while saving the insular apparatus of the rest of the pancreas. Of great importance is the fact that occlusion of the pancreatic duct considerably reduces the number of severe postoperative complications connected with the stump treatment as well as pancreatonecrosis, incompetence of anastomoses, and the formation of a pancreatic fistula.

Polyurethane adhesives in neurosurgery. KL-3 has been widely used in the following surgical situations [421]: recovery of potency of cerebral arteries with saccular aneurysms that cannot be eliminated from the circulation; plasty of cerebral fistulas in the anterior cranial fossa; the closing of defects in the dura mater; plasty of trepanation holes; fixation of bone fragments in craniotomic defects; and sealing of extra- and intracranial anastomoses.

Also of great importance is the method of managing cerebral tumors, glioblastomas in particular, by concurrent use of KL-3 composition and a cytostatic drug, methotrexate [421].

Use of KL-3 in urology. KL-3 adhesive ensures good hemostasis and reliable sealing in the management of various lesions of the kidney of rabbits [464]: linear (longitudinal and transverse) incised wounds that penetrate into the pelvic-caliceal system; stab wounds; cutting off the kidney's pole and subsequent rejoining by the adhesive method; perforating wounds made by a punch in the form of a metal tube of 0.5 cm diameter; and severe crush wounds.

The functional activity of kidneys glued together was determined by excretory urography, radioisotopic renography, and magnetic resonance imaging; function recovered 1–2 months postoperatively [462, 463].

The KL-3 composition was also used in a clinic to seal surgical lesions of the kidney following resection of its lower segment for multiple concretions in the pelvic-caliceal system and to consolidate the sutures following plastic operations performed in connection with hydronephrosis [465].

KL-3 adhesive was used in a clinic [466, 467] in managing 101 patients with various operative interventions on the kidneys and urinary tract: closing lesions on the kidneys; nephrotomy; nephrolithotomy; enucleation of tumorlike nodes of the kidneys; plane and wedgelike resections of the kidneys; sealing anastomoses in the intestinal plasty of ureters; sealing the wall of the urinary bladder following adenectomy and prostatectomy; sealing the sutures in urethroplasty for hypostasis and traumatic damage; and in plasty of resected vesicovaginal fistulas.

A combination of suturing with synthetic thread in an atraumatic needle and the use of KL-3 adhesive was employed to close lesions of the kidneys. The renal defect in nephrolithotomy was treated with a continuous suture before application of KL-3 adhesive to the wall of the renal parenchyma. The ligation of large intraorgan vessels (more than 7 mm in diameter) has been performed and appeared to be beneficial when the adhesive joining the renal tissue was followed with suturing of the defect of the renal fibrous capsule, which allowed the wound to be sealed and isolated the adhesive from adjacent tissues.

KL-3 adhesive was used for patients with terminal renal insufficiency and renal transplants for the conservative and surgical treatment of postoperative complications [467]. Spontaneous rupture of the transplanted kidney was eliminated in 21 cases by means of additional sealing with glued flaps of aponeurotic tissue.

The technique for closing urinary fistulas was as follows: KL-3 adhesive was applied over the first row of sutures, and then the second row of sutures was laid over the adhesive film. The porous adhesive film did not deform the sutures in any of 11 operations and reliably sealed the wound [468].

Filling of urinary point-fistulas (no longer than 0.3–0.5 mm) was done for 18 patients. KL-3 composition was supplemented with antibacterial drugs as the filler. The choice of antibacterial drug was determined beforehand on the basis of the patient's antibioticogram. The technique consists in administering the polymer mass via an external hole of the fistula using a polyvinyl chloride catheter.

For closing parenchymatous renal hemorrhages, a needleless injector was used [468]. This method consists in obtaining a hemostatic infiltrate on the wound surface, which is achieved by introducing easily polymerized cyanoacrylate adhesives into the organ parenchyma using a needleless injector. A thin adhesive film that is formed serves as a reliable sealant. Over time, the polymeric film undergoes biodegradation and is replaced with regenerative connective tissue.

Maxillofacial surgery. Fixation of mucous-periosteal grafts to the palate during uranostaphyloplasty was carried out with a use of KL-3 adhesive in 58 children with congenital cleft palate [469]. The following advantages of the adhesive method were noted:

1. The preoperative period is shortened by 2–3 days, by avoidance of the need to make up a protective palatal plate.
2. The patient does not experience any discomfort connected with obtaining the cast and bearing the plate during the postoperative period.
3. The postoperative wound care is simplified considerably.
4. The good clinical effect is combined with cost-saving considerations.

9.4 Cyanoacrylate Adhesive

Cyanoacrylate adhesives fix various materials quickly and securely and are now widely used. Their only drawback is low water resistance of adhesive joints, especially when bonding metals, glass, and ceramics. For example, the adhesion strength when gluing metal reduces by 80% after a month in water. To increase water resistance we proposed to modify cyanoacrylate adhesive with an oligomer that cures much less rapidly than cyanoacrylate. Internal stresses in the seal were sharply reduced, and adhesion strength reached 364 MPa when metal was glued. This strength displayed only 3% decrease after the glued joint was exposed to salt water for one year.

"Tsvet" adhesive thus obtained may also be used for gluing in water or oil media and at any ambient temperature. It is widely used when instruments and sensors are to be glued to aircraft and watercraft hulls and when bonding polyethylene to fluoroplastics.

9.5 Use of Polymer Compositions for Nuclear Energy Applications

One of the basic problems of the nuclear power industry is avoidance of environmental pollution, which is directly related to the necessity of reliable burial of radioactive waste. One type of waste comprises ion-exchange resins used to purify process waters and render them free of radioactive cations. The problem of disposal of used radioactive ion-exchange resins is rather difficult and we have suggested using Silor composition for this purpose. This composition envelops the swollen resin granules. This envelope is resistant to radiation exposure, and a 150 Mrad dose produces no alterations in it. It reliably eliminates the possibility of escape of water and salts from the granules. In addition, crown-ethers included into the Silor act as traps for cations if they should diffuse into the envelope volume. After the envelope is formed, the granules are mixed with cement paste to form tubes that are stored in repositories.

Silor is also used to impregnate the interior cavities of tanks containing radioactive water. This impregnation strengthens the concrete, improves its radiation resistance, and eliminates the possibility of radioactive cations penetrating into the body of the concrete.

9.6 Quality Enhancement for Articles Made of Porous Materials

Impregnation of tiles, roofing slate, figured sidewalk elements, gypsum items, and wood increases their strength, avoids various types of corrosion including those of chemical and biological origin, and improves the appearance. For example, impregnation of the upper surface of a concrete highway prolongs its service life significantly and removes the need for asphalt coating; and impregnation of wood shuttering for concreting may increase its service life several times; impregnation of tiles prevents soaking; and so on.

9.7 Brick and Concrete Paints

The use of inorganic binding materials and components based on organic binding acrylic, predominantly as dispersions in water or dissolved in organic solvents, led to first-generation brick and concrete paints. The main drawback of these coatings is comparatively short service life.

Silor-F, a brick and concrete paint of the third generation, displays a major difference from other paints in not only forming attractive and strong films on building facades but also impregnating into concrete, wood, plaster, brick, etc., wall materials and interacting chemically with the material, eliminating the boundary layer between wall surface and paint layer.

Certification tests undertaken by the Construction, Architecture and Geodesy University of Sofia, Bulgaria and NIIZhB of Moscow, Russia showed that concrete strength after application of the paints is increased 3–8 times under fluctuating and vibration loads. Increase in wall surface strength is of the most value when softening of the wall material occurs. Thus, when Silor-F is applied onto sand only, a material exhibiting strength in excess of 500 kgf/cm^2 is obtainable.

Poor water resistance of wall surfaces (e.g., paint) is a basic factor leading to destruction of facade coatings. Water getting into a wall surface layer beneath a paint layer and freezing there results in delamination. Apart from this, at a borderline the concentration of internal stresses from shrinkage, temperature, or moisture and the appearance of fungi, lichens, and micoorganisms is always observed. Silor-F penetrating into a wall practically eliminates the opportunity for these negative developments. As was shown by numerous tests, operational usage of paint to obtain 0.2–5 mm of wall impregnation is quite sufficient to obtain a positive result.

Notwithstanding that Silor-F painting of facades is performed to a significant depth, paint consumption negligibly exceeds that when a conventional paint is used. Thus, even when wall porosity is 5% and provided that total filling of pores by paint is achieved, 50 g of paint per 1 m^2 of facade will be required to impregnate a wall to a depth of 1 mm. Furthermore, the majority of the as-applied weight of conventional paint constitutes water or volatile solvents lost on drying; Silor-F paint, being based on low-viscosity oligomers, can have 100% dry residue and therefore no weight decrease on curing.

Cured Silor-F paint is characterized by high strength and chemical stability, providing for prolonged facade coating service life. Thus, the paint withstands the impact of salt, acids, alkali solutions, solvents, and oil products and may be used for covering interior wall surfaces of industrial premises affected by aggressive gases, and painted surfaces may be allowed contact with aggressive liquids. When Silor-F is applied to walls contaminated with fungi or lichens, the organisms are completely destroyed and no recurrence is possible. Silor-F is not flammable and does not support combustion.

The paint is available in several modifications. Painted surfaces may be matt or glossy, hydrophobic or hydrophilic; impregnated wall surface layers may be porous or smooth. Gloss painting onto a smooth impregnated wall may be used when a wall is in contact with water or splashes; water absorption will be zero. Matt hydrophobic paint prevents the infiltration of walls by salts, offers rain resistance and a permanent fresh appearance, and at the same time provides no obstacle for water vapor to penetrate the wall when outer-layer pores are not filled with paint and their surface is not hydrophobicized. Hydrophilic paint accelerates drying of the wall in damp or wet room interiors.

The paint consists of two components. After the components are mixed the solution may last for several days, which allows the paint to be applied onto surfaces by any conventional means.

9.8 Manufacture of Floors

There is large demand for so-called dust-free floors. These floors may be manufactured by pouring polymer compositions onto a concrete floor. They are easily cleaned of dirt, feature attractive appearance and strength, and form no dust particles in operation. For these reasons they are widely used at enterprises producing nuclear power, pharmaceuticals, and food products, as well as in offices, etc., although their cost is estimated at $50–100 per square meter.

We have developed know-how for manufacturing floors similar in quality significantly more cheaply. Techniques include embedding colored stones or chips into cement–sand tie placed onto a floor body. After the ties are dried off, a Silor composition is applied onto the stones. This composition impregnates, strengthens, and seals the upper part of the tie, and stones and chips become securely bonded in. Consumption of composition may be only 100–200 g per square meter. A wear-resistant polyurethane composition of 0.1–0.2 mm thickness is applied to the floor. In this way, polymer consumption in floor manufacturing may be reduced 3–5 times.

9.9 Manufacture of Heat Insulation Panels

One of the most important tasks in civil engineering is increasing buildings' heat resistance. The majority of current methods for heat saving in buildings use mineral wool or heat-insulating polymeric materials. These methods have a number of drawbacks, including the hygroscopic properties and inadequate service life of mineral wool, high costs, vulnerability to biological impacts, and high toxicity of gases evolved when polymeric heat-insulation materials burn.

We have developed techniques for manufacturing heat insulation materials from mineral sources, i.e., perlite and vermiculite. These materials may be pressed with the addition of inorganic glue containing specific additives introduced to prevent its suction by porous materials. In this way, glue consumption may be reduced by 2–5% of the weight of the heat insulation material. Heat insulation material may be made as bricks, plates, envelopes, etc. and the outer heat insulation layer impregnated with the Silor. The composition may be coated to imitate a desirable stone look. This processing may prevent soaking of the insulation, increase its strength and attractive appearance, and facilitate gluing of the product onto building facades.

For example, a brick made of swollen perlite wastes has the characteristics listed below:

- 40 MPa compression strength
- 0.79% water absorption
- 0.048 W/m·K heat conductivity
- Over 35 cycles of freeze–thaw resistance

This brick may be used not only to increase heat resistance of buildings, engineering facilities, tanks, etc., but also as an independent construction material.

9.10 Strengthening and Sealing of Rocks

Methods for the control of adhesive properties described herein have been used successfully in creating polymer compositions intended for strengthening and sealing rocks. These compositions were used for the first time to construct the North Muy tunnel of 19 km length on the Baikal–Amur railroad (Russia). This tunnel cuts through granite massifs in a zone of stretching of the earth's crust. This causes formation of an abundance of cracks: about 100 cracks were found along the tunnel axis, mostly of length of 1000 m. The cracks were filled with sedimentary rock in soil-like form, and stratum pressures reached dozens of atmospheres. We were approached to develop a polymer composition for injection into cracks to strengthen sedimentary rocks sufficiently to withstand these stratum pressures after the tunnel was advanced. A similar situation was faced when the Leningrad subway tunnel was engineered. This was attempted by freezing rocks with liquid nitrogen, resulting in numerous accidents, including fatal ones.

The task set was solved by introduction of a composition based on a polymer–surfactant mixture. This composition ensured displacement of water from rocks and their impregnation, with a subsequent curing process. About 100 tonnes of composition was used.

Regularities measured were used to develop a composition intended to strengthen rocks for mining purposes, e.g., in coal production. Compositions used currently comprise polyisocyanate and polyol mixed before pumping into the rocks. Usually the pump-in procedure is performed using 150 atmospheres pressure and tiny cracks are opened in rocks for the composition to penetrate into and glue after curing.

Current compositions are still not able to strengthen water-saturated rocks, because they do not wet such rocks, which are usually clay shale in water. In addition, polyisocyanate reacts not only with the polyols but also with water, and water destroys the equimolar nature of the system and causes a sharp decrease in molecular weight of the polymer formed. Moreover, sorption separation of components occurs as the composition moves along narrow cracks.

Epoxy, polyester, and polyurethane resins were used as the basis for our compositions; all of which display the ability to reliably strengthen rocks containing water, thus demonstrating the possibility of using the techniques developed to adjust polymer composite properties irrespective of their content.

References

[1] I. Ludinov, A. Voronov, *Dokl. Akad. Nauk Ukraine*, **4**, 93 (1995).
[2] G. Fourche, *Polym. Eng. Sci.*, **35**, 957 (1995).
[3] J. Joraelachvili, D. Tabor, *Proc. Roy. Soc.*, **331**, 19 (1972).
[4] A. Kornev, in *Glues, Mending Plastics and Metals*, LDNTP, Leningrad, p. 51 (1979).
[5] J. Bikerman, *Uspekhi Khimii*, **16**, 1431 (1972).
[6] S. Spring, in *Cleaning of Metal Surfaces*, Mir, Moscow, p. 349 (1966).
[7] A. Berlin, in *Aspects of Polymers Adhesion*, Khimiya, Moscow, p. 390 (1974).
[8] E. Ovchinnikov, V. Struk, in *2nd Intl. Symp. on Theory and Applications of Plasma Chemistry* (ISTAPS–95) (1995).
[9] L. Barish, *J. Appl. Polym. Sci.*, **6**, 617 (1962).
[10] J. Bikerman, *J. Appl. Chem.*, **11**, 81 (1961).
[11] N. Demarouette, M. Kamal, *Polym. Eng. Sci.*, **34**, 1823 (1994).
[12] V. Denisenko, R. Veselovsky, in *Advantages in the Field of Creation and Applying of Glues*, MDNTP, Moscow (1983).
[13] F. Fabulyak, J. Lipatov, *Questions Chem. Tech.*, **32**, 3 (1974).
[14] P. Klegt, in *Polymers Manufacturing*, Khimiya, Moscow, p. 254 (1965).
[15] B. Kamensky, *Kauchuk i Rezina*, **8**, 35 (1964).
[16] J. Lipatov, A. Filipovich, R. Veselovsky, *Dokl. Akad. Nauk USSR*, **275**, 118 (1984).
[17] E. Trostyanskaya, in *Fillers for Polymers*, MDNTP, Moscow, p. 3 (1969).
[18] Yu. Lipatov, in *Polymer Reinforcement*, ChemTec Publishing, Canada, p. 406 (1994).
[19] R. Veselovsky, V. Kestelman, *Int. J. Polym. Mater.*, **23**, 139 (1994).
[20] R. Veselovsky, A. Filipovich, *Vysokomol. Soed., B*, **27**, 497 (1985).
[21] F. Fabulyak, in *Molecular Flexibility of Polymers in the Border Layers*, Naukova Dumka, Kiev, p. 144 (1983).
[22] T. Lipatova, in *Composite Polymer Materials*, Naukova Dumka, Kiev, p. 37 (1974).
[23] L. Sergeeva, Yu. Lipatov, *Kauchuk i Rezina*, **5**, 11 (1995).
[24] E. Morosova, in *Structure and Properties of the Boundary Layers of Polymers*, Naukova Dumka, Kiev, p. 221 (1972).
[25] V. Berstain, in *Thermodynamical and Structural Properties of the Boundary Layers of Polymers*, Naukova Dumka, Kiev, p. 66 (1976).
[26] E. Trostyanskaya, A. Poymanov, *Mech. Polym.*, **1**, 26 (1965).
[27] V. Berstain, V. Nikitin, *Plasticheskiye Massy*, **10**, 30 (1963).
[28] E. Trostyanskaya, *Plasticheskiye Massy*, **11**, 67 (1965).
[29] A. Nufury, T. Lipatova, R. Veselovsky, in *Physical Chemistry of Polymer Compositions*, Naukova Dumka, Kiev, p. 25 (1974).
[30] D. Balar, in *Chemistry of Co-ordinate Compounds*, Inostr. Literatura, Moscow, p. 154 (1960).
[31] W. Zisman, H. Fox, E. Hare, *J. Phys. Chem.*, **59**, 1097 (1955).

[32] V. Babich, Yu. Lipatov, in *Composite Polymer Materials*, Naukova Dumka, Kiev, p. 175 (1975).
[33] V. Zelenov, Y. Krasnov, *Plasticheskiye Massy*, **9**, 54 (1976).
[34] Y. Lobanov, in *Adhesion of Polymers*, Publ. Acad. Sci. USSR, p. 1979 (1963).
[35] A. Petrova, *Plasticheskiye Massy*, **9**, 55 (1971).
[36] Yu. Lipatov, B. Kargin, *J. Phyz. Khymii*, **32**, 131 (1958).
[37] V. Gul, *Vysokomol. Soed.*, **32**, 2077 (1963).
[38] V. Gromov, *J. Phys. Khymii*, **32**, 2077 (1963).
[39] J. Huntsberger, *J. Polym. Sci.*, **A1**, 1339 (1963).
[40] E. Hamurcu, B. Baysal, in *35 IUPAC Congress*, Istanbul, Abstr. 2, Sec. 4–6, p. 822 (1995).
[41] E. Tremein, *Plast. Technol.*, **9**, 11 (1963).
[42] N. Shonhorn, *Khymiya i Technol. Polym.*, **8**, 138 (1964).
[43] G. Voronkov, E. Lasskaya, *Prikl. Khymiya*, **38**, 1430 (1965).
[44] F. Borisova, *Plasticheskiye Massy*, **11**, 57 (1968).
[45] T. Lotmentseva, in *Thermodynamic and Structural Properties of the Boundary Layers of Polymers*, Naukova Dumka, Kiev, p. 146 (1976).
[46] V. Spitsin, *Dokl. Akad. Nauk USSR*, **177**, 1138 (1967).
[47] A. Lyubimov, in *NMR in Polymers*, Nauka, Moscow (1966).
[48] E. Amelina, R. Yusupov, *Colloid J.*, **37**, 332 (1975).
[49] E. Amelina, R. Yusupov, *Colloid J.*, **36**, 931 (1974).
[50] Y. Taubman, S. Tolstaya, *Colloid J.*, **26**, 356 (1964).
[51] S. Tolstaya, S. Shabanova, in *Application of Surfactants in Coating Industry*, Khimiya, Moscow, p. 141 (1976).
[52] S. Tolstaya, S. Michaylova, in *Macromolecules in Interface*, Naukova Dumka, Kiev, p. 78 (1971).
[53] S. Mikhaylova, I. Kuleshova, in *Structure and Properties of the Boundary Layers of Polymers*, Naukova Dumka, Kiev, p. 113 (1972).
[54] Y. Sinitsina, R. Gadjieva, *Lakokras. Mater.*, **3**, 22 (1965).
[55] T. Shtiupel, in *Synthetic Detergents*, Goskhymizdat, Moscow, p. 448 (1960).
[56] V. Rudoy, V. Ogarev, *Colloid J.*, **37**, 401 (1975).
[57] L. Floyd, in *Polyamides*, Goskhymizdat, Moscow, p. 132 (1960).
[58] V. Tikhomirova, in *Polymeric Coatings in the Atomic Industry*, Atomizdat, Moscow, p. 35 (1965).
[59] P. Zubov, L. Sukhareva, *Colloid J.*, **26**, 454 (1964).
[60] A. Freydin, I. Sokolnikova, in *Structure and Properties of the Boundary Layers of Polymers*, Naukova Dumka, Kiev, p. 233 (1972).
[61] S. Hoiland, H. Hoiland, in *13th IUPAC Conf. Chemical Thermodynamics, 25th AFCAT Conf.*, Clermont-Ferrand, p. 165 (1994).
[62] A. Yakovlev, O. Ruban, in *Heterogenic Polymeric Materials*, Naukova Dumka, Kiev, p. 23 (1973).
[63] P. Rebinder, *J. Vsesoyuzn. Khymicheskogo Obschestva im. Mendeleeva*, **11**, 362 (1966).
[64] Z. Markina, *Colloid J.*, **26**, 76 (1964).
[65] A. Taubman, in *2nd Conf. USSR on Colloid Chemistry*, Moscow, p. 52 (1952).
[66] A. Taubman, *Dokl. Acad. Nauk USSR*, **74**, 759 (1950).
[67] A. Shwarts, D. Perry, in *Surfactants and Detergents*, Inostr. Literatura, Moscow, p. 540 (1960).
[68] K. Shinova, T. Nakagava, in *Colloid Surfactants*, Mir, Moscow, p. 32 (1966).
[69] A. Petrov, G. Nikitin, *Masloboyno-Zhirov. Promyshlennost*, **8**, 12 (1960).
[70] G. Simakova, *Colloid J.*, **6**, 893 (1969).
[71] M. Kusakov, A. Koshevnik, *J. Phyz. Khymii*, **27**, 1887 (1953).
[72] V. Zisman, *Khymiya i Technol. Polymerov*, **11**, 107 (1964).
[73] T. Amphiteatrova, *Colloid J.*, **27**, 489 (1965).
[74] P. Zubov, L. Sukhareva, *Colloid J.*, **38**, 643 (1976).
[75] L. Sukhareva, L. Krylova, *Colloid J.*, **34**, 268 (1972).
[76] R. Veselovsky, V. Kestelman, *Intl. J. Polym. Mater.*, **23**, 117 (1994).
[77] V. Myshko, R. Veselovsky, *Vysokomolek. Soed., B*, **13**, 114 (1971).

[78] I. Vlasova, I. Chudina, *Plasticheskiye Massy*, **2**, 14 (1962).
[79] C. Hare, Polyurethanes, in *J. Prof. Coat. Linings*, **12**, 71 (1994).
[80] R. Veselovsky, B. Kuporev, *Compos. Polym. Mater.*, **19**, 25 (1983).
[81] P. Flory, in *Principles of Polymer Chemistry*, Cornell Univerity Press, New York, p. 270 (1953).
[82] K. Mittal, in *Adhesion Science and Technology*, Plenum Press, New York, p. 129 (1975).
[83] B. Damaskin, in *Adsorption of Organic Compositions on Electrodes*, Nauka, Moscow, p. 260 (1968).
[84] R. Veselovsky, G. Shapoval, *Ukrain. Khymich. J.*, **37**, 554 (1971).
[85] V. Gul, N. Dvoretskaya, *Mech. Polym.*, **4**, 625 (1966).
[86] V. Gul, N. Dvoretskaya, *Dokl. Acad. Nauk USSR*, **172**, 637 (1967).
[87] V. Gul, N. Dvoretskaya, in *Physics of Composites Strength*, Leningrad, p. 96 (1979).
[88] V. Basin, *Mech. Polym.*, **4**, 731 (1971).
[89] V. Basin, G. Artemova, *Mech. Polym.*, **4**, 731 (1971).
[90] Yu. Lipatov, V. Babich, *Dokl. Acad. Nauk USSR*, **239**, 371 (1978).
[91] V. Babich, in *New Methods for Investigation of Polymers*, Naukova Dumka, Kiev, p. 118 (1975).
[92] V. Moshev, *Ukrain. Polym. J.*, **1**, 267 (1992).
[93] R. Veselovsky, Y. Znachkov, *Dokl. Acad. Nauk Ukraine*, **A4**, 373 (1977).
[94] Y. Malinsky, *Uspekhi Khimii*, **39**, 1511 (1970).
[95] L. Maders, G. Zeltserman, *Vysokomol. Soed., A*, **17**, 551 (1975).
[96] V. Myshko, R. Veselovsky, *Synth. Phys. Chem. Polym.*, **10**, 77 (1974).
[97] Yu. Lipatov, R. Veselovsky, *Synth. Phys. Chem. Polym.*, **9**, 123 (1971).
[98] Yu. Lipatov, R. Veselovsky, *Ukrain. Khym. J.*, **39**, 465 (1973).
[99] A. Sholokhov, *Plasticheskiye Massy*, **5**, 72 (1965).
[100] A. Chistyakov, M. Suchareva, *Plasticheskiye Massy*, **5**, 53 (1967).
[101] S. Davies, Patent Great Britain 9309943.0 (2277890) 14.05.93, Publ. 16.11.94.
[102] Yu. Lipatov, V. Myshko, *Vysokomol. Soed., A*, **16**, 1148 (1974).
[103] Yu. Lipatov, *Eur. Conf. EURADN '94*, **272**, 14 (1994).
[104] S. Lee, S. Kit, *Polym. Eng. Sci.*, **33**, 598 (1993).
[105] E. Shchukin, *XVI Intl. Conf. Surfactants*, Khimiya, Moscow, vol. 2, pp. 1–15 (1985).
[106] S. Krause, in *Polymer Mixtures*, Mir, Moscow, p. 26 (1981).
[107] J. Gibbs, in *Collected Works*, Yale University Press, New Haven, p. 360 (1948).
[108] A. Tager, L. Kolmakova, *Vysokomol. Soed., A*, **22**, 483 (1980).
[109] G. Allen, G. Gee, J. Nicholson, *Polymer*, **2**, 8 (1961).
[110] A. Tager, *Vysokomol. Soed., A*, **10**, 1659 (1977).
[111] A. Farooque, D. Deshpande, *Eur. Polym. J.*, **28**, 1597 (1992).
[112] L. Yee, M. Maxwell, *J. Macromol. Sci.*, **B17**, 543 (1980).
[113] D. Patterson, *Macromolecules*, **4**, 690 (1978).
[114] L. Sperling, in *Interpenetrating Polymer Networks*, Mir, Moscow, p. 328 (1984).
[115] S. Omel'chenko, V. Matyushov, *Vysokomol. Soed., B*, **11**, 7 (1968).
[116] L. Nilsen, in *Mechanical Properties of Polymer and Polymer Compositions*, Khimiya, Moscow, p. 310 (1978).
[117] A. Komolinova, Y. Zuev, *Vysokomol. Soed., B*, **20**, 149 (1978).
[118] A. Tinny, in *Strength and Failure of Polymers under Influence of Liquid Media*, Naukova Dumka, Kiev, p. 206 (1975).
[119] Yu. Lipatov, R. Veselovsky, in *VIII Kongress o Lepeni Kovov "Intermetalbond '81,"* Bratislava, p. 37 (1981).
[120] S. Wiederhorn, A. Shorb, *J. Appl. Phys.*, **39**, 1569 (1968).
[121] I. Galinga, B. Vegemund, Patent USSR 227211, Publ. 14.12.1966.
[122] R. Veselovsky, B. Lyashenko, *Zavodskaya Labor.*, **43**, 1510 (1977).
[123] B. Lyashenko, R. Veselovsky, in *Advanced Glues and Bonding Plastics and Metals*, LDNTP, Leningrad, p. 13 (1971).
[124] R. Veselovsky, S. Koryagin, in Avt. Svid. USSR 529201, publ. 25.09.76.
[125] R. Veselovsky, K. Zabela, in Avt. Svid. USSR 529340, publ. 25.09.76.

[126] K. Backnell, in *Reinforced Plastics*, Khimiya, Leningrad, p. 328 (1981).
[127] D. Verchero, J. Pascault, *J. Appl. Polym. Sci.*, **43**, 293 (1991).
[128] A. Meeks, *Polymer*, **15**, 675 (1974).
[129] N. Paul, D. Richards, *Polymer*, **18**, 945 (1977).
[130] A. Yamanaka, A. Kaji, K. Kan, in *4th SPSJ Intl. Polymer Conf. "New Developments in Polymer Science and Technology,"* Yokohama, Tokyo, p. 113 (1992).
[131] Ashida Tadashi, *J. Adhes. Soc. Jpn.*, **31**, 250 (1995).
[132] J. Koenig, in *Abstr.: MACROAKRON' 94: 35th IUPAC Symp. Macromolecules*, Akron, Ohio, p. 778 (1994).
[133] R. Drake, W. McCarthy, *Rubber World*, **159**, 51 (1968).
[134] K. Koto, *Polymer*, **8**, 33 (1967).
[135] G. Roginskaya, V. Volkov, *Vysokomol. Soed., A,* **21**, 2111 (1979).
[136] L. Manzione, J. Gillham, *J. Appl. Polym. Sci.*, **26**, 907 (1981).
[137] E. Merz, G. Claver, M. Baer, *J. Polym. Sci.*, **22**, 326 (1965).
[138] J. Duly, R. Pethrick, *Polymer*, **22**, 37 (1981).
[139] Yu. Lipatov, in *Physical Chemistry of Filled Polymers*, Khimiya, Moscow, p. 304 (1977).
[140] I. Perepechko, in *Methods for Investigation of Polymers*, Khimiya, Moscow, p. 226 (1973).
[141] R. Veselovsky, Yu. Lipatov, *Dokl. Akad. Nauk USSR*, **248**, 915 (1979).
[142] A. Nesterov, R. Veselovsky, *Compos. Polym. Mater.*, **7**, 11 (1980).
[143] G. Vysotskaya, R. Veselovsky, in *Reactive Polymers*, NIITEHIM, Moscow, p. 74 (1983).
[144] R. Veselovsky, Y. Kochergin, *Ukrain. Khim. J.*, **49**, 325 (1983).
[145] R. Bagnetri, R. Pearson, in *MACROAKRON '94: 35th IUPAC Intl. Symp. Macromolecules*, Akron, Ohio, p. 1026 (1994).
[146] T. Kulik, Y. Kochergin, *Compos. Polym. Mater.*, **19**, 11 (1983).
[147] Y. Kochergin, T. Kulik, *Compos. Polym. Mater.*, **19**, 37 (1983).
[148] R. Veselovsky, A. Nesterov, *Compos. Polym. Mater.*, **20**, 50 (1984).
[149] Y. Kochergin, G. Vysotskaya, *Compos. Polym. Mater.*, **25**, 42 (1985).
[150] V. Vakula, L. Pritykin, in *Physical Chemistry Adhesion of Polymers*, Khimiya, Moscow, p. 224 (1984).
[151] E. Lebedev, Yu. Lipatov, *Compos. Polym. Mater.*, **3**, 18 (1979).
[152] R. Veselovsky, T. Lipatova, *Dokl. Akad. Nauk USSR*, **233**, 1146 (1977).
[153] R. Veselovsky, in *New Methods of Manufacturing and Investigation of Polymers*, Naukova Dumka, Kiev, p. 78 (1978).
[154] A. Mel'nichenko, R. Veselovsky, in *The Problems of Polymer Composite Materials*, Naukova Dumka, Kiev, p. 53 (1979).
[155] R. Veselovsky, Yu. Lipatov, *Dokl. Akad. Nauk USSR*, **248**, 915 (1979).
[156] Yu. Lipatov, V. Shilov, *Dokl. Akad. Nauk USSR,* **270**, 912 (1983).
[157] O. Vakshteyn, in *Diffraction of X-rays on Chain Molecules*, Izdat. Akad. Nauk USSR, Moscow, p. 372 (1963).
[158] Sato Toso, *Finish and Paints*, **6**, 52 (1994).
[159] P. Zubov, L. Suchareva, *Lakokras. Mater.*, **6**, 28 (1963).
[160] A. Sanzharovsky, in *Physical and Mechanical Properties of Polymer and Paint Coatings*, Khimiya, Moscow, p. 183 (1978).
[161] A. Freydin, A. Sholokhova, *Mech. Polym.*, **2**, 240 (1966).
[162] P. Zubov, L. Suchareva, *Colloid J.*, **38**, 643 (1976).
[163] P. Zubov, L. Lepilkina, *Colloid J.*, **5**, 564 (1961).
[164] P. Zubov, L. Suchareva, *Mech. Polym.*, **3**, 67 (1965).
[165] P. Zubov, L. Suchareva, *Vysokomol. Soed., B,* **14**, 103 (1972).
[166] Y. Kolekov, V. Solomko, in *Thermodynamical and Structural Properties of Boundary Layers of Polymers*, Naukova Dumka, Kiev, p. 74 (1976).
[167] F. Jones, P. Jacobs, in *5th Intl. Conf. Fibre Reinforced Composites*, Newcastle-upon-Tyne, 1992: FRC '92, London, p. 23 (1992).
[168] V. Vinogradov, *Plasticheskiye Massy*, **11**, 51 (1976).

[169] V. Solomko, I. Usov, in *Modification of Properties of Polymers*, Naukova Dumka, Kiev, p. 25 (1965).
[170] Yu. Lipatov, in *Physical Chemistry of Filled Polymers*, Khimiya, Moscow, p. 304 (1977).
[171] V. Tikhomirov, in *Physical Chemical Base Fabrication of Manufacture*, Legkaya Industr., Moscow, p. 73 (1969).
[172] P. Seliverstov, Y. Polyakov, *Plasticheskiye Massy*, **1**, 47 (1967).
[173] B. Dolezhal, in *Corrosion of Plastics and Rubbers*, Khimiya, Moscow, p. 75 (1964).
[174] P. Zubov, M. Kadyrov, *Mech. Polym.*, **2**, 123 (1968).
[175] V. Myshko, S. Garf, in *New Methods of Investigation of Polymers*, Naukova Dumka, Kiev, p. 154 (1975).
[176] V. Myshko, S. Garf, T. Lipatova, *Synth. Phys. Chem. Polym.*, **16**, 77 (1975).
[177] V. Belyi, N. Egorenkov, Y. Pleskachevsky, in *Adhesion of Polymers to Materials*, Nauka i Tekhnika, Minsk, p. 3 (1971).
[178] M. Kobrin, in *Determination of Internal Strength*, Mashinostroenie, Moscow, p. 196 (1965).
[179] L. Sedov, in *Mechanics of Media*, Nauka, Moscow, p. 492 (1970).
[180] D. Argiris, in *Advanced Methods of Matrix*, Inostr. Literatura, Moscow, p. 240 (1968).
[181] G. Kron, in *Tensor Analysis of Nets*, Sovets. Radio, Moscow, p. 720 (1978).
[182] E. Trostyanskaya, *Plasticheskiye Massy*, **6**, 50 (1976).
[183] P. Zubov, *Colloid J.*, **38**, 643 (1976).
[184] L. Sukhareva, *Colloid J.*, **20**, 266 (1967).
[185] P. Zubov, in *Adhesion of Polymers*, Khimiya, Moscow, p. 85 (1963).
[186] L. Sukhareva, *Lakokras. Mater.*, **2**, 26 (1968).
[187] P. Zubov, *Colloid J.*, **25**, 438 (1963).
[188] L. Sukhareva, *Colloid J.*, **25**, 743 (1963).
[189] A. Narinskaya, L. Sukhareva, *Vysokomol. Soed., A*, **9**, 2571 (1967).
[190] L. Sukhareva, Doctoral thesis, Academy of Sciences, USSR, 1968.
[191] P. Zubov, *Vestnik Akad. Nauk USSR*, **12**, 32 (1963).
[192] L. Sukhareva, P. Zubov, in *Polymers, Symp.*, Baku, p. 120 (1972).
[193] A. Berlin, Avt. Svid. USSR 129328, publ. 1960.
[194] V. Malyanov, *Mech. Polym.*, **4**, 734 (1968).
[195] L. Sukhareva, A. Zemtsov, *Colloid J.*, **36**, 992 (1974).
[196] L. Sukhareva, L. Krylova, *Colloid J.*, **34**, 268 (1972).
[197] V. Vinogradov, in *Plastics*, Khimiya, Moscow, p. 46 (1974).
[198] Yu. Lipatov, L. Sergeeva, in *Synthesis and Properties of Interpenetrating Networks*, Special Issue, *Uspekhi Khimii*, **45**, 138 (1970).
[199] Yu. Lipatov, L. Sergeeva, in *Interpenetrating Polymer Networks*, Naukova Dumka, Kiev, p. 170 (1979).
[200] E. Morozova, in *Formation of Polymer Films*, Khimiya, Moscow, p. 84 (1971).
[201] P. Rebinder, in *Physical Chemistry of Flotation*, Metalurgizdat, Moscow, p. 133 (1933).
[202] G. Babalyan, in *Application of Surfactants in the Oil Industry*, Gostoptech, Moscow, p. 30 (1961).
[203] V. Kulikov, *Lakokras. Mater.*, **4**, 37 (1977).
[204] V. Kulikov, *Lakokras. Mater.*, **1**, 85 (1977).
[205] E. Klenikov, E. Kameper, *Colloid J.*, **27**, 828 (1965).
[206] E. Klenikov, *Colloid J.*, **34**, 68 (1972).
[207] E. Morozova, in *Formation of Polymer Films*, Khimiya, Moscow, p. 84 (1971).
[208] A. Ryabov, L. Smirnov, *Khimiya i Khim. Tech. Gorky*, **2**, 78 (1970).
[209] A. Koretsky, *Izves. Sibirs. Otdel. Akad. Nauk USSR, Khimiya*, **4**, 32 (1972).
[210] A. Freydin, Doctoral thesis, Moscow, 1972.
[211] A. Freydin, K. Chanskiy, in *The Technology of Manufacturing of Glued Goods*, Gosstroyizdat, Moscow, p. 45 (1966).
[212] E. Brostow, C. Roger, in *Failure of Plastics*, Hanse, Munich, vol. XXII, p. 486 (1986).

[213] V. Belyi, N. Egorenkov, Yu. Pleskachevsky, *Adhesion of Polymers to Metals*, Nauka i Tekhnika, Minsk, p. 255 (1971).
[214] U. Basaron, B. Deryagin, in *Investigation in the Field of Surface Forces*, Izd. Akad. Nauk USSR, Moscow, p. 122 (1967).
[215] G. Voronkov, *J. Prikl. Khimii*, **38**, 1483 (1965).
[216] B. Kuporev, R. Veselovsky, *Compos. Polym. Mater.*, **19**, 25 (1983).
[217] A. Freydin, in *Strength of Glued Materials*, Khimiya, Moscow, p. 256 (1971).
[218] M. Orshakovsky, *Lakokras. Mater.*, **5**, 52 (1965).
[219] A. Tinny, in *Phys. Chem. Mech. Mater.*, **4**, 312 (1969).
[220] A. Tinny, in *Strength and Failure of Polymers under Influence of Liquids*, Naukova Dumka, Kiev, p. 206 (1975).
[221] E. Sinevitch, A. Ryzhkov, *Plasticheskiye Massy*, **2**, 23 (1978).
[222] N. Pertsov, E. Sinevitch, N. Ivanova, *Plasticheskiye Massy*, **1**, 25 (1978).
[223] R. Veselovsky, in *Physicochemistry of Multicomponent Polymeric Systems*, Naukova Dumka, Kiev, vol. 1, p. 250 (1986).
[224] T. Lipatova, Sh. Vengerovskaya, A. Feinerman, L. Sheinina, *J. Polym. Sci.*, **21**, 2085 (1983).
[225] T.E. Lipatova, S.A. Zubko, *Vysokomol. Soed. A*, **12**, 1555 (1970).
[226] L. Gallacher, J.A. Battelcheim, *J. Polym. Sci.*, **58**, 697 (1962).
[227] A.E. Nesterov, T.E. Lipatova, F. Dusek, L. Pelzbauer, M. Houska, J. Hradill, Yu. Lipatov, *Angew. Makromol. Chem.*, **52**, 39 (1976).
[228] V.A. Ogarev, *Colloid. Polym. Sci.*, **252**, 582 (1974).
[229] V.A. Ogarev, V.V. Arslanov, A.A. Trapeznikov, *Kolloid Zhurn.*, **34**(3), 372 (1972).
[230] T. Lipatova, Sh. Vengerovskaya, L. Sheinina, T. Khramova, *Vysokomol. Soed. A*, **27**(1), 174 (1985).
[231] J. Spevacek, P. Schneider, *Makromol. Chem.*, **176**(11), 310 (1975).
[232] J.J. Jasper, B.L. Housman, *J. Phys. Chem*, **69**(1), 310 (1965).
[233] P.A. Rebinder, in *Uspekhi Kolloidnoi Khimii*, Nauka, Moscow, p. 9 (1973).
[234] G. Nemethy, H.A. Sheraga, *J. Phys. Chem.*, **66**(10), 1773 (1962).
[235] A.B. Taubman, S.A. Nikitina, *Dokl. Akad. Nauk SSSR*, **135**, 1179 (1960).
[236] M.A. Markevich, B.L. Rytov, L.V. Vladimirov, D.P. Shashkin, P.A. Shiryaev, A.G. Solovyev, *Vysokomol. Soed. A*, **28**(8), 1595 (1986).
[237] V.A. Zakupra, *Methods of Analysis and Control in Production of Surface Active Substances*, Khimiya, Moscow, p. 367 (1977).
[238] R.A. Veselovsky, L.S. Sheinina, Sh.G. Vengerovskaya, A.Yu. Filipovich, *Ukrain. Khim. Zhurn.*, **54**(7), 759 (1988).
[239] P.P. Kushch, B.A. Komarov, B.A. Rozenberg, *Vysokomol. Soed. A*, **21**(8), 1697 (1979).
[240] A.A. Shvartz, *Anionic Polymerization Carbanions, Living Polymers and Processes with Electron Transfer*, Mir, Moscow, 1971, p. 669.
[241] J. Brandt, G. Aglinton, *Application of Spectroscopy in Organic Chemistry*, Mir, Moscow, p. 279 (1967).
[242] N.P. Shusherina, T.I. Likhomanova, E.V. Adamskaya, *Khim. Geterocyk. Soed., B*, **1**, 72 (1978).
[243] A.M. Noshov, in VINITI 28.07.75 #3503-75, p. 16 (1975).
[244] A. Trapeznikov, T. Zatsepina, T. Gracheva, R. Shcherbakova, V. Ogarev, *Dokl. Akad. Nauk SSSR*, **160**, 174 (1965).
[245] A. Nesterov, Yu. Lipatov, *High-molecular Compounds,* **15**(11), 2601–2606 (1973).
[246] S. Zubko, A. Nesterov, Sh. Vengerovskaya, L. Sheynina, *High-molecular Compounds,* **17**(5), 1054–1057 (1975).
[247] A. Feinerman, Yu. Lipatov, *Device for Surface Tension Determination*, Inform. Letter No. 6, ICHM Acad. Sci. Ukrain. SSR, 1969.
[248] J. Jasper, L. Houseman, *J. Phys. Chem.*, **69**(1), 310–314 (1965).
[249] T. Lipatova, Sh. Vengerovskaya, L. Sheinina, *High-molecular Compounds,* **27**(1), 174–179 (1985).
[250] T. Lipatova, Sh. Vengerovskaya, A. Feinerman, L. Sheinina, *J. Polym. Sci., Polym. Chem. Ed.*, **21**, 2085–2094 (1983).
[251] T. Novikova, T. Lipatova, Yu. Nizel'sky, *Rep. Acad. Sci. Ukraine B,* **8**, 45–48.

[252] Dow offers surfactants. Dow Chemical Co., *Urethanes Technol.*, **13**(6), 36 (1996–1997).
[253] D. Kardashov, A. Petrova, *Polymer Adhesives*, Khimiya, Moscow, p. 256 (1983).
[254] H. Brandle, *Converter*, **17**(9), 12–14 (1980).
[255] Yu. Lipatov, *Physical Chemistry of the Filled Polymers*, Khimiya, Moscow, p. 304 (1977).
[256] G. Zdorikova, S. Blagova, L. Samoylenko, *et al.*, Avt. Svid. 749864 USSR, CI 08 L 27/06.
[257] Yu. Kercha, Yu. Lipatov, *Uspekhi Khimiyi*, **46**(2), 336–366 (1977).
[258] P. Write, L. Cumming, *Polyurethane Elastomers*, Khimiya, Leningrad, p. 304 (1973).
[259] V. Grigorieva, S. Baturin, S. Entelis, *High-molecular Compounds,* **14**(7), 1345–1349 (1972).
[260] A. Degtiareva, L. Markovskaya, N. Voloshina, R. Veselovsky, *Compos. Polym. Mater.*, **3**(34), 33–36 (1987).
[261] B. Kupor'ev, Yu. Kuksin, R. Veselovsky, *Compos. Polym. Mater.*, **2**(1), 24–29 (1979).
[262] E. Terman, L. Kalinina, N. Seraya, L. Suhkareva, *Colloid J.*, **37**(3), 601–603 (1975).
[263] A. Freydin, *Adhesive Joints Durability and Strength*, Khimiya, Moscow, p. 272 (1981).
[264] L. Aristovskaya, P. Babyevsky, S. Vlasov, *et al.*, *Practical Work on Polymer Material Science*, Khimiya, Moscow, p. 255 (1980).
[265] L. Kopusov, Thesis Abstract, 02.00.06, Kiev, p. 16 (1976).
[266] F. Sherman, *Rheology of Emulsions*, Khimiya, p. 217 (1972).
[267] D. Fridrihsberg, *Colloid Chemistry Course,* Khimiya, Leningrad, p. 358 (1964). Year
[268] S. Paik, N. Scheider, *Macromolecules*, **8**(11), 68–73 (1975).
[269] R. Bonart, J. Morbitzer, H. Rinke, *Kolloid Z.*, **240**(5), 39–45 (1970).
[270] I. Kuleznev, *Polymer Mixtures*, Khimiya, Moscow, p. 304 (1980).
[271] N. Bespalov, N. Konovalenko, *Multi-component Systems on Polymer Base*, Khimiya, Leningrad, p. 88 (1981).
[272] Yu. Godovsky, *Polymer Thermal Physics*, Khimiya, Moscow, p. 280 (1982).
[273] J. Adamson, *Physical Chemistry of Surfaces*, Mir, Moscow, p. 568 (1979).
[274] Yu. Trapeznikov, T. Zatsepina, I. Gracheva, *et al.*, *Rep. Acad. Sci. USSR*, **160**(1), 174–177 (1965).
[275] V. Ogarev, V. Arslanov, Yu. Trapeznikov, *Colloid J.*, **34**(3), 372–378 (1972).
[276] Yu. Ol'khov, S. Baturin, S. Entelis, *High-molecular Compounds,* **19**(6), 1307–1311 (1977).
[277] T. Tanaka, T. Yokogama, J. Yamaguchi, *J. Polym. Sci., A-1*, **6**(8), 2137–2152 (1968).
[278] G. Kozlova, V. Zharkov, *High-molecular Compounds*, *B*, **20**(4), 285–286 (1978).
[279] G. Kozlova, B. Naymark, V. Zharkov, *High-molecular Compounds,* **17**(6), 1277–1281 (1975).
[280] T. Lipatova, L. Sheynina, L. Vladimirova, *et al.*, *High-molecular Compounds,* **29**(4), 747–752 (1987).
[281] A. Askadsky, L. Kolmakova, G. Gaprh, *et al.*, *High-molecular Compounds,* **19**(5), 1004–1008 (1977).
[282] J. Rabek, *Experimental Methods in Polymer Chemistry*, Mir, Moscow, vol. 1, p. 382 (1983).
[283] *Investigation of Small Polymer Additives Influence*, Report on R&D Works, ICHM Acad. Sci. Ukrain. SSR, Kiev, p. 77 (1990).
[284] D. Gee, *Macromolecule Chemistry*, Mir, Moscow, p. 137 (1948).
[285] D. Mangray, *Macromol. Chem.*, *B*, **65**(1), 2946 (1963).
[286] D.X. Saunders, K.K. Frisch, *Polyurethane Chemistry*, Khimiya, Moscow, p. 470 (1968).
[287] A.K. Zhitinkina, O.G. Tarakanov, N.A. Tolstykh *et al.*, *Polym. Synth. Phys. Chem.*, **6**(21), 3–14 (1977).

[288] R.P. Tiger, L.M. Sarynina, S.G. Entelis, *Uspekhi Khimii*, **41**(9), 1672–1695 (1972).
[289] L.A. Onosova, Diss. Cand. Chem. Sci., Moscow, p. 22 (1974).
[290] S.S. Medved', N.A. Belova, L.I. Gracheva, *et al.*, *Plasticheskiye Massy*, (3), 711 (1986).
[291] S.S. Yufit, *Inter-phase Catalysis Mechanism*, Nauka, Moscow, p. 263 (1984).
[292] T.V. Kozlova, N.A. Kozlov, V.V. Zharkov, *J. Phys. Chem.*, **45**(8), 2110–2111 (1971).
[293] L.A. Bakalo, R.A. Veselovsky, N.L. Zbanatskaya, *et al.*, *Plasticheskiye Massy*, **5**(3), 34–35 (1986).
[294] L.I. Tarushina, F.O. Pozdnyakova, *Polymer Spectral Analysis*, Khimiya, Leningrad, p. 248 (1986).
[295] L.L. Shevchenko, *Uspekhi Khimii*, **32**(4), 457–462 (1963).
[296] V.V. Zharkov, L.I. Kopusov, T.V. Kozlova, *Plasticheskiye Massy*, (12), 41–45 (1983).
[297] A.K. Zhitinkina, N.A. Shibanova, O.G. Tarakanov, *Uspekhi Khimii*, **6**(11), 1866–1899 (1985).
[298] T.E. Pavlova, A.A. Fedorov, I.S. Pominov, *et al.*, *J. Appl. Spectrosc.*, 13, 544–546 (1970).
[299] V.D. Mayorov, N.V. Librovich, *J. Phys. Chem.*, **50**(11), 2817–2820 (1976).
[300] K.B. Yatsymirsky, *Coordination Chemistry Problems*, Naukova Dumka, Kiev, p. 180 (1977).
[301] K.B. Yatsymirsky, Ya.D. Lampeka, *Physical Chemistry of Metal Complexes with Macrocyclic Ligands*, Naukova Dumka, Kiev, p. 256 (1985).
[302] V.A. Barabanov, S.L. Davydova, *Vysokomol. Soed., A*, **24**(5), 899–926 (1982).
[303] K.D. Pedersen, H.K. Frensdorf, *Uspekhi Khimii*, **42**(3), 492–510 (1973).
[304] N.K. Ivanov, I.P. Maximova, T.G. Guryeva, *J. Appl. Chem.*, **48**(4), 873–875 (1975).
[305] M.A. Matveyev, A.I. Rabukhin, *J. Appl. Chem.*, **35**(6), 1254–1259 (1962).
[306] R.K. Ailer, *Silica Chemistry*, Mir, Moscow, vols. 1–2 (1982).
[307] *Guide for Injection Works under Tunnel Construction in Complex Engineering Conditions*, Ministry for Transport Construction, All-Union Transport Construction Research Institute, Moscow (1983).
[308] S.L. Davydova, V.A. Barabanov, *Koordinatsionnaya Khimiya*, **6**(6), 823–855 (1980).
[309] N.A. Tsarenko, V.V. Yakshin, N.G. Zhukova, *et al.*, *Rep. Acad. Sci. USSR*, **258**(2), 366–368 (1981).
[310] L.A. Yanovskaya, S.S. Yufit, *Organic Synthesis in Two-phase Systems*, Khimiya, Moscow, p. 184 (1982).
[311] R.P. Tiger, I.G. Badayeva, S.P. Bondarenko, *et al.*, *High-molecular Compounds*, **19**(2), 419–427 (1977).
[312] Y. Iwakura, K. Uno, V. Kobayashi, *J. Polym. Sci.*, **48**(6) (1983).
[313] R.A. Veselovsky, S.S. Ischenko, T.I. Novikova, N.L. Zbanatskaya, *J. Appl. Chem.*, **49**(8), 1872–1876 (1988).
[314] Z.N. Medved', N.A. Starikova, A.K. Zhitinkina, *et al.*, *Polymer Synthesis and Physical Chemistry*, Naukova Dumka, Kiev, No. 18, pp. 39–43 (1976).
[315] V.V. Strelko, *Colloid J.*, **32**(3), 430–436 (1970).
[316] Ya.I. Lankin, S.I. Kontorovich, E.D. Shchukin, *J. Colloid Chem.*, **42**(4), 653–655 (1980).
[317] D.I. Mendeleyev, *Selected Lectures on Chemistry*, Vysshaya Shkola, Moscow, p. 352 (1968).
[318] N.K. Ivanov, A.M. Arbuzov, N.V. Vorontsova, *et al.*, *J. Appl. Chem.*, **49**(9), 1897–1900 (1976).
[319] N.K. Ivanov, A.M. Arbuzov, I.P. Maximova, *J. Appl. Chem.*, **51**(3), 572–577 (1978).
[320] U. Mason (ed.), *Physical Acoustics*, Mir, Moscow, p. 419 (1969).
[321] A.K. Zhitinkina, N.A. Tolstykh, L.V. Turetsky, *et al.*, Avt. Svid. 727659 USSR (1980).
[322] A.K. Zhitinkina, O.G. Tarakanov, in *All-Union Conference on Polyurethane Chemistry and Physical Chemistry*, Naukova Dumka, Kiev, pp. 17–18 (1975).
[323] A.A. Tager, *Polymer Physical Chemistry*, Khimiya, Moscow, p. 544 (1978).

[324] L. Nilsen, *Mechanical Properties of Polymers and Polymer Compositions* (transl. from English), Khimiya, Moscow, p. 310 (1978).
[325] V.F. Rossovitsky, V.V. Shifrin, in *Physical Methods of Polymer Investigation*, Naukova Dumka, Kiev, p. 82 (1982).
[326] I.G. Ganeyev, V.I. Pakhomov, *Dokl. Akad. Nauk SSSR*, **191**(2), 351–354 (1970).
[327] F. Fabuljak, Yu. Lipatov, *Vysokomol. Soedin., A*, **12**(4), 738–752 (1970).
[328] Yu. Lipatov, F. Fabuljak, *Vysokomol. Soedin., A*, **11**(4), 708–716 (1969).
[329] J. Bikerman, *Uspekhi Khimii*, **41**(8), 1431–1464 (1972).
[330] Yu. Lipatov, Yu. Feinerman, O. Anokhin, *Dokl. Akad. Nauk SSSR*, **231**(2), 381–384 (1976).
[331] P. Write, A. Cumming, *Polyurethane Elastomers*, Khimiya, Leningrad, p. 304 (1973).
[332] V. Yakushin, O. Strukov, *Uspekhi Khimii*, **41**(8), 1504–1535 (1972).
[333] I. Dekhant, R. Dantz, V. Kimmer, R. Shmolke, *IR-Spectroscopy of Polymers*, Khimiya, Moscow, p. 472 (1976).
[334] Yu. Lipatov, F. Fabuljak, in *Surface Phenomena in Polymers*, Naukova Dumka, Kiev, pp. 15–19 (1970).
[335] F. Fabuljak, Yu. Lipatov, S. Suslo, D. Vovchuk, *Vysokomol. Soedin., B*, **22**(4), 282–286 (1980).
[336] T. Lipatova, V. Ivashchenko, *Synth. Phys. Chem. Polym.*, **6**, 73–76 (1970).
[337] T. Lipatova, L. Sheinina, *Vysokomol. Soedin., B*, (1), 44–47 (1976).
[338] A. Berlin, T. Keffeli, V. Korolev, *Polyetheracrylics*, Nauka, Moscow, p. 372 (1967).
[339] Yu. Lipatov, in *Composite Polymeric Materials*, Naukova Dumka, Kiev, pp. 75–82 (1975).
[340] T. Sogolova, N. Bessonova, *Vysokomol. Soedin., B*, **18**(7), 528–537 (1976)
[341] M. Buleva, J. Petkanebin, H. Sonntag, S. Stoylov, *Colloid Polym. Sci.*, **257**(3), 324–327 (1979).
[342] S. Kontorovich, V. Kononenko, N. Shchukin, *Colloid. J.*, **43**(5), 980–981 (1981).
[343] S. Kontorovich, V. Kononenko, E. Shchukin, *Colloid. J.*, **39**(3), 569–570 (1977).
[344] Yu. Lipatov, V. Privalko, V. Shumsky, *Vysokomol. Soedin. B*, **15**(9), 2106–2109 (1973).
[345] T.I. Smith, P.A. Bruce, *J. Colloid Interface Sci.*, **72**(1), 13–26 (1979).
[346] J. Saunders, K. Frish, *Chemistry of Polyurethanes*, Khimiya, Moscow, p. 270 (1968).
[347] Yu. Lipatov, Yu. Kercha, L. Sergeeva, *Structure and Properties of Polyurethanes*, Naukova Dumka, Kiev, p. 280 (1970).
[348] Yu. Kercha, *Physical Chemistry of Polyurethanes*, Naukova Dumka, Kiev, p. 224 (1979).
[349] S. Glough, N. Schneider, *J. Macromol. Sci. Phys.*, **2**(4), 553–566 (1968).
[350] R. Bonart, L. Morbitzer, G. Hentze, *J. Macromol. Sci. Phys.*, **3**(2), 337–356 (1969).
[351] Yu. Lipatov, T. Gritsenko, F. Fabuljak, *Dokl. Akad. Nauk SSSR*, **234**(2), 375–378 (1977).
[352] Yu. Kercha, Yu. Lipatov, S. Krafchik, *et al.*, *Vysokomol. Soedin. A*, **17**(7), 1596–1599 (1975).
[353] R.W. Seymour, G.M. Estes, S.L. Cooper, *Macromolecules*, **3**(5), 579–583 (1970).
[354] R.W. Seymour, S.L. Cooper, *Macromolecules*, **6**(10), 48–53 (1973).
[355] V. Zharkov, L. Kopusov, T. Kozlova, *Plasticheskiye Massy*, (12), 41–45 (1981).
[356] T. Kozlova, V. Zharkov, in *Surface Phenomena in Polymers*, Naukova Dumka, Kiev, pp. 51–61 (1976).
[357] V. Khranovskii, L. Gul'ko, *J. Macromol. Sci. Phys.*, **22**(4), 497–508 (1983).
[358] R. Veselovsky, V. Mishko, Yu. Lipatov, *Ukrain. Chem. J.*, **39**(5), 465–471 (1973).
[359] R. Veselovsky, T. Lipatova, Yu. Lipatov, *Dokl. Akad. Nauk SSSR*, **233**(6), 1146–1149 (1977).
[360] Yu. Znachkov, R. Veselovsky, B. Lyashenko, K. Zabela, in *Transport and Storage of Oil and Petroleum*, VNIIOENG, Moscow, No.12, pp. 17–19 (1976).
[361] B. Deryagin, N. Krotova, V. Smilga, *Adhesion of Solids*, Nauka, Moscow, p. 280 (1973).

[362] M. Syomin, I. Skoriy, in *Reliability and Durability of Elements of Machines*, Saratov University, pp. 22–31 (1975).
[363] V. Kukovjakin, I. Skoriy, *Bulletin of the Machine Industry*, (4), 41–44 (1972).
[364] F. Belyankin, V. Yatzenko, G. Margolin, *Durability and Deformability of Glass-Reinforced Plastics under Biaxial Squeezing*, Naukova Dumka, Kiev, p. 156 (1971).
[365] G. Pisarenko, V. Agarev, A. Kvitka, V. Popkov, E. Umansky, *Resistance of Materials*, Vishcha Shkola, Kiev, p. 672 (1973).
[366] I. Kragelsky, M. Mikhin, K. Ljapin, N. Dobichin, *Factory Laboratory*, (7), 852–854 (1970).
[367] R. Mises, *Z. Angew. Math. Mech.*, p. 3 (1928).
[368] I. Goldenblatt (ed.), *Glass-Reinforced Plastics Resistance*, Mashinostroyenie, Moscow, p. 303 (1968).
[369] L. Fisher, *Modern Plast.*, **37**(10), 120, 122, 127–128 (1960).
[370] L. W. Hu, J.J. Marin, *Appl. Mech.*, **271**(6) (1955).
[371] P.B. Norris, Forest Prod. Labor. Rep. (1950).
[372] F. Warren, *Trans. ASME*, **76**(4) (1953).
[373] J.J. Marin, *Aeronaut. Sci.*, **29**(4) (1957).
[374] K. Zakharov, *Plastics*, **8**, 59–62 (1961).
[375] A. Malmeister, V. Tamuzh, G. Teters, *Resistance of Rigid Polymeric Materials*, Zinatne, Riga, p. 398 (1967).
[376] I. Goldenblatt, V. Kopnov, *Proc. Acad. Sci. USSR, Mechanics*, **6**, 77–83 (1965).
[377] E. Ashkenazi, *Anisotropy of Machine-Building Materials*, Mashinostroyenie, Leningrad, p. 111 (1969).
[378] Yu. Tarnopolsky, A. Skudra, *Structural Durability and Deformability of Glass-Reinforced Plastics*, Zinatne, Riga, p. 260 (1966).
[379] A. Antipov, M. Morozov, *Works of the Nikolaev Shipbuilding Institute*, **98**, 50–57 (1975).
[380] A. Antipov, M. Morozov, *Works of the Nikolaev Shipbuilding Institute*, **98**, 58–64 (1975).
[381] M. Morozov, *Works of the Nikolaev Shipbuilding Institute*, **98**, 103–108 (1975).
[382] V. Nikolaev, V. Perevozchikov, *Problems of Durability*, (11), 62–64 (1974).
[383] G. Pisarenko, A. Lebedev, *Strength of Materials to Deformation and Shattering under Composite Tension*, Naukova Dumka, Kiev, p. 210 (1969).
[384] P. Balandin, *Bulletin of the Engineers and Technicians*, No. 1 (1937).
[385] E. Ashkenazi, *Durability of Anisotropic Wood and Synthetic Materials*, Lesnaya Promyshlennost, Moscow, 1966.
[386] P. Ogibalov, Yu. Suvorova, *Mechanics of Reinforced Plastics*, MGU Publisher, Moscow, p. 480 (1965).
[387] A. Malkin (ed.), *Mechanical Characteristics of Rigid Polymers*, Khimiya, Moscow, p. 358 (1975).
[388] J. Friedman, *Mechanical Characteristics of Metals*, Oborongiz, Moscow, p. 556 (1952).
[389] A. Botkin, *Proceedings of the Research Institute of Hydraulic Engineering*, **36**, 205–236 (1940).
[390] U. Yagn, *Bulletin of the Engineers and Technicians*, No. 6 (1931).
[391] I. Miroliubov, *Works of the Leningrad Technological Institute*, No. 25 (1953).
[392] B. Deryagin, *Proc. Acad. Sci. USSR*, **280** (1963)
[393] B. Deryagin, V. Lazarev, *Proc. Acad. Sci. USSR*, **106** (1949).
[394] P.W. Brigman, *J. Appl. Phys.*, **8**, 328 (1937).
[395] A. Rabinovich, M. Shtarkov, E. Dmitrieva, *Works of the Moscow Physical and Engineering Institute*, Explorations in Mechanics and Applied Mathematics, **I** (1958).
[396] V. Nalimov, N. Chernova, *Statistical Methods of Scheduling of Extreme Experiments*, Nauka, Moscow, p. 340 (1965).
[397] V. Nalimov (ed.), *New Ideas in Scheduling Experiments*, Nauka, Moscow, p. 334 (1969).
[398] V. Nalimov, *Mendeleyev' Soc. J.*, (1), 3–4 (1980).

[399] M. Katz, *Automation of Chemical Manufactures*, (2), 13–19 (1982).
[400] M. Katz, *Development of Methods of Identification and Suboptimization*, Doctoral thesis, Kharkov, p. 40 (1992).
[401] M. Katz, *Automation of Chemical Manufactures*, (2), 5–8 (1980).
[402] M. Katz, *Chem. Technol.*, (10), 55–57 (1988).
[403] Yu.M. Bazhenov, *Organic Concrete*, Stroyizdat, Moscow (1983).
[404] E. Tazawa, S. Kobayashi, Properties and Applications of Polymer Impregnated Cementitious Materials, in *Polymers in Concrete,* Publ. SP-30, American Concrete Institute, Detroit (1973).
[405] Eric Baer (ed.), *Engineering Design for Plastics*, Reinhold, New York (1960).
[406] R.A. Veselovsky, *Digest of Physical Chemistry of Multicomponent Polymer Systems*, Vol. 1, Naukova Dumka, Kiev, p. 375 (1986).
[407] R. Veselovsky, K. Veselovsky, V. Kestelman, *Intl. J. Polym. Mater.*, **23**, 117–138 (1994).
[408] Yu.C. Lipatov, A.Yu. Filipovitch, R.A. Veselovsky, *Dokl. Acad. Nauk USSR*, **275**(1), 118–121 (1984).
[409] I.G. Manets, R.A. Veselovsky, *Polymeric Composite Materials in Mining*, Nedra, Moscow, p. 235 (1988).
[410] R.A. Veselovsky, *Control of Adhesion Strength of Polymers*, Naukova Dumka, Kiev, p. 176 (1988).
[411] R.A. Veselovsky, V.V. Efanova, *Mech. Compos. Mater.*, **30**(1), 3–11 (1994).
[412] R.A. Veselovsky, V.N. Kestelman, *Adhesion Strength of Polymers*, Swissbonding, 1994.
[413] A.W. Adamson, *Physical Chemistry of Surfaces*, Wiley, New York, p. 134 (1987).
[414] M.T. Bryk, *Khimia i Technol. Vysokomol. Soed.*, **4**, 142–184 (1973).
[415] Yu.N. Savvin, N.D. Zverev, E.P. Nikolaeva, *J. Tech. Phys.*, **60**, 138–140 (1990).
[416] J. Klivava, *Phys. Stat. Sol. (B)*, **134**, 411–457 (1986).
[417] R.A. Veselovsky, A.Yu. Filipovich, V.A. Khranovskyj, *Vysokomol. Soed., B*, **27**, 497–500 (1985).
[418] Yu.S. Lipatov, A.Yu. Filipovich, R.A. Veselovsky, *Proc. Acad. Sci. USSR,* **275**, 118–121 (1984).
[419] P.W. Lipatova, G.A. Pkhakadze, *Use of Polymers in Surgery,* Naukova Dumka, Kiev, p. 129 (1977).
[420] T.E. Lipatova, G.A. Pkhakadze, *Medical Adhesives,* Naukova Dumka, Kiev, p. 44 (1979).
[421] T.E. Lipatova, G.A. Pkhakadze, *Polymers in Endoprosthetics*, Naukova Dumka, Kiev, p. 160 (1983).
[422] G.A. Pkhakadze, *The Morphological and Biochemical Aspects of Polymeric Biodestruction*, Naukova Dumka, Kiev, p. 159 (1986).
[423] T.E. Lipatova, O.E. Novikova, G.A. Pkhakadze, *Proc. Acad. Sci. Ukraine B*, 810–814 (1974).
[424] T.E. Lipatova, G.A. Pkhakadze, *Vestnik Acad. Sci. Ukraine,* **12**, 22–30 (1980).
[425] G.A. Pkhakadze, T.L. Tereshchenko, V.P Yatsenko, A.K. Kolomiytsev, *Polymers in Medicine*, Vroslav, **2**, 105–112 (1977).
[426] V.P. Yatsenko, *et al.*, *Current Problems in Modern Pathophysiology*, Naukova Dumka, Kiev, pp. 425–427 (1981).
[427] G.A. Pkhakadze, T.T. Alekseyeva, *Biodestructible Polymeric Materials*, Naukova Dumka, Kiev, pp. 12–19 (1982).
[428] G.A. Pkhakadze, L.V. Shchukina, R.A. Veselovsky, T.E. Lipatova, *Clin. Surg.*, **9**, 60–62 (1973).
[429] M.M. Kovalev, T.L Tereshchenko., E.I. Suslov, G.A. Pkhakadze, *Polymers in Medicine*, Vroslav, **14**, 277–283 (1974).
[430] T.E. Lipatova, V.A. Mironov, G.A. Pkhakadze, Avt. Svid. USSR No. 566832 (1977).
[431] T.E. Lipatova, R.A. Veselovsky, G.A Pkhakadze, Avt. Svid. USSR No. 884705 (1975).
[432] T.E. Lipatova., R.A. Veselovsky, G.A. Pkhakadze, USA Patent No. 4057535 (1977).
[433] G.A. Pkhakadze, L.V. Shchukina, T.L. Tereshchenko, *Ukrain. Biochem. J.*, 25–33 (1978).

[434] G.A. Pkhakadze, G.V. Burenko, S.V. Komisaranko, *Ukrain. Biochem. J.*, 703–708 (1970).
[435] G.A. Pkhakadze, *Proc. Acad. Sci. Ukr. SSR, B*, **7B**, 656–659 (1972).
[436] N.O. Mezhiborskaya, G.V. Burenko, G.A. Pkhakadze, *Proc. Acad. Sci. Ukrain. SSR*, 271–274 (1971).
[437] G.A. Pkhakadze, T.L. Tereshchenko, in *The Morphological and Biochemical Aspects of Polymeric Biodestruction*, Naukova Dumka, Kiev, p. 101 (1986).
[438] G.A. Pkhakadze, L.V. Shchukina, T.E. Lipatova, *Ukrain. Biochem. J.*, 568–572 (1976).
[439] A.I. Snegirev, G.V. Burenko, G.A. Pkhakadze, in *The Morphological and Biochemical Aspects of Polymeric Biodestruction*, Naukova Dumka, Kiev, p. 31 (1986).
[440] I.M. Kebuladze, M.M. Kovalev, G.V. Burenko, in *Polymers in Medicine*, Naukova Dumka, Kiev, p. 83 (1976).
[441] I.M. Kebuladze, M.M. Kovalev, T.E. Lipatova, *et al.*, *Sabchota Med.*, **5**, 53–55 (1973).
[442] G.A. Pkhakadze, L.V. Shchukina, T.E. Lipatova, *et al.*, *Proc. 29th Congress of Surgeons*, Zdorov'ya, Kiev, pp. 222–223 (1974).
[443] M.M. Kovalev, G.V. Burenko, I.M. Kebuladze, *et al.*, *Polymers in Medicine,* Vroslav, **2**, 117–123 (1973).
[444] M.M. Kovalev, I.M. Kebuladze, V.A. Mironov, in *The Problems of Producing Medical Drugs and Polymeric Materials for Medicine*, Naukova Dumka, Kiev, pp. 76–78 (1979).
[445] V.P. Roy, I.M. Kebuladze, *Clin. Surg.*, **2**, 45–47 (1979).
[446] G.G. Gorodenko, I.M. Slepukha, V.R. Belkin, *et al.*, *Thoracic Surgery*, **5**, 57–62 (1974).
[447] V.I. Titarenko, G.G. Gorodenko, I.M. Slepukha, *et al.*, in *Polymers in Medicine*, Naukova Dumka, Kiev, p. 41 (1976).
[448] V.Ya. Djugostran, *Thoracic Surgery,* **5**, 58–60 (1987).
[449] R.V. Mikulyak, *Clin. Surg.*, **1**, 25–27 (1984).
[450] R.V. Mikulyak, Ya.Ya. Bondar, I.K. Loyko, V.I. Polovchik, *Clin. Surg.,* **1**, 71–72 (1985).
[451] Yu.N. Mokhnyuk, A.I. Poyda, Avt. Svid. USSR No. 990199 (1983).
[452] S.G. Vainshtein, M.N. Zhukovski, Z.I. Shust, *et al.*, Avt. Svid USSR No. 1172541 (1985).
[453] A.I. Poyda, *Clin. Surg.*, **2**, 41–42 (1982).
[454] S.G. Vainshtein, M.N. Zhukovski, Z.I. Shust, *Clin. Surg.*, **10**, 67–68 (1985).
[455] S.G. Vainshtein, M.N. Zhukovski, Z.I. Shust, Ya.N. Fedorov, *Kazan' Med. J.*, **6**, 403–404 (1986).
[456] Yu.Z. Livshits, I.S. Dyachuk, L.S. Belyansky, Z.I. Murtazaev, *Clin. Surg.*, **9**, 67–68 (1987).
[457] S.A. Shalimov, V.S. Zemskov, A.P. Podpryatov, *et al.*, *Clin. Surg.*, **6**, 47–50 (1982).
[458] S.G. Shtofin, *Surgery*, **2**, 126–127 (1987).
[459] V.S. Zemskov, Yu.Z. Livshits, G.I. Litvinenko, *et al.*, *Clin. Surg.*, **11**, 8–11 (1981).
[460] A.A. Shalimov, V.S. Zemskov, S.A. Shalimov, *et al.*, *Clin. Surg.*, **11**, 1–7 (1981).
[461] T.L. Tereshchenko, in *Polymers in Medicine*, Naukova Dumka, Kiev, p. 53 (1976).
[462] T.E. Lipatova, G.A. Pkhakadze, T.L. Tereshchenko, *Vojenske Zdravotnickelisty (CSSR)*, **1**, 25–29 (1976).
[463] P.A. Malinovsky, V.A. Mironov, in *The Problems of Producing Medical Drugs and Polymeric Materials for Medicine*, Naukova Dumka, Kiev, pp. 76–78 (1979).
[464] B.S. Gekman, P.A. Malinovsky, G.A. Pkhakadze, *Urol. Nephrol.*, 560–564 (1981).
[465] *Proc. 31st All-Union Congress of Surgeons, FAM*, Tashkent, p. 186 (1986).
[466] A.A. Polyakov, *10th All-Union Congress on Transplantation of Organs*, Kiev, pp. 70–73 (1985).
[467] A.A. Polyakov, V.A. Tkhor, *Transplantation of Kidneys, Thesis of III All-Union Symposium*, Riga, pp. 35–36 (1984).
[468] L.V. Khar'kov, *Stomatology*, **5**, 53–55 (1984).
[469] T.E. Lipatova, Yu. Lipatov, *Mendeleyev' Soc. J.*, 483–496 (1985).

Index

About the Authors

ROMAN A. VESELOVSKY, D.SC., is a polymers expert at the Institute of Macromolecular Chemistry of the National Academy of Sciences of the Ukraine, Academician of the Ukrainian Technological Academy, and a member of the International Scientific Board "Swissbonding," based in Switzerland. He has published 6 books and 220 journal articles and holds 120 patents, all related to materials science.

VLADIMIR KESTELMAN, D.SC., is an internationally respected expert on the mechanics and technology of composite materials and is president of KVN International, a worldwide consulting firm. An Academician of the Ukrainian Technological Academy, he is a former professor at Moscow Aviation Institute and the author of *Electrophysical Phenomena in the Tribology of Polymers*, *Electrets in Engineering: Fundamentals and Applications*, and other books. He has published 26 books and 520 journal articles and holds 75 patents.